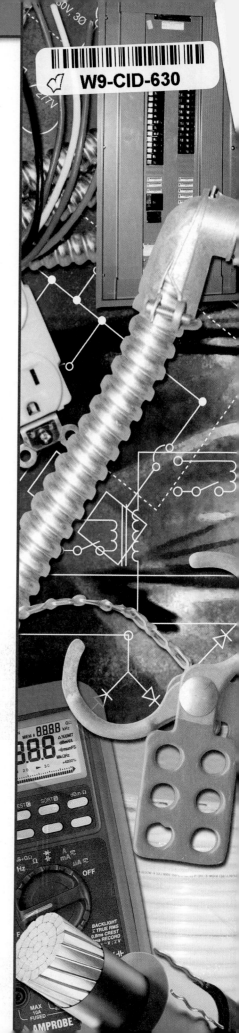

# Electrical

## *Level Two*

### Trainee Guide
2011 *NEC*® Revision

**Prentice Hall**
Boston   Columbus   Indianapolis   New York   San Francisco   Upper Saddle River
Amsterdam   Cape Town   Dubai   London   Madrid   Milan   Munich   Paris   Montreal   Toronto
Delhi   Mexico City   São Paulo   Sydney   Hong Kong   Seoul   Singapore   Taipei   Tokyo

**National Center for Construction Education and Research**
*President:* Don Whyte
*Director of Product Development:* Daniele Stacey
*Electrical Project Manager:* Daniele Stacey
*Production Manager:* Tim Davis

*Quality Assurance Coordinator:* Debie Ness
*Desktop Publishing Coordinator:* James McKay
*Production Specialist:* Laura Wright
*Editors:* Chris Wilson, Debie Ness

**Writing and development services provided by Topaz Publications, Liverpool, NY**
*Lead Writer/Project Manager:* Veronica Westfall
*Desktop Publisher:* Joanne Hart
*Art Director:* Megan Paye

*Permissions Editors:* Andrea LaBarge, Tonia Burke
*Writers:* Thomas Burke, Gerald Shannon, Nancy Brown,
Charles Rogers, Anna Meade

**Pearson Education, Inc.**
*Editorial Director:* Vernon R. Anthony
*Senior Product Manager:* Lori Cowen
*Senior Managing Editor:* JoEllen Gohr
*Senior Project Manager:* Stephen C. Robb
*AV Project Manager:* Janet Portisch
*Operations Supervisor:* Deidra M. Skahill
*Art Director:* Jayne Conte
*Cover Photo:* Tim Davis

*Director of Marketing:* David Gesell
*Executive Marketing Manager:* Derril Trakalo
*Senior Marketing Coordinator:* Alicia Wozniak
*Full-Service Project Management*: Michael B. Kopf,
S4Carlisle Publishing Services
*Composition:* S4Carlisle Publishing Services
*Printer/Binder:* Courier/Kendallville, Inc.
*Cover Printer:* Lehigh-Phoenix Color/Hagerstown
*Text Fonts:* Palatino and Univers

Credits and acknowledgments for content borrowed from other sources and reproduced, with permission, in this textbook appear at the end of each module.

10 9 8 7 6 5 4 3 2 1

**Prentice Hall**
is an imprint of

www.pearsonhighered.com

Perfect bound  ISBN-13: 978-0-13-256953-8
ISBN-10:  0-13-256953-1
Case bound  ISBN-13: 978-0-13-257123-4
ISBN-10:  0-13-257123-4

# Preface

## To the Trainee

Electricity powers the applications that make our daily lives more productive and efficient. The demand for electricity has led to vast job opportunities in the electrical field. Electricians constitute one of the largest construction occupations in the United States, and they are among the highest-paid workers in the construction industry. According to the U.S. Bureau of Labor Statistics, job opportunities for electricians are expected to be very good as the demand for skilled craftspeople is projected to outpace the supply of trained electricians.

Electricians install electrical systems in structures. They install wiring and other electrical components, such as circuit breaker panels, switches, and light fixtures. Electricians follow blueprints, the *National Electrical Code®*, and state and local codes. They use specialized tools and testing equipment, such as ammeters, ohmmeters, and voltmeters. Electricians learn their trade through craft and apprenticeship programs. These programs provide classroom instruction and on-the-job learning with experienced electricians.

We wish you success in your second year of training in the electrical craft, and we hope that you will continue your training beyond this textbook. There are more than 700,000 people employed in electrical work in the United States, and, as most of them can tell you, there are many opportunities awaiting those with the skills and desire to move forward in the construction industry.

## New with *Electrical Level Two*

NCCER and Pearson are pleased to present a fresh format for this edition of *Electrical Level Two*. This seventh edition has been updated to meet the 2011 *National Electrical Code®* and includes revisions to the module examinations. The workbook questions, now Supplemental Exercises, have been incorporated into the textbook so that they're contained in one convenient location for easy reference.

The opening page of each of the eleven modules in this textbook features photos of award-winning construction projects from Associated Builders and Contractors, Inc., and The Associated General Contractors of America. Check out these awesome projects and see where a career as an electrician could take you.

We wish you success as you progress through this training program. If you have any comments on how NCCER might improve upon this textbook, please complete the User Update form located at the back of each module and send it to us. We will always consider and respond to input from our customers.

We invite you to visit the NCCER website at **www.nccer.org** for information on the latest product releases and training, as well as online versions of the *Cornerstone* newsletter and Pearson's Contren® product catalog.

Your feedback is welcome. You may email your comments to **curriculum@nccer.org** or send general comments and inquiries to **info@nccer.org**.

## Contren® Learning Series

The National Center for Construction Education and Research (NCCER) is a not-for-profit 501(c)(3) education foundation established in 1996 by the world's largest and most progressive construction companies and national construction associations. It was founded to address the severe workforce shortage facing the industry and to develop a standardized training process and curricula. Today, NCCER is supported by hundreds of leading construction and maintenance companies, manufacturers, and national associations. The Contren® Learning Series was developed by NCCER in partnership with Pearson Education, Inc., the world's largest educational publisher.

Some features of NCCER's Contren® Learning Series are as follows:

- An industry-proven record of success
- Curricula developed by the industry for the industry
- National standardization providing portability of learned job skills and educational credits
- Compliance with the Office of Apprenticeship requirements for related classroom training (*CFR 29:29*)
- Well-illustrated, up-to-date, and practical information

NCCER also maintains a National Registry that provides transcripts, certificates, and wallet cards to individuals who have successfully completed modules of NCCER's Contren® Learning Series. *Training programs must be delivered by an NCCER Accredited Training Sponsor in order to receive these credentials.*

# Special Features

In an effort to provide a comprehensive, user-friendly training resource, we have incorporated many different features for your use. Whether you are a visual or hands-on learner, this book will provide you with the proper tools to get started in the electrical industry.

## Introduction

This page is found at the beginning of each module and lists the Objectives, Performance Tasks, Trade Terms, and Required Trainee Materials for that module. The Objectives list the skills and knowledge you will need in order to complete the module successfully. The Performance Tasks give you an opportunity to apply your knowledge to the real-world duties that electricians perform. The list of Trade Terms identifies important terms you will need to know by the end of the module. Required Trainee Materials list the materials and supplies needed for the module.

## On Site

On Site features provide a head start for those entering the electrical field by presenting technical tips and professional practices from master electricians in a variety of disciplines. The On Site features often include real-life scenarios similar to those you might encounter on the job site.

## Color Illustrations and Photographs

Full-color illustrations and photographs are used throughout each module to provide vivid detail. These figures highlight important concepts from the text and provide clarity for complex instructions. Each figure reference is denoted in the text in *italic type* for easy reference.

*Figure 2* Pushing down on the bender to complete the bend.

## Notes, Cautions, and Warnings

Safety features are set off from the main text in highlighted boxes and are organized into three categories based on the potential danger of the issue being addressed. Notes simply provide additional information on the topic area. Cautions alert you of a danger that does not present potential injury but may cause damage to equipment. Warnings stress a potentially dangerous situation that may cause injury to you or a co-worker.

## Going Green

Going Green looks at ways to preserve the environment, save energy, and make good choices regarding the health of the planet. Through the introduction of new construction practices and products, you will see how the "greening of America" has already taken root.

## Case History

Case History features emphasize the importance of safety by citing examples of the costly (and often devastating) consequences of ignoring *National Electrical Code*® or OSHA regulations.

## What's wrong with this picture?

What's wrong with this picture? features include photos of actual code violations for identification and encourage you to approach each installation with a critical eye.

*What's wrong with this picture?*

108SA14.EPS

## Think About It

Think About It features use "What if?" questions to help you apply theory to real-world experiences and put your ideas into action.

### Think About It

#### Residential Wiring

Make a mental wiring tour of your home. Picture several rooms, including the kitchen and utility/laundry room. How is each device connected to the power source, and what is the probable amperage and overcurrent protection? What other devices might or might not be included in the circuit? How many branch circuits serve each room? Later, examine the panelboard and exposed wiring to see how accurately you identified the branch circuits.

## Step-by-Step Instructions

Step-by-step instructions are used throughout to guide you through technical procedures and tasks from start to finish. These steps show you not only how to perform a task but also how to do it safely and efficiently.

are energized until you have verified that the circuit is de-energized. This is called a live-dead-live test. Follow these steps to verify that a circuit is de-energized:

*Step 1* Ensure that the circuit is properly tagged and locked out *(CFR 1910.333/1926.417)*.

*Step 2* Verify the test instrument operation on a known source using the appropriately rated tester.

*Step 3* Using the test instrument, check the circuit to be de-energized. The voltage should be zero.

*Step 4* Verify the test instrument operation, once again on a known power source.

## Trade Terms

Each module presents a list of Trade Terms that are discussed within the text and defined in the Glossary at the end of the module. These terms are denoted in the text with **blue bold type** upon their first occurrence. To make searches for key information easier, a comprehensive Glossary of Trade Terms from all modules is located at the back of this book.

Electricity is all about cause and effect. The presence of voltage (volts) in a closed circuit will cause current (amps) to flow. The more voltage you apply, the more current will flow. However, the amount of current flow is also determined by how much resistance (ohms) the load offers to the flow of current. In order to convert electrical energy into work, the load consumes energy. The amount of energy a device consumes is called power, and is expressed in watts (W). Volts (V), amps, ohms, and watts are related in such a way that if any one of them changes, the others are proportionally affected. This relationship can be seen using basic math principles that you will learn in this module. You will also learn how electricity is produced and how test instruments are used to measure electricity.

## Review Questions

Review Questions are provided to reinforce the knowledge you have gained. This makes them a useful tool for measuring what you have learned.

# Contren® Curricula

NCCER's training programs comprise more than 80 construction, maintenance, pipeline, and utility areas and include skills assessments, safety training, and management education.

Boilermaking
Cabinetmaking
Carpentry
Concrete Finishing
Construction Craft Laborer
Construction Technology
Core Curriculum:
    Introductory Craft Skills
Drywall
Electrical
Electronic Systems Technician
Heating, Ventilating, and
    Air Conditioning
Heavy Equipment Operations
Highway/Heavy Construction
Hydroblasting
Industrial Coating and Lining
    Application Specialist
Industrial Maintenance
    Electrical and
    Instrumentation Technician
Industrial Maintenance
    Mechanic
Instrumentation
Insulating
Ironworking
Masonry
Millwright
Mobile Crane Operations
Painting
Painting, Industrial
Pipefitting
Pipelayer
Plumbing
Reinforcing Ironwork
Rigging
Scaffolding
Sheet Metal
Signal Person
Site Layout
Sprinkler Fitting
Tower Crane Operator
Welding

## Green/Sustainable Construction

Building Auditor
Fundamentals of
    Weatherization
Introduction to Weatherization
Sustainable Construction
    Supervisor
Weatherization Crew Chief
Weatherization Technician
Your Role in the Green
    Environment

## Energy

Introduction to the Power
    Industry
Introduction to Solar
    Photovoltaics
Introduction to Wind Energy
Power Industry Fundamentals
Power Generation Maintenance
    Electrician
Power Generation I&C
    Maintenance Technician
Power Generation Maintenance
    Mechanic
Power Line Worker
Solar Photovoltaic Systems
    Installer

## Pipeline

Control Center Operations,
    Liquid
Corrosion Control
Electrical and Instrumentation
Field Operations, Liquid
Field Operations, Gas
Maintenance
Mechanical

## Safety

Field Safety
Safety Orientation
Safety Technology

## Management

Fundamentals of Crew
    Leadership
Project Management
Project Supervision

## Supplemental Titles

Applied Construction Math
Careers in Construction
Tools for Success

## Spanish Translations

Basic Rigging
    (Principios Básicos de
    Maniobras)
Carpentry Fundamentals
    (Introducción a la
    Carpintería, Nivel Uno)
Carpentry Forms
    (Formas para Carpintería,
    Nivel Trés)
Concrete Finishing, Level One
    (Acabado de Concreto,
    Nivel Uno)
Core Curriculum:
    Introductory Craft Skills
    (Currículo Básico:
    Habilidades Introductorias
    del Oficio)
Drywall, Level One
    (Paneles de Yeso, Nivel Uno)
Electrical, Level One
    (Electricidad, Nivel Uno)
Field Safety
    (Seguridad de Campo)
Insulating, Level One
    (Aislamiento, Nivel Uno)
Ironworking, Level One
    (Herrería, Nivel Uno)
Masonry, Level One
    (Albañilería, Nivel Uno)
Pipefitting, Level One
    (Instalación de Tubería
    Industrial, Nivel Uno)
Reinforcing Ironwork, Level One
    (Herreria de Refuerzo,
    Nivel Uno)
Safety Orientation
    (Orientación de Seguridad)
Scaffolding
    (Andamios)
Sprinkler Fitting, Level One
    (Instalación de Rociadores,
    Nivel Uno)

# Acknowledgments

This curriculum was revised as a result of the farsightedness and leadership of the following sponsors:

Associated Builders and Contractors of Iowa
Beacon Electrical Contractors
Cianbro Corporation
Construction Industry Training Council
Dvorak, LLC
Lamphear Electric
Lee College
Madison Comprehensive High School

Pumba Electric, LLC
Putnam Career and Technical Center
Rust Constructors Inc.
TIC – The Industrial Company
Tri-City Electrical Contractors, Inc.
Vector Electric and Controls Inc.
Willmar Electric Service Corporation

This curriculum would not exist were it not for the dedication and unselfish energy of those volunteers who served on the Authoring Team. A sincere thanks is extended to the following:

Kelly Allen
Clarence "Ed" Cockrell
Tim Dean
Tim Ely
Al Hamilton

Dan Lamphear
Leonard "Skip" Layne
L. J. LeBlanc
David Lewis
Scott Mitchell

Jim Mitchem
Christine Porter
Michael J. Powers
Raymond Saldivar
Wayne Stratton

*A final note:* This book is the result of a collaborative effort involving the production, editorial, and development staff at Pearson Education, Inc., and the National Center for Construction Education and Research. Thanks to all of the dedicated people involved in the many stages of this project.

# NCCER Partners

American Fire Sprinkler Association
Associated Builders and Contractors, Inc.
Associated General Contractors of America
Association for Career and Technical Education
Association for Skilled and Technical Sciences
Carolinas AGC, Inc.
Carolinas Electrical Contractors Association
Center for the Improvement of Construction
    Management and Processes
Construction Industry Institute
Construction Users Roundtable
Construction Workforce Development Center
Design Build Institute of America
Merit Contractors Association of Canada
Metal Building Manufacturers Association
NACE International
National Association of Minority Contractors
National Association of Women in Construction
National Insulation Association
National Ready Mixed Concrete Association
National Technical Honor Society
National Utility Contractors Association
NAWIC Education Foundation
North American Technician Excellence

Painting & Decorating Contractors of America
Portland Cement Association
SkillsUSA
Steel Erectors Association of America
The Manufacturers Institute
U.S. Army Corps of Engineers
University of Florida, M. E. Rinker School of
    Building Construction
Women Construction Owners & Executives, USA

# Contents

**Note:** *NFPA 70®*, *National Electrical Code®*, and *NEC®* are registered trademarks of the National Fire Protection Association, Inc., Quincy, MA 02269. All *National Electrical Code®* and *NEC®* references in this textbook refer to the 2011 edition of the *National Electrical Code®*.

## 26209-11

# Grounding and Bonding

Focuses on the purpose of grounding and bonding electrical systems. Thoroughly covers *NEC*® requirements. (15 Hours)

## 26210-11

# Circuit Breakers and Fuses

Describes fuses and circuit breakers along with their practical applications. Also covers sizing. (12.5 Hours)

## 26211-11

# Control Systems and Fundamental Concepts

Gives basic descriptions of various types of contactors and relays along with their practical applications. (12.5 Hours)

# Glossary

# Index

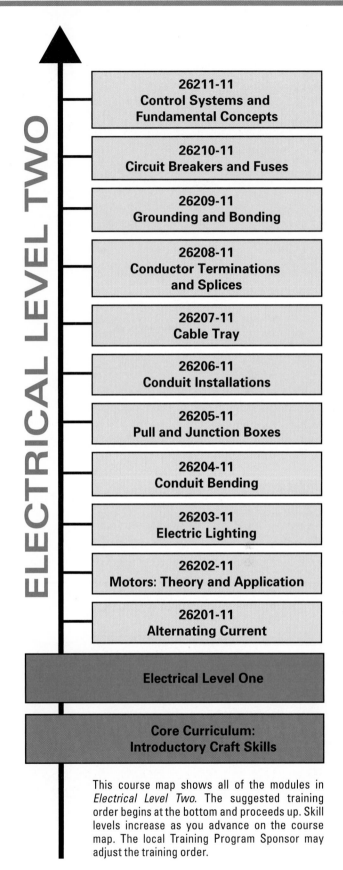

ELECTRICAL LEVEL TWO

**26211-11**
**Control Systems and**
**Fundamental Concepts**

**26210-11**
**Circuit Breakers and Fuses**

**26209-11**
**Grounding and Bonding**

**26208-11**
**Conductor Terminations**
**and Splices**

**26207-11**
**Cable Tray**

**26206-11**
**Conduit Installations**

**26205-11**
**Pull and Junction Boxes**

**26204-11**
**Conduit Bending**

**26203-11**
**Electric Lighting**

**26202-11**
**Motors: Theory and Application**

**26201-11**
**Alternating Current**

**Electrical Level One**

**Core Curriculum:**
**Introductory Craft Skills**

This course map shows all of the modules in *Electrical Level Two*. The suggested training order begins at the bottom and proceeds up. Skill levels increase as you advance on the course map. The local Training Program Sponsor may adjust the training order.

# Alternating Current

**Pawnee Health Center**

W. W. Enterprises won an Excellence in Construction award from ABC in the Commercial-Electrical division for its work on the Pawnee Health Center in Pawnee, Oklahoma, which serves the Native American Pawnee Nation.

## 26201-11

Trainees with successful module completions may be eligible for credentialing through NCCER's National Registry. To learn more, go to **www.nccer.org** or contact us at **1.888.622.3720.** Our website has information on the latest product releases and training, as well as online versions of our *Cornerstone* newsletter and Pearson's Contren® product catalog.

Your feedback is welcome. You may email your comments to **curriculum@nccer.org,** send general comments and inquiries to **info@nccer.org,** or use the User Update form at the back of this module.

 V.1 7/11

# Alternating Current

## Objectives

When you have completed this module, you will be able to do the following:

1. Calculate the peak and effective voltage or current values for an AC waveform.
2. Calculate the phase relationship between two AC waveforms.
3. Describe the voltage and current phase relationship in a resistive AC circuit.
4. Describe the voltage and current transients that occur in an inductive circuit.
5. Define inductive reactance and state how it is affected by frequency.
6. Describe the voltage and current transients that occur in a capacitive circuit.
7. Define capacitive reactance and state how it is affected by frequency.
8. Explain the relationship between voltage and current in the following types of AC circuits:
   - RL circuit
   - RC circuit
   - LC circuit
   - RLC circuit
9. Explain the following terms as they relate to AC circuits:
   - True power
   - Apparent power
   - Reactive power
   - Power factor
10. Explain basic transformer action.

## Performance Tasks

This is a knowledge-based module. There are no Performance Tasks.

## Trade Terms

| | | |
|---|---|---|
| Capacitance | Inductance | Reactance |
| Frequency | Micro | Root-mean-square (rms) |
| Hertz (Hz) | Peak voltage | Self-inductance |
| Impedance | Radian | |

## Required Trainee Materials

1. Pencil and paper
2. Appropriate personal protective equipment
3. Copy of the latest edition of the *National Electrical Code®*

Note: *NFPA 70®*, *National Electrical Code®*, and *NEC®* are registered trademarks of the National Fire Protection Association, Inc., Quincy, MA 02269. All *National Electrical Code®* and *NEC®* references in this module refer to the 2011 edition of the *National Electrical Code®*.

## Contents

*Topics to be presented in this module include:*

## Figures and Tables ———————————

## 1.0.0 INTRODUCTION

Alternating current (AC) and its associated voltage reverses between positive and negative polarities and varies in amplitude with time. One complete waveform or cycle includes a complete set of variations, with two alternations in polarity. Many sources of voltage change direction with time and produce a resultant waveform. The most common AC waveform is the sine wave.

## 2.0.0 SINE WAVE GENERATION

To understand how the alternating current sine wave is generated, some of the basic principles learned in magnetism should be reviewed. Two principles form the basis of all electromagnetic phenomena:

* An electric current in a conductor creates a magnetic field that surrounds the conductor.
* Relative motion between a conductor and a magnetic field, when at least one component of that relative motion is in a direction that is perpendicular to the direction of the field, creates a voltage in the conductor.

*Figure 1* shows how these principles are applied to generate an AC waveform in a simple one-loop rotary generator. The conductor loop rotates through the magnetic field to generate the induced AC voltage across its open terminals. The magnetic flux shown here is vertical.

There are three factors affecting the magnitude of voltage developed by a conductor through a magnetic field: the strength of the magnetic field; the length of the conductor; and the rate at which the conductor cuts directly across or perpendicular to the magnetic field.

Assuming that the strength of the magnetic field and the length of the conductor making the loop are both constant, the voltage produced will vary depending on the rate at which the loop cuts directly across the magnetic field.

The rate at which the conductor cuts the magnetic field depends on the speed of the generator in revolutions per minute (rpm) and the angle at which the conductor is traveling through the field. If the generator is operated at a constant rpm, the voltage produced at any moment will depend on the angle at which the conductor is cutting the field at that instant.

In *Figure 2*, the magnetic field is shown as parallel lines called lines of flux. These lines always go from the north to south poles in a generator. The motion of the conductor is shown by the large arrow.

Assuming the speed of the conductor is constant, as the angle between the flux and the conductor motion increases, the number of flux

201F01.EPS

*Figure 1*   Conductor moving across a magnetic field.

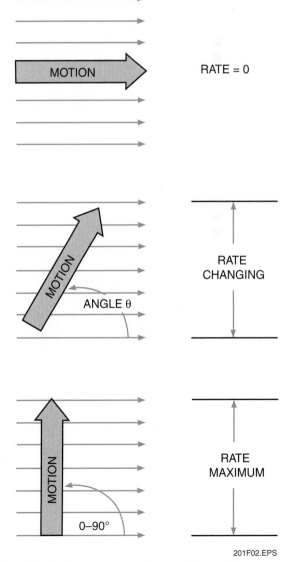

201F02.EPS

*Figure 2*   Angle versus rate of cutting lines of flux.

# Why Do Power Companies Generate and Distribute AC Power Instead of DC Power?

The transformer is the key. Power plants generate and distribute AC power because it permits the use of transformers, which makes power delivery more economical. Transformers used at generation plants step the AC voltage up, which decreases the current. Decreased current allows smaller-sized wires to be used for the power transmission lines. Smaller wire is less expensive and easier to support over the long distances that the power must travel from the generation plant to remotely located substations. At the substations, transformers are again used to step AC voltages back down to a level suitable for distribution to homes and businesses.

There is no such thing as a DC transformer. This means DC power would have to be transmitted at low voltages and high currents over very large-sized wires, making the process very uneconomical. When DC is required for special applications, the AC voltage may be converted to DC voltage by using rectifiers, which make the change electrically, or by using AC motor–DC generator sets, which make the change mechanically.

lines cut in a given time (the rate) increases. When the conductor is moving parallel to the lines of flux (angle of 0°), it is not cutting any of them, and the voltage will be zero.

The angle between the lines of flux and the motion of the conductor is called θ (theta). The magnitude of the voltage produced will be proportional to the sine of the angle. Sine is a trigonometric function. Each angle has a sine value that never changes.

The sine of 0° is 0. It increases to a maximum of 1 at 90°. From 90° to 180°, the sine decreases back to 0. From 180° to 270°, the sine decreases to −1. Then from 270° to 360° (back to 0°), the sine increases to its original 0.

Because voltage is proportional to the sine of the angle, as the loop goes 360° around the circle the voltage will increase from 0 to its maximum at 90°, back to 0 at 180°, down to its maximum negative value at 270°, and back up to 0 at 360°, as shown in *Figure 3*.

Notice that at 180° the polarity reverses. This is because the conductor has turned completely around and is now cutting the lines of flux in the opposite direction. This can be shown using the left-hand rule for generators. The curve shown in *Figure 3* is called a sine wave because its shape is generated by the trigonometric function sine. The value of voltage at any point along the sine wave can be calculated if the angle and the maximum obtainable voltage ($E_{max}$) are known.

The formula used is:

$$E = E_{max} \sin \theta$$

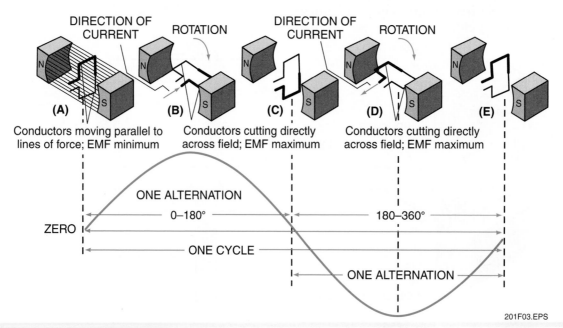

*Figure 3*   One cycle of alternating voltage.

*Where:*

E = voltage induced

$E_{max}$ = maximum induced voltage

θ = angle at which the voltage is induced

Using the above formula, the values of voltage anywhere along the sine wave in *Figure 3* can be calculated. Sine values can be found using either a scientific calculator or trigonometric tables. With an $E_{max}$ of 10 volts (V), the following values are calculated as examples:

θ = 0°, sine = 0
E = $E_{max}$sine θ
E = (10V)(0)
E = 0V

θ = 45°, sine = 0.707
E = $E_{max}$sine θ
E = (10V)(0.707)
E = 7.07V

θ = 90°, sine = 1.0
E = $E_{max}$sine θ
E = (10V)(1.0)
E = 10V

θ = 135°, sine = 0.707
E = $E_{max}$sine θ
E = (10V)(0.707)
E = 7.07V

θ = 180°, sine = 0
E = $E_{max}$sine θ
E = (10V)(0)
E = 0V

θ = 225°, sine = −0.707
E = $E_{max}$sine θ
E = (10V)(−0.707)
E = −7.07V

θ = 270°, sine = −1.0
E = $E_{max}$sine θ
E = (10V)(−1.0)
E = −10V

θ = 315°, sine = −0.707
E = $E_{max}$sine θ
E = (10V)(−0.707)
E = −7.07V

## 3.0.0 SINE WAVE TERMINOLOGY

There are a number of AC voltage terms that are specific to sine waves. The following sections discuss some of these terms, including frequency, wavelength, peak value, average value, and effective value.

### 3.1.0 Frequency

The frequency of a waveform is the number of times per second an identical pattern repeats itself. Each time the waveform changes from zero to a peak value and back to zero is called an alternation. Two alternations form one cycle. The number of cycles per second is the frequency. The unit of frequency is hertz (Hz). One hertz equals one cycle per second (cps).

For example, let us determine the frequency of the waveform shown in *Figure 4*.

In one-half second, the basic sine wave is repeated five times. Therefore, the frequency (f) is:

$$f = \frac{5 \text{ cycles}}{0.5 \text{ second}} = 10 \text{ cycles per second (Hz)}$$

### 3.1.1 Period

The period of a waveform is the time (t) required to complete one cycle. The period is the inverse of frequency:

$$t = \frac{1}{f}$$

*Where:*

t = period (seconds)

f = frequency (Hz or cps)

For example, let us determine the period of the waveform in *Figure 4*. If there are five cycles in one-half second, then the frequency for one cycle is 10 cps (5 ÷ 0.5 = 10). Therefore, the period is:

$$t = \frac{1}{cps}$$

$$t = \frac{1}{10} = 0.1 \text{ second}$$

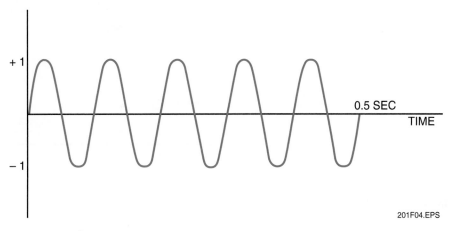

201F04.EPS

*Figure 4*   Frequency measurement.

## 3.2.0 Wavelength

The wavelength or λ (lambda) is the distance traveled by a waveform during one period. Since electricity travels at the speed of light (186,000 miles/second or 300,000 kilometers/second), the wavelength of electrical waveforms equals the product of the period and the speed of light (c):

$$\lambda = tc$$

*or:*

$$\lambda = \frac{c}{f}$$

*Where:*

λ = wavelength (meters)
t = period (seconds)
c = speed of light (meters/second)
f = frequency (Hz or cps)

## 3.3.0 Peak Value

The peak value is the maximum value of voltage ($V_M$) or current ($I_M$). For example, specifying that a sine wave has a **peak voltage** of 170V applies to either the positive or the negative peak. To include both peak amplitudes, the peak-to-peak (p–p) value may be specified. In the above example, the peak-to-peak value is 340V, double the peak value of 170V, because the positive and negative peaks are symmetrical. However, the two opposite peak values cannot occur at the same time. Furthermore, in some waveforms the two peaks are not equal. The positive peak value and peak-to-peak value of a sine wave are shown in *Figure 5*.

## 3.4.0 Average Value

The average value is calculated from all the values in a sine wave for one alternation or half cycle. The half cycle is used for the average because over a full cycle the average value is zero, which is useless for comparison purposes. If the sine values for all angles up to 180° in one alternation are added and then divided by the number of values, this average equals 0.637.

Since the peak value of the sine is 1 and the average equals 0.637, the average value can be calculated as follows:

**Average value = 0.637 × peak value**

For example, with a peak of 170V, the average value is 0.637 × 170V, which equals approximately 108V. *Figure 5* shows where the average value would fall on a sine wave.

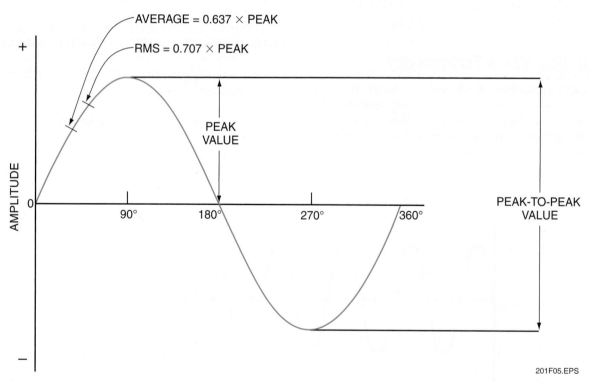

*Figure 5*  Amplitude values for a sine wave.

# Frequency

The frequency of the utility power generated in the United States is normally 60Hz. In some European countries and elsewhere, utility power is often generated at a frequency of 50Hz. Which of these frequencies (60Hz or 50Hz) has the shortest period?

### 3.5.0 Root-Mean-Square or Effective Value

Meters used in AC circuits indicate a value called the effective value. The effective value is the value of the AC current or voltage wave that indicates the same energy transfer as an equivalent direct current (DC) or voltage.

The direct comparison between DC and AC is in the heating effect of the two currents. Heat produced by current is a function of current amplitude only and is independent of current direction. Thus, heat is produced by both alternations of the AC wave, although the current changes direction during each alternation.

In a DC circuit, the current maintains a steady amplitude. Therefore, the heat produced is steady and is equal to $I^2R$. In an AC circuit, the current is continuously changing; periodically high, periodically low, and periodically zero. To produce the same amount of heat from AC as from an equivalent amount of DC, the instantaneous value of the AC must at times exceed the DC value.

By averaging the heating effects of all the instantaneous values during one cycle of alternating current, it is possible to find the average heat produced by the AC current during the cycle. The amount of DC required to produce that heat will be equal to the effective value of the AC.

The most common method of specifying the amount of a sine wave of voltage or current is by stating its value at 45°, which is 70.7% of the peak. This is its root-mean-square (rms) value. Therefore:

Value of rms = 0.707 × peak value

For example, with a peak of 170V, the rms value is 0.707 × 170, or approximately 120V. This is the voltage of the commercial AC power line, which is always given in rms value.

## 4.0.0 AC PHASE RELATIONSHIPS

In AC systems, phase is involved in two ways: the location of a point on a voltage or current wave with respect to the starting point of the wave, or with respect to some corresponding point on the same wave. In the case of two waves of the same frequency, it is the time at which an event of one takes place with respect to a similar event of the other.

Often, the event is the starting of the waves at zero or the points at which the waves reach their maximum values. When two waves are compared in this manner, there is a phase lead or lag of one with respect to the other unless they are alternating in unison, in which case they are said to be in phase.

### 4.1.0 Phase Angle

Suppose that a generator started its cycle at 90° where maximum voltage output is produced instead of starting at the point of zero output. The two output voltage waves are shown in *Figure 6*. Each is the same waveform of alternating voltage, but wave B starts at the maximum value while wave A starts at zero. The complete cycle of wave B through 360° takes it back to the maximum value from which it started.

Wave A starts and finishes its cycle at zero. With respect to time, wave B is ahead of wave A in its values of generated voltage. The amount it leads in time equals one quarter revolution, which is 90°. This angular difference is the phase angle between waves B and A. Wave B leads wave A by the phase angle of 90°.

The 90° phase angle between waves B and A is maintained throughout the complete cycle and in all successive cycles as long as they both have the same frequency. At any instant in time, wave B has the value that A will have 90° later. For instance, at 180°, wave A is at zero, but B is already at its negative maximum value, the point where wave A will be later at 270°.

To compare the phase angle between two waves, both waves must have the same frequency. Otherwise, the relative phase keeps changing. Both waves must also have sine wave variations, because this is the only kind of waveform that is measured in angular units of time. The amplitudes can be different for the two waves. The phases of two voltages, two currents, or a current with a voltage can be compared.

### 4.2.0 Phase Angle Diagrams

To compare AC phases, it is much more convenient to use vector diagrams corresponding to the voltage and current waveforms, as shown in *Figure 6*. $V_A$ and $V_B$ represent the vector quantities corresponding to the generator voltage.

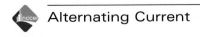

A vector is a quantity that has magnitude and direction. The length of the arrow indicates the magnitude of the alternating voltage in rms, peak, or any AC value as long as the same measure is used for all the vectors. The angle of the arrow with respect to the horizontal axis indicates the phase angle.

In *Figure 6*, the vector $V_A$ represents the voltage wave A, with a phase angle of 0°. This angle can be considered as the plane of the loop in the rotary generator where it starts with zero output voltage. The vector $V_B$ is vertical to show the phase angle of 90° for this voltage wave, corresponding to the vertical generator loop at the start of its cycle. The angle between the two vectors is the phase angle.

The symbol for a phase angle is θ (theta). In *Figure 7*, θ = 0°. *Figure 7* shows the waveforms and vector diagram of two waves that are in phase but have different amplitudes.

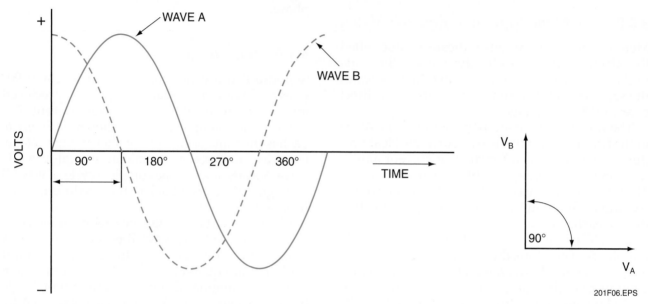

*Figure 6*  Voltage waveforms 90° out of phase.

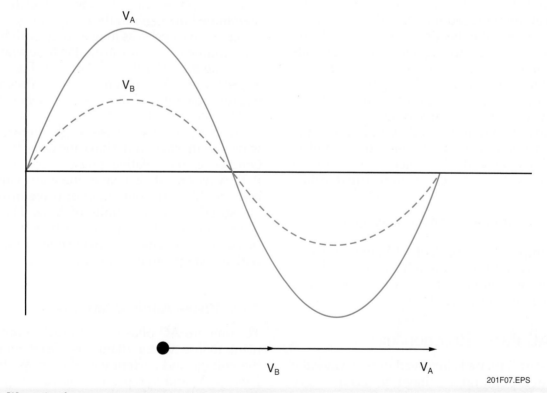

*Figure 7*  Waves in phase.

NCCER — *Contren® Learning Series*  26201-11

## Left-Hand Rule for Generators

Hand rules for generators and motors give direction to the basic principles of induction. For a generator, if you move a conductor through a magnetic field made up of flux lines, you will induce an EMF, which drives current through a conductor. The left-hand rule for generators will help you determine which direction the current will flow in the conductor. It states that if you hold the thumb, first, and middle fingers of the left hand at right angles to one another with the first finger pointing in the flux direction (from the north pole to the south pole), and the thumb pointing in the direction of motion of the conductor, the middle finger will point in the direction of the induced voltage (EMF). The polarity of the EMF determines the direction in which current will flow as a result of this induced EMF. The left-hand rule for generators is also called Fleming's first rule.

201SA01.EPS

## 5.0.0 Nonsinusoidal Waveforms

The sine wave is the basic waveform for AC variations for several reasons. This waveform is produced by a rotary generator, as the output is proportional to the angle of rotation. Because of its derivation from circular motion, any sine wave can be analyzed in angular measure, either in degrees from 0° to 360° or in radians from 0 to $2\pi$ radians.

In many electronic applications, however, other waveshapes are important. Any waveform that is not a sine (or cosine) wave is a nonsinusoidal

# RMS Amplitude

The root-mean-square (rms) value, also called the effective value, is the value assigned to an alternating voltage or current that results in the same power dissipation in a given resistance as DC voltage or current of the same numerical value. This is illustrated below. As shown, 120 (peak) VAC will not produce the same light (350 lumens versus 500 lumens) as 120VDC from a 60W lamp. In order to produce the same light (500 lumens), 120V rms must be applied to the lamp. This requires that the applied sinusoidal AC waveform have a peak voltage of about 170V (170V × 0.707V = 120V).

201SA02.EPS

waveform. Common examples are the square wave and sawtooth wave in *Figure 8*.

With nonsinusoidal waveforms for either voltage or current, there are important differences and similarities to consider. Note the following comparisons with sine waves:

- In all cases, the cycle is measured between two points having the same amplitude and varying in the same direction. The period is the time for one cycle.
- Peak amplitude is measured from the zero axis to the maximum positive or negative value.

However, peak-to-peak amplitude is better for measuring nonsinusoidal waveshapes because they can have asymmetrical peaks, as with the rectangular wave in *Figure 8*.

- The rms value 0.707 of peak applies only to sine waves, as this factor is derived from the sine values in the angular measure used only for the sine waveform.
- Phase angles apply only to sine waves, as angular measure is used only for sine waves. Note that the phase angle is indicated only on the sine wave of *Figure 8*.

Figure 8  AC waveforms.

201F08.EPS

Think About It

## Phase Angles

Why is the phase angle 90° in *Figure 6* and 0° in *Figure 7*? Why is the vector diagram in *Figure 7* shown as a straight line?

## 6.0.0 RESISTANCE IN AC CIRCUITS

An AC circuit has an AC voltage source. Note the circular symbol with the sine wave inside it shown in *Figure 9*. It is used for any source of sine

wave alternating voltage. This voltage connected across an external load resistance produces alternating current of the same waveform, frequency, and phase as the applied voltage.

According to Ohm's law, current (I) equals voltage (E) divided by resistance (R). When E is an rms value, I is also an rms value. For any instantaneous value of E during the cycle, the value of I is for the corresponding instant of time.

In an AC circuit with only resistance, the current variations are in phase with the applied voltage, as shown in *Figure 9*. This in-phase relationship between E and I means that such an AC circuit can be analyzed by the same methods used for DC circuits since there is no phase angle to

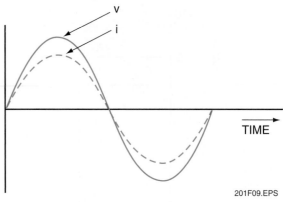

**Figure 9**  Resistive AC circuit.

consider. Components that have only resistance include resistors, the filaments for incandescent light bulbs, and vacuum tube heaters.

In purely resistive AC circuits, the voltage, current, and resistance are related by Ohm's law because the voltage and current are in phase.

$$I = \frac{E}{R}$$

Unless otherwise noted, the calculations in AC circuits are generally in rms values. For example, in *Figure 9*, the 120V applied across the 10Ω resistance $R_L$ produces an rms current of 12A. This is determined as follows:

$$I = \frac{E}{R_L} = \frac{120V}{10\Omega} = 12A$$

Furthermore, the rms power (true power) dissipation is I²R or:

$$P = (12A)^2 \times 10\Omega = 1{,}440W$$

*Figure 10* shows the relationship between voltage and current in purely resistive AC circuits. The voltage and current are in phase, their cycles

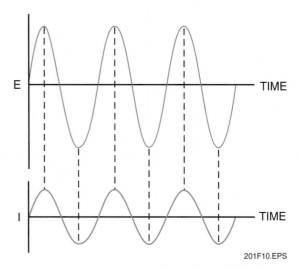

**Figure 10**  Voltage and current in a resistive AC circuit.

begin and end at the same time, and their peaks occur at the same time.

The value of the voltage shown in *Figure 10* depends on the applied voltage to the circuit. The value of the current depends on the applied voltage and the amount of resistance. If resistance is changed, it will affect only the magnitude of the current.

The total resistance in any AC circuit, whether it is a series, parallel, or series-parallel circuit, is calculated using the same rules that were learned and applied to DC circuits with resistance. Power computations are discussed later in this module.

### 7.0.0 INDUCTANCE IN AC CIRCUITS

Inductance is the characteristic of an electrical circuit that opposes the change of current flow. It is the result of the expanding and collapsing field caused by the changing current. This moving flux cuts across the conductor that is providing the current, producing induced voltage in the wire itself. Furthermore, any other conductor in the field, whether carrying current or not, is also

---

## *What's wrong with this picture?*

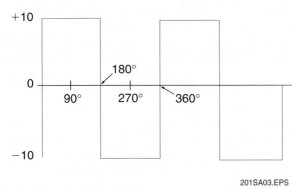

---

NCCER — *Contren® Learning Series*  26201-11

cut by the varying flux and has induced voltage. This induced current opposes the current flow that generated it.

In DC circuits, a change must be initiated in the circuit to cause inductance. The current must change to provide motion of the flux. A steady DC of 10A cannot produce any induced voltage as long as the current value is constant. A current of 1A changing to 2A does induce voltage. Also, the faster the current changes, the higher the induced voltage becomes, because when the flux moves at a higher speed it can induce more voltage.

However, in an AC circuit the current is continuously changing and producing induced voltage. Lower frequencies of AC require more inductance to produce the same amount of induced voltage as a higher frequency current. The current can have any waveform as long as the amplitude is changing.

The ability of a conductor to induce voltage in itself when the current changes is its **self-inductance**, or simply inductance. The symbol for inductance is L and its unit is the henry (H). One henry is the amount of inductance that allows one volt to be induced when the current changes at the rate of one ampere per second.

### 7.1.0 Factors Affecting Inductance

An inductor is a coil of wire that may be wound on a core of metal or paper, or it may be self-supporting. It may consist of turns of wire placed side by side to form a layer of wire over the core or coil form. The inductance of a coil or inductor depends on its physical construction. Some of the factors affecting inductance are:

- *Number of turns* – The greater the number of turns, the greater the inductance. In addition, the spacing of the turns on a coil also affects inductance. A coil that has widely-spaced turns has a lower inductance than one that has the same number of more closely-spaced turns. The reason for this higher inductance is that the closely-wound turns produce a more concentrated magnetic field, causing the coil to exhibit a greater inductance.
- *Coil diameter* – The inductance increases directly as the cross-sectional area of the coil increases.
- *Length of the core* – When the length of the core is decreased, the turn spacing is decreased, increasing the inductance of the coil.
- *Core material* – The core of the coil can be either a magnetic material (such as iron) or a non-magnetic material (such as paper or air). Coils wound on a magnetic core produce a stronger magnetic field than those with non-magnetic cores, giving them higher values of inductance. Air-core coils are used where small values of inductance are required.
- *Winding the coil in layers* – The more layers used to form a coil, the greater the effect the magnetic field has on the conductor. Layering a coil can increase the inductance.

Factors affecting the inductance of a coil can be seen in *Figure 11*.

| | LOW INDUCTANCE | HIGH INDUCTANCE |
|---|---|---|
| NUMBER OF TURNS | FEW TURNS | MANY TURNS |
| COIL DIAMETER | NARROW COIL | WIDE COIL |
| CORE LENGTH | LONG CORE | SHORT CORE |
| CORE MATERIAL | NON-MAGNETIC CORE | MAGNETIC CORE |

201F11.EPS

*Figure 11*   Factors affecting the inductance of a coil.

# Inductance in an AC Circuit

Can you name three commonly used electrical devices that insert inductance into an AC circuit?

## 7.2.0 Voltage and Current in an Inductive AC Circuit

The self-induced voltage across an inductance L is produced by a change in current with respect to time ($\Delta i/\Delta t$) and can be stated as:

$$V_L = L\frac{\Delta i}{\Delta t}$$

*Where:*

$$\Delta = \text{change}$$
$$V_L = \text{volts}$$
$$L = \text{henrys}$$
$$\Delta i/\Delta t = \text{amperes per second}$$

This gives the voltage in terms of how much magnetic flux is cut per second. When the magnetic flux associated with the current varies the same as I, this formula gives the same results for calculating induced voltage. Remember that the induced voltage across the coil is actually the result of inducing electrons to move in the conductor, so there is also an induced current.

For example, what is the self-induced voltage $V_L$ across a 4h inductance produced by a current change of 12A per second?

$$V_L = L\frac{\Delta i}{\Delta t}$$
$$V_L = 4h \times \frac{12A}{1}$$
$$V_L = 4 \times 12$$
$$V_L = 48V$$

The current through a 200 microhenry (µh) inductor changes from 0 to 200 milliamps (mA) in 2 microseconds (µsec). (The prefix micro [µ] means one-millionth.) What is the $V_L$?

$$V_L = L\frac{\Delta i}{\Delta t}$$
$$V_L = \left(200 \times 10^{-6}\right)\frac{200 \times 10^{-3}}{2 \times 10^{-6}}$$
$$V_L = 20V$$

The induced voltage is an actual voltage that can be measured, although $V_L$ is produced only while the current is changing. When $\Delta i/\Delta t$ is present for only a short time, $V_L$ is in the form of a voltage pulse. With a sine wave current that is always changing, $V_L$ is a sinusoidal voltage that is 90° out of phase with $I_L$.

The current that flows in an inductor is induced by the changing magnetic field that surrounds the inductor. This changing magnetic field is produced by an AC voltage source that is applied to the inductor. The magnitude and polarity of the induced current depend on the field strength, direction, and rate at which the field cuts the inductor windings. The overall effect is that the current is out of phase and lags the applied voltage by 90°.

At 270° in *Figure 12*, the applied electromotive force (EMF) is zero, but it is increasing in the positive direction at its greatest rate of change. Likewise, electron flow due to the applied EMF is also increasing at its greatest rate. As the electron flow increases, it produces a magnetic field that is building with it. The lines of flux cut the conductor as they move outward from it with the expanding field.

As the lines of flux cut the conductor, they induce a current into it. The induced current is at its maximum value because the lines of flux are expanding outward through the conductor at their greatest rate. The direction of the induced current is in opposition to the force that generated it. Therefore, at 270° the applied voltage is zero and is increasing to a positive value, while the current is at its maximum negative value.

At 0° in *Figure 12*, the applied voltage is at its maximum positive value, but its rate of change is zero. Therefore, the field it produces is no longer expanding and is not cutting the conductor. Because there is no relative motion between the field and conductor, no current is induced. Therefore, at 0° voltage is at its maximum positive value, while current is zero.

At 90° in *Figure 12*, voltage is once again zero, but this time it is decreasing toward negative at its greatest rate of change. Because the applied voltage is decreasing, the magnetic field is collapsing inward on the conductor. This has the effect of reversing the direction of motion between the field and conductor that existed at 0°.

Therefore, the current will flow in a direction opposite of what it was at 0°. Also, because the applied voltage is decreasing at its greatest rate, the field is collapsing at its greatest rate. This causes the flux to cut the conductor at the greatest rate, causing the induced current magnitude to be maximum. At 90°, the applied voltage is zero

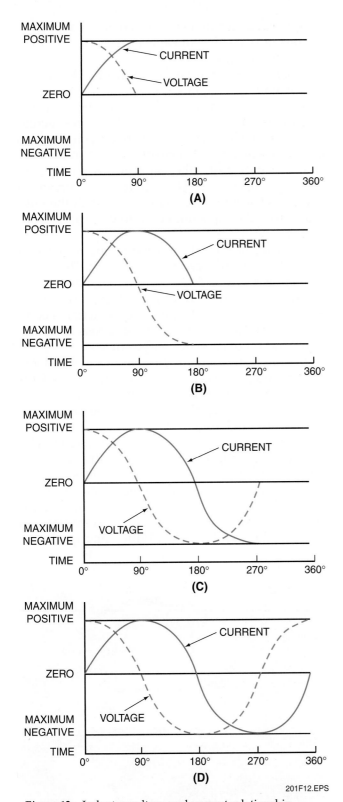

*Figure 12*   Inductor voltage and current relationship.

decreasing toward negative, while the current is maximum positive.

At 180° in *Figure 12*, the applied voltage is at its maximum negative value, but just as at 0°, its rate of change is zero. At 180°, therefore, current will be zero. This explanation shows that the voltage peaks positive first, then 90° later the current peaks positive. Current thus lags the applied voltage in an inductor by 90°. This can easily be remembered using the phrase "ELI the ICE man." ELI represents voltage (E), inductance (L), and current (I). In an inductor, the voltage leads the current just as the letter E leads or comes before the letter I. The word ICE will be explained in the section on capacitance.

## 7.3.0 Inductive Reactance

The opposing force that an inductor presents to the flow of alternating current cannot be called resistance since it is not the result of friction within a conductor. The name given to this force is inductive reactance because it is the reaction of the inductor to alternating current. Inductive reactance is measured in ohms and its symbol is $X_L$.

Remember that the induced voltage in a conductor is proportional to the rate at which magnetic lines of force cut the conductor. The greater the rate or higher the frequency, the greater the counter-electromotive force (CEMF). Also, the induced voltage increases with an increase in inductance; the more turns, the greater the CEMF. Reactance then increases with an increase of frequency and with an increase in inductance. The formula for inductive reactance is as follows:

$$X_L = 2\pi fL$$

*Where:*

$X_L$ = inductive reactance in ohms

$2\pi$ = a constant in which the Greek letter pi (π) represents 3.14 and 2 × pi = 6.28

f = frequency of the alternating current in hertz

L = inductance in henrys

For example, if f is equal to 60Hz and L is equal to 20h, find $X_L$:

$$X_L = 2\pi fL$$
$$X_L = 6.28 \times 60Hz \times 20h$$
$$X_L = 7,536\Omega$$

Once calculated, the value of $X_L$ is used like resistance in a form of Ohm's law:

$$I = \frac{E}{X_L}$$

*Where:*

I = effective current (amps)
E = effective voltage (volts)
$X_L$ = inductive reactance (ohms)

Unlike a resistor, there is no power dissipation in an ideal inductor. An inductor limits current, but it uses no net energy since the energy required to build up the field in the inductor is given back to the circuit when the field collapses.

## 8.0.0 CAPACITANCE

A capacitor is a device that stores an electric charge in a dielectric material. Capacitance is the ability to store a charge. In storing a charge, a capacitor opposes a change in voltage. *Figure 13* shows a simple capacitor in a circuit, schematic representations of two types of capacitors, and a photo of common capacitors.

*Figure 14(A)* shows a capacitor in a DC circuit. When voltage is applied, the capacitor begins to charge, as shown in *Figure 14(B)*. The charging continues until the potential difference across the capacitor is equal to the applied voltage. This charging current is transient or temporary since it flows only until the capacitor is charged to the applied voltage. Then there is no current in the circuit. *Figure 14(C)* shows this with the voltage across the capacitor equal to the battery voltage or 10V.

The capacitor can be discharged by connecting a conducting path across the dielectric. The stored charge across the dielectric provides the potential difference to produce a discharge current, as shown in *Figure 14(D)*. Once the capacitor is completely discharged, the voltage across it equals zero, and there is no discharge current.

In a capacitive circuit, the charge and discharge current must always be in opposite directions. Current flows in one direction to charge the capacitor and in the opposite direction when the capacitor is allowed to discharge.

Current will flow in a capacitive circuit with AC voltage applied because of the capacitor charge and discharge current. There is no current through the dielectric, which is an insulator. While the capacitor is being charged by increasing applied voltage, the charging current flows in one direction to the plates. While the capacitor is

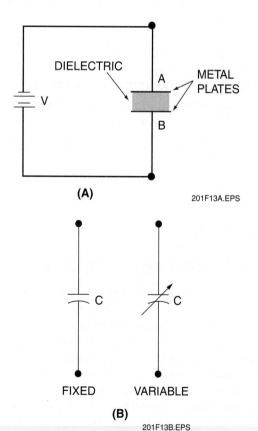

(A)

201F13A.EPS

C

C

FIXED          VARIABLE

(B)

201F13B.EPS

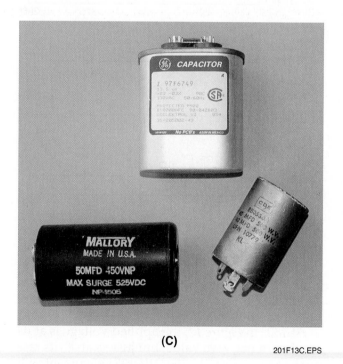

(C)

201F13C.EPS

*Figure 13*  Capacitors.

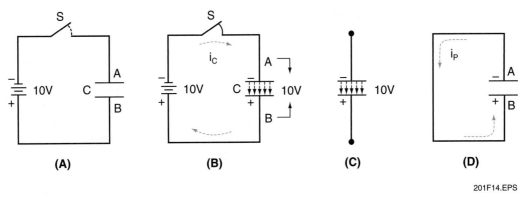

*Figure 14*  Charging and discharging a capacitor.

201F14.EPS

discharging as the applied voltage decreases, the discharge current flows in the reverse direction. With alternating voltage applied, the capacitor alternately charges and discharges.

First, the capacitor is charged in one polarity, and then it discharges; next, the capacitor is charged in the opposite polarity, and then it discharges again. The cycles of charge and discharge current provide alternating current in the circuit at the same frequency as the applied voltage. The amount of capacitance in the circuit will determine how much current is allowed to flow.

Capacitance is measured in farads (F), where one farad is the capacitance when one coulomb is stored in the dielectric with a potential difference of one volt. Smaller values are measured in microfarads (µF). A small capacitance will allow less charge and discharge current to flow than a larger capacitance. The smaller capacitor has more opposition to alternating current, because less current flows with the same applied voltage.

In summary, capacitance exhibits the following characteristics:

- DC is blocked by a capacitor. Once charged, no current will flow in the circuit.
- AC flows in a capacitive circuit with AC voltage applied.
- A smaller capacitance allows less current.

### 8.1.0 Factors Affecting Capacitance

A capacitor consists of two conductors separated by an insulating material called a dielectric. There are many types and sizes of capacitors with different dielectric materials. The capacitance of a capacitor is determined by three factors:

- *Area of the plates* – The initial charge displacement on a set of capacitor plates is related to the number of free electrons in each plate. Larger plates will produce a greater capacitance than smaller ones. Therefore, the capacitance of a

capacitor varies directly with the area of the plates. For example, if the area of the plates is doubled, the capacitance is doubled. If the size of the plates is reduced by 50%, the capacitance would also be reduced by 50%.

- *Distance between plates* – As two capacitor plates are brought closer together, more electrons will move away from the positively charged plate and move into the negatively charged plate. This is because the mutual attraction between the opposite charges on the plates increases as the plates move closer together. This added movement of charge is an increase in the capacitance of the capacitor. In a capacitor composed of two plates of equal area, the capacitance varies inversely with the distance between the plates. For example, if the distance between the plates is decreased by one-half, the capacitance will be doubled. If the distance between the plates is doubled, the capacitance would be one-half as great.
- *Dielectric permittivity* – Another factor that determines the value of capacitance is the permittivity of the dielectric. The dielectric is the material between the capacitor plates in which the electric field appears. Relative permittivity expresses the ratio of the electric field strength in a dielectric to that in a vacuum. Permittivity has nothing to do with the dielectric strength of the medium or the breakdown voltage. An insulating material that will withstand a higher applied voltage than some other substance does not always have a higher dielectric permittivity. Many insulating materials have a greater dielectric permittivity than air. For a given applied voltage, a greater attraction exists between the opposite charges on the capacitor plates, and an electric field can be set up more easily than when the dielectric is air. The capacitance of the capacitor is increased when the permittivity of the dielectric is increased if all the other parameters remain unchanged.

# Capacitance

The concept of capacitance, like many electrical quantities, is often hard to visualize or understand. A comparison with a balloon may help to make this concept clearer. Electrical capacitance has a charging effect similar to blowing up a balloon and holding it closed. The expansion capacity of the balloon can be changed by changing the thickness of the balloon walls. A balloon with thick walls will expand less (have less capacity) than one with thin walls. This is like a small 10µF capacitor that has less capacity and will charge less than a larger 100µF capacitor.

## 8.2.0 Calculating Equivalent Capacitance

Connecting capacitors in parallel is equivalent to adding the plate areas. Therefore, the total capacitance is the sum of the individual capacitances, as illustrated in *Figure 15*.

A 10µF capacitor in parallel with a 5µF capacitor, for example, provides a 15µF capacitance for the parallel combination. The voltage is the same across the parallel capacitors. Note that adding parallel capacitance is opposite to the case of inductances in parallel and resistances in parallel.

Connecting capacitances in series is equivalent to increasing the thickness of the dielectric. Therefore, the combined capacitance is less than the smallest individual value. The combined equivalent capacitance is calculated by the reciprocal formula, as shown in *Figure 16*.

Capacitors connected in series are combined like resistors in parallel. Any of the shortcut calculations for the reciprocal formula apply. For

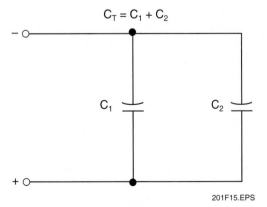

$$C_T = C_1 + C_2$$

*Figure 15*   Capacitors in parallel.

$$C_T = \frac{1}{\frac{1}{C_1} + \frac{1}{C_2}}$$

*Figure 16*   Capacitors in series.

example, the combined capacitance of two equal capacitances of 10µF in series is 5µF.

Capacitors are used in series to provide a higher voltage breakdown rating for the combination. For instance, each of three equal capacitances in series has one-third the applied voltage.

In series, the voltage across each capacitor is inversely proportional to its capacitance. The smaller capacitance has the larger proportion of the applied voltage. The reason is that the series capacitances all have the same charge because they are in one current path. With equal charge, a smaller capacitance has a greater potential difference.

## 8.3.0 Capacitor Specifications

This specifies the maximum potential difference that can be applied across the plates without puncturing the dielectric.

### 8.3.1 Voltage Rating

Usually, the voltage rating is for temperatures up to about 60°C. High temperatures result in a lower voltage rating. Voltage ratings for general-purpose paper, mica, and ceramic capacitors are typically 200V to 500V. Ceramic capacitors with ratings of 1 to 5kV are also available.

Electrolytic capacitors are commonly used in 25V, 150V, and 450V ratings. In addition, 6V and 10V electrolytic capacitors are often used in transistor circuits. For applications where a lower voltage rating is permissible, more capacitance can be obtained in a smaller physical size.

The potential difference across the capacitor depends on the applied voltage and is not necessarily equal to the voltage rating. A voltage rating higher than the potential difference applied across the capacitor provides a safety factor for long life in service. With electrolytic capacitors, however, the actual capacitor voltage should be close to the rated voltage to produce the oxide film that provides the specified capacitance.

# Capacitance

Suppose you had a motor with a bad 30μF starting capacitor and no 30μF direct replacement capacitor was available. As a temporary measure, you are authorized to substitute two equal-value capacitors in its place. What size capacitors (μF) should be used if you are connecting them in parallel?

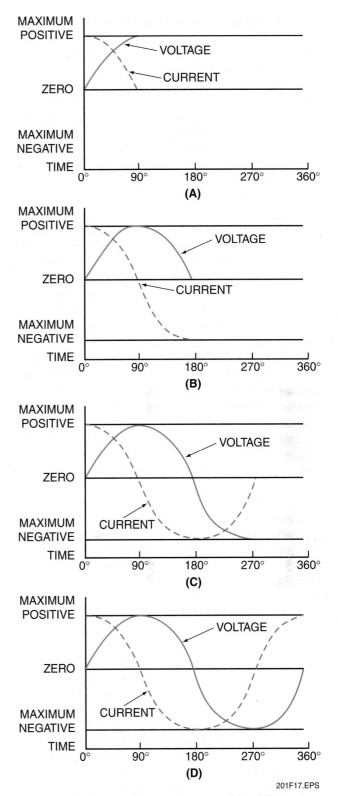

**Figure 17** Voltage and current in a capacitive AC circuit.

201F17.EPS

The voltage ratings are for applied DC voltage. The breakdown rating is lower for AC voltage because of the internal heat produced by continuous charge and discharge.

## 8.3.2 Leak Resistance

Consider a capacitor charged by a DC voltage source. After the charging voltage is removed, a perfect capacitor would keep its charge indefinitely. After a long period of time, however, the charge will be neutralized by a small leakage current through the dielectric and across the insulated case between terminals, because there is no perfect insulator. For paper, ceramic, and mica capacitors, the leakage current is very slight, or inversely, the leakage resistance is very high. For paper, ceramic, or mica capacitors, $R_l$ is 100MΩ or more. However, electrolytic capacitors may have a leakage resistance of 0.5MΩ or less.

## 8.4.0 Voltage and Current in a Capacitive AC Circuit

In a capacitive circuit driven by an AC voltage source, the voltage is continuously changing. Thus, the charge on the capacitor is also continuously changing. The four parts of *Figure 17* show the variation of the alternating voltage and current in a capacitive circuit for each quarter of one cycle.

The solid line represents the voltage across the capacitor, and the dotted line represents the current. The line running through the center is the zero or reference point for both the voltage and the current. The bottom line marks off the time of the cycle in terms of electric degrees. Assume that the AC voltage has been acting on the capacitor for some time before the time represented by the starting point of the sine wave.

At the beginning of the first quarter-cycle (0° to 90°), the voltage has just passed through zero and is increasing in the positive direction. Since the

zero point is the steepest part of the sine wave, the voltage is changing at its greatest rate.

The charge on a capacitor varies directly with the voltage; therefore, the charge on the capacitor is also changing at its greatest rate at the beginning of the first quarter-cycle. In other words, the greatest number of electrons are moving off one plate and onto the other plate. Thus, the capacitor current is at its maximum value.

As the voltage proceeds toward maximum at 90°, its rate of change becomes lower and lower, making the current decrease toward zero. At 90°, the voltage across the capacitor is maximum, the capacitor is fully charged, and there is no further movement of electrons from plate to plate. That is why the current at 90° is zero.

At the end of the first quarter-cycle, the alternating voltage stops increasing in the positive direction and starts to decrease. It is still a positive voltage; but to the capacitor, the decrease in voltage means that the plate that has just accumulated an excess of electrons must lose some electrons. The current flow must reverse its direction. The second part of the figure shows the current curve to be below the zero line (negative current direction) during the second quarter-cycle (90° to 180°).

At 180°, the voltage has dropped to zero. This means that, for a brief instant, the electrons are equally distributed between the two plates; the current is maximum because the rate of change of voltage is maximum.

Just after 180°, the voltage has reversed polarity and starts building to its maximum negative peak, which is reached at the end of the third quarter-cycle (180° to 270°). During the third quarter-cycle, the rate of voltage change gradually decreases as the charge builds to a maximum at 270°. At this point, the capacitor is fully charged and carries the full impressed voltage. Because the capacitor is fully charged, there is no further exchange of electrons and the current flow is zero at this point. The conditions are exactly the same as at the end of the first quarter-cycle (90°), but the polarity is reversed.

Just after 270°, the impressed voltage once again starts to decrease, and the capacitor must lose electrons from the negative plate. It must discharge, starting at a minimum rate of flow and rising to a maximum. This discharging action continues through the last quarter-cycle (270° to 360°) until the impressed voltage has reached zero. The beginning of the entire cycle is 360°, and everything starts over again.

In *Figure 17*, note that the current always arrives at a certain point in the cycle 90° ahead of the voltage because of the charging and discharging action. This voltage-current phase relationship in a capacitive circuit is exactly opposite to that in an inductive circuit. The current through a capacitor leads the voltage across the capacitor by 90°. A convenient way to remember this is the phrase "ELI the ICE man" (ELI refers to inductors, as

previously explained). ICE pertains to capacitors as follows:

$$I = current$$
$$C = capacitor$$
$$E = voltage$$

In capacitors (C), current (I) leads voltage (E) by 90°.

It is important to realize that the current and voltage are both going through their individual cycles at the same time during the period the AC voltage is impressed. The current does not go through part of its cycle (charging or discharging) and then stop and wait for the voltage to catch up. The amplitude and polarity of the voltage and the amplitude and direction of the current are continually changing.

Their positions, with respect to each other and to the zero line at any electrical instant or any degree between 0° and 360°, can be seen by reading upward from the time-degree line. The current swing from the positive peak at 0° to the negative peak at 180° is not a measure of the number of electrons or the charge on the plates. It is a picture of the direction and strength of the current in relation to the polarity and strength of the voltage appearing across the plates.

## 8.5.0 Capacitive Reactance

Capacitors offer a very real opposition to current flow. This opposition arises from the fact that, at a given voltage and frequency, the number of electrons that go back and forth from plate to plate is limited by the storage ability (capacitance) of the capacitor. As the capacitance is increased, a greater number of electrons changes plates every cycle. Since current is a measure of the number of electrons passing a given point in a given time, the current is increased.

Increasing the frequency will also decrease the opposition offered by a capacitor. This occurs because the number of electrons that the capacitor is capable of handling at a given voltage will change plates more often. As a result, more electrons will pass a given point in a given time (greater current flow). The opposition that a capacitor offers to AC is therefore inversely proportional to frequency and capacitance. This opposition is called capacitive reactance. Capacitive reactance decreases with increasing frequency or, for a given frequency, the capacitive reactance decreases with increasing capacitance. The symbol for capacitive reactance is $X_C$. The formula is:

$$X_C = \frac{1}{2\pi fC}$$

*Where:*

$X_c$ = capacitive reactance in ohms

f = frequency in hertz

C = capacitance in farads

$2\pi$ = 6.28 (2 × 3.14)

For example, what is the capacitive reactance of a 0.05µF capacitor in a circuit whose frequency is 1 megahertz?

$$X_c = \frac{1}{2\pi fC} = \frac{1}{(6.28)(10^6 \text{ hertz})(5 \times 10^{-8} \text{ farads})}$$

$$X_c = \frac{1}{3.14 \times 10^{-1}} = \frac{1}{0.314} = 3.18 \text{ ohms}$$

The capacitive reactance of a 0.05µF capacitor operated at a frequency of 1 megahertz is 3.18 ohms. Suppose this same capacitor is operated at a lower frequency of 1,500 hertz instead of 1 megahertz. What is the capacitive reactance now? Substituting where $1{,}500 = 1.5 \times 10^3$ hertz:

$$X_c = \frac{1}{2\pi fC} = \frac{1}{(6.28)(1.5 \times 10^3 \text{ hertz})(5 \times 10^{-8} \text{ farads})}$$

$$X_c = \frac{1}{4.71 \times 10^{-4}} = 2{,}123 \text{ ohms}$$

---

---

Note a very interesting point from these two examples. As frequency is decreased from 1 megahertz to 1,500 hertz, the capacitive reactance increases from 3.18 ohms to 2,123 ohms. Capacitive reactance increases as the frequency decreases.

## 9.0.0 LC AND RLC CIRCUITS

AC circuits often contain inductors, capacitors, and/or resistors connected in series or parallel combinations. When this is done, it is important to determine the resulting phase relationship between the applied voltage and the current in the circuit. The simplest method of combining factors that have different phase relationships is vector addition with the trigonometric functions. Each quantity is represented as a vector, and the resultant vector and phase angle are then calculated.

In purely resistive circuits, the voltage and current are in phase. In inductive circuits, the voltage leads the current by 90°. In capacitive circuits, the current leads the voltage by 90°. *Figure 18* shows the phase relationships of these components used in AC circuits. Recall that these characteristics are summarized by the phrase "ELI the ICE man."

ELI = *E* Leads *I* (inductive)

ICE = *I* Capacitive (leads) *E*

The impedance Z of a circuit is defined as the total opposition to current flow. The magnitude of the impedance Z is given by the following equation in a series circuit:

$$Z = \sqrt{R^2 + X^2}$$

*Where:*

Z = impedance (ohms)

R = resistance (ohms)

X = net reactance (ohms)

The current through a resistance is always in phase with the voltage applied to it; thus resistance is shown along the 0° axis. The voltage across an inductor leads the current by 90°; thus inductive reactance is shown along the 90° axis. The voltage across a capacitor lags the current by 90°; thus capacitive reactance is shown along the −90° axis. The net reactance is the difference between the inductive reactance and the capacitive reactance:

X = net reactance (ohms)

$X_L$ = inductive reactance (ohms)

$X_C$ = capacitive reactance (ohms)

---

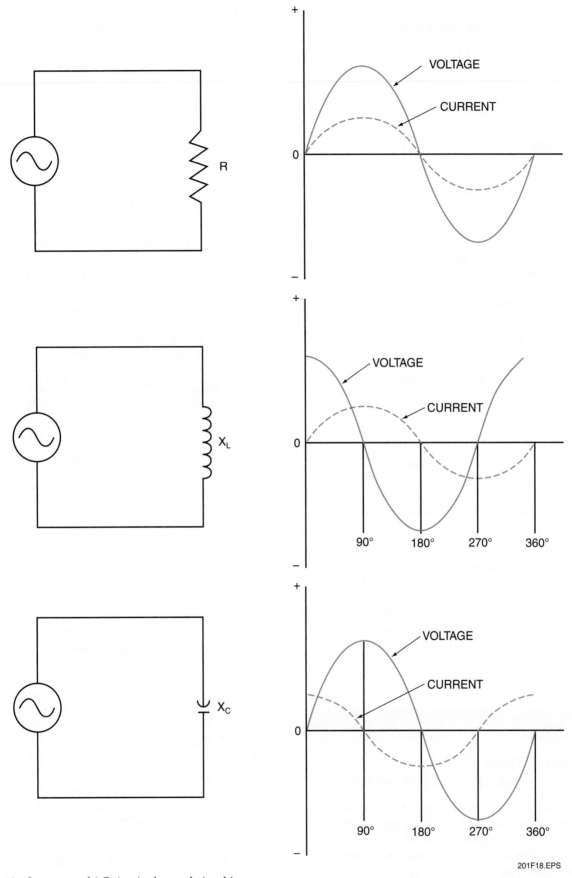

*Figure 18*  Summary of AC circuit phase relationships.

201F18.EPS

The impedance Z is the vector sum of the resistance R and the net reactance X. The angle, called the phase angle, gives the phase relationship between the applied voltage and current.

## 9.1.0 RL Circuits

RL circuits combine resistors and inductors in a series, parallel, or series-parallel configuration. In a pure inductive circuit, the current lags the voltage by an angle of 90°. In a circuit containing both resistance and inductance, the current will lag the voltage by some angle between zero and 90°.

### 9.1.1 Series RL Circuit

*Figure 19* shows a series RL circuit. Since it is a series circuit, the current is the same in all portions of the loop. Using the values shown, the circuit will be analyzed for unknown values such as $X_L$, Z, I, $E_L$, and $E_R$.

The solution would be worked as follows:

*Step 1* Compute the value of $X_L$.

$$X_L = 2\pi fL$$
$$X_L = 6.28 \times 100 \times 4 = 2,512 \text{ ohms}$$

*Step 2* Draw vectors R and $X_L$ as shown in *Figure 19*. R is drawn horizontally because the circuit current and voltage across R are in phase. It therefore becomes the reference line from which other angles are measured. $X_L$ is drawn upward at 90° from R because voltage across $X_L$ leads circuit current through R.

*Step 3* Compute the value of circuit impedance Z, which is equal to the vector sum of $X_L$ and R.

$$\tan = \frac{X_L}{R} = \frac{2,512}{1,500} = 1.67$$
$$\text{arctan } 1.67 = 59.1°$$
$$\cos 59.1° = 0.5135$$

Find Z using the cosine function:

$$\cos = \frac{R}{Z}$$
$$Z = \frac{R}{\cos}$$
$$Z = \frac{1,500}{0.5135} = 2,921 \text{ ohms}$$

*Step 4* Compute the circuit current using Ohm's law for AC circuits.

$$I = \frac{E}{Z} = \frac{100V}{2,921\Omega} = 0.034A$$

*Step 5* Compute voltage drops in the circuit.

$$E_L = IX_L = 0.034 \times 2,512 = 85 \text{ volts}$$
$$E_R = IR = 0.034 \times 1,500 = 51 \text{ volts}$$

Note that the voltage drops across the resistor and inductor do not equal the supply voltage because they must be added vectorially. This would be done as follows (because of rounding, numbers are not exact):

$$\tan = \frac{E_L}{E_R} = \frac{85}{51} = 1.67$$
$$\text{arctan } 1.67 = 59.1°$$
$$\cos = \frac{E_R}{E_Z} \qquad E_Z = \frac{E_R}{\cos}$$
$$E_Z = \frac{51}{0.5135} = \text{approx. } 100V = E_S$$

*Figure 19* Series RL circuit and vector diagram.

# Vector Analysis

When using vector analysis, the horizontal line is the in-phase value and the vertical line pointing up represents the leading value. The vertical line pointing down represents the lagging value.

In this inductive circuit, the current lags the applied voltage by an angle equal to 59.1°.

**NOTE**

You could also use the Pythagorean theorem (discussed later) to find this answer.

*Figure 20* shows another series RL circuit, its associated waveforms, and vector diagrams. This circuit is used to summarize the characteristics of a series RL circuit:

- The current I flows through all the series components.
- The voltage across $X_L$, labeled $V_L$, can be considered an $IX_L$ voltage drop, just as $V_R$ is used for an IR voltage drop.
- The current I through $X_L$ must lag $V_L$ by 90°, as this is the angle between current through an inductance and its self-induced voltage.
- The current I through R and its IR voltage drop have the same phase. There is no reactance to

sine wave current in any resistance. Therefore, I and IR have the same phase, or this phase angle is 0°.
- $V_T$ is the vector sum of the two out-of-phase voltages $V_R$ and $V_L$.
- Circuit current I lags $V_T$ by the phase angle.
- Circuit impedance is the vector sum of R and $X_L$.

In a series circuit, the higher the value of $X_L$ compared with R, the more inductive the circuit is. This means there is more voltage drop across the inductive reactance, and the phase angle increases toward 90°. The series current lags the applied generator voltage.

Several combinations of $X_L$ and R in series are listed in *Table 1* with their resultant impedance and phase angle. Note that a ratio of 10:1 or more for $X_L/R$ means that the circuit is practically all inductive. The phase angle of 84.3° is only slightly less than 90° for the ratio of 10:1, and the total impedance Z is approximately equal to $X_L$. The voltage drop across $X_L$ in the series circuit will be equal to the applied voltage, with almost none across R.

**Table 1** Series R and $X_L$ Combinations

| R (Ω) | $X_L$ (Ω) | Z (Ω) (Approx.) | Phase Angle (θ) (°) |
|---|---|---|---|
| 1 | 10 | $\sqrt{101} = 10$ | 84.3° |
| 10 | 10 | $\sqrt{200} = 14$ | 45° |
| 10 | 1 | $\sqrt{101} = 10$ | 5.7° |

*Figure 20* Series RL circuit with waveforms and vector diagram.

At the opposite extreme, when R is 10 times as large as $X_L$, the series circuit is mainly resistive. The phase angle of 5.7° means the current has almost the same phase as the applied voltage, the total impedance Z is approximately equal to R, and the voltage drop across R is practically equal to the applied voltage, with almost none across $X_L$.

### 9.1.2 Parallel RL Circuit

In a parallel RL circuit, the resistance and inductance are connected in parallel across a voltage source. Such a circuit thus has a resistive branch and an inductive branch.

The 90° phase angle must be considered for each of the branch currents, instead of voltage drops in a series circuit. Remember that any series circuit has different voltage drops, but one common current. A parallel circuit has different branch currents, but one common voltage.

In the parallel circuit in *Figure 21*, the applied voltage $V_A$ is the same across $X_L$, R, and the generator, since they are all in parallel. There cannot be any phase difference between these voltages. Each branch, however, has its individual current. For the resistive branch $I_R = V_A/R$; in the inductive branch $I_L = V_A/X_L$.

The resistive branch current $I_R$ has the same phase as the generator voltage $V_A$. The inductive branch current $I_L$ lags $V_A$, however, because the current in an inductance lags the voltage across it by 90°.

The total line current, therefore, consists of $I_R$ and $I_L$, which are 90° out of phase with each other. The phasor sum of $I_R$ and $I_L$ equals the total line current $I_T$. These phase relations are shown by the waveforms and vectors in *Figure 21*. $I_T$ will lag $V_A$ by some phase angle that results from the vector addition of $I_R$ and $I_L$.

The impedance of a parallel RL circuit is the total opposition to current flow by the R of the resistive branch and the $X_L$ of the inductive branch. Since $X_L$ and R are vector quantities, they must be added vectorially.

If the line current and the applied voltage are known, Z can also be calculated by the equation:

$$Z = \frac{V_A}{I_{Line}}$$

The Z of a parallel RL circuit is always less than the R or $X_L$ of any one branch. The branch of a parallel RL circuit that offers the most opposition to current flow has the lesser effect on the phase angle of the current.

Several combinations of $X_L$ and R in parallel are listed in *Table 2*. When $X_L$ is 10 times R, the parallel circuit is practically resistive because there is little inductive current in the line. The small value of $I_L$ results from the high $X_L$. The total impedance

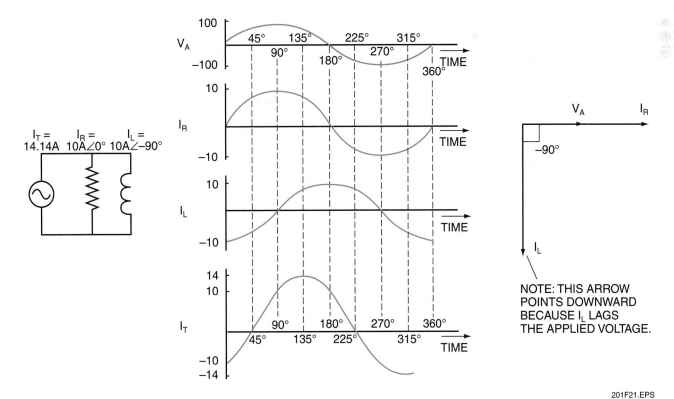

*Figure 21*  Parallel RL circuit with waveforms and vector diagram.

**Table 2**  Parallel R and X_L Combinations

| R ($\Omega$) | X$_L$ ($\Omega$) | I$_R$ (A) | I$_L$ (A) | I$_T$ (A) (Approx.) | Z$_T$ = V$_A$/I$_T$ ($\Omega$) | Phase Angle ($\theta$)$_I$ (°) |
|---|---|---|---|---|---|---|
| 1 | 10 | 10 | 1 | $\sqrt{101}$ = 10 | 1 | −5.7° |
| 10 | 10 | 1 | 1 | $\sqrt{2}$ = 1.4 | 7.07 | −45° |
| 10 | 1 | 1 | 10 | $\sqrt{101}$ = 10 | 1 | −84.3° |

of the parallel circuit is approximately equal to the resistance then, since the high value of X$_L$ in a parallel branch has little effect. The phase angle of −5.7° is practically 0° because almost all the line current is resistive.

As X$_L$ becomes smaller, it provides more inductive current in the main line. When X$_L$ is $\frac{1}{10}$R, practically all the line current is the I$_L$ component. Then, the parallel circuit is practically all inductive, with a total impedance practically equal to X$_L$. The phase angle of −84.3° is almost −90° because the line current is mostly inductive. Note that these conditions are opposite from the case of X$_L$ and R in series.

### 9.2.0 RC Circuits

In a circuit containing resistance only, the current and voltage are in phase. In a circuit of pure capacitance, the current leads the voltage by an angle of 90°. In a circuit that has both resistance and capacitance, the current will lead the voltage by some angle between 0° and 90°.

### 9.2.1 Series RC Circuit

*Figure 22A* shows a series RC circuit with resistance R in series with capacitive reactance X$_C$. Current I is the same in X$_C$ and R since they are in series. Each has its own series voltage drop, equal to IR for the resistance and IX$_C$ for the reactance.

In *Figure 22B*, the phasor is shown horizontal as the reference phase, because I is the same

throughout the series circuit. The resistive voltage drop IR has the same phase as I. The capacitor voltage IX$_C$ must be 90° clockwise from I and IR, as the capacitive voltage lags. Note that the IX$_C$ phasor is downward, exactly opposite from an IX$_L$ phasor, because of the opposite phase angle.

If the capacitive reactance alone is considered, its voltage drop lags the series current I by 90°. The IR voltage has the same phase as I, however, because resistance provides no phase shift. Therefore, R and X$_C$ combined in series must be added by vectors because they are 90° out of phase with each other, as shown in *Figure 22C*.

As with inductive reactance, $\theta$ (theta) is the phase angle between the generator voltage and its series current. As shown in *Figure 22B* and *Figure 22C*, $\theta$ can be calculated from the voltage or impedance triangle.

With series X$_C$ the phase angle is negative, clockwise from the zero reference angle of I because the X$_C$ voltage lags its current. To indicate the negative phase angle, this 90° phasor points downward from the horizontal reference, instead of upward as with the series inductive reactance.

In series, the higher the X$_C$ compared with R, the more capacitive the circuit. There is more voltage drop across the capacitive reactance, and the phase angle increases toward −90°. The series X$_C$ always makes the current lead the applied voltage. With all X$_C$ and no R, the entire applied voltage is across X$_C$ and equals −90°. Several combinations of X$_C$ and R in series are listed in *Table 3*.

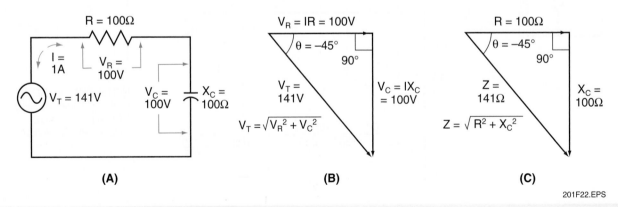

**(A)**          **(B)**          **(C)**

201F22.EPS

*Figure 22*  Series RC circuit with vector diagrams.

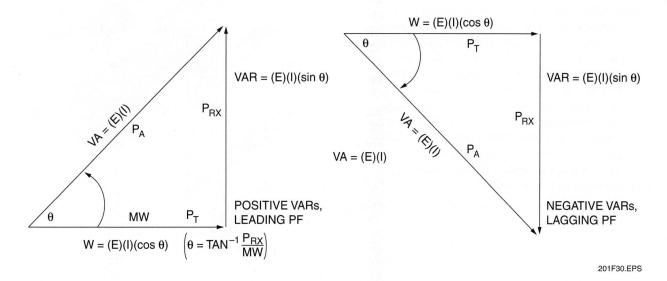

$$P_A = \sqrt{(P_T{}^2) + (P_{RX}{}^2)}$$

$$P_T = \sqrt{(P_A{}^2) - (P_{RX}{}^2)}$$

$$P_{RX} = \sqrt{(P_A{}^2) + (P_T{}^2)}$$

201F30.EPS

*Figure 30*  Power triangle.

# 11.0.0 TRANSFORMERS

A transformer is a device that transfers electrical energy from one circuit to another by electromagnetic induction (transformer action). The electrical energy is always transferred without a change in frequency, but may involve changes in the magnitudes of voltage and current. Because a transformer works on the principle of electromagnetic induction, it must be used with an input source voltage that varies in amplitude.

## 11.1.0 Transformer Construction

*Figure 31* shows the basic components of a transformer. In its most basic form, a transformer consists of:

- A primary coil or winding
- A secondary coil or winding
- A core that supports the coils or windings

A simple transformer action is shown in *Figure 32*. The primary winding is connected to a 60Hz AC voltage source. The magnetic field or flux builds up (expands) and collapses (contracts)

around the primary winding. The expanding and contracting magnetic field around the primary winding cuts the secondary winding and induces an alternating voltage into the winding. This voltage causes AC to flow through the load. The voltage may be stepped up or down depending on the design of the primary and secondary windings.

*Figure 31*  Basic components of a transformer.

# Power Factor

In power distribution circuits, it is desirable to achieve a power factor approaching a value of 1 in order to obtain the most efficient transfer of power. In AC circuits where there are large inductive loads such as in motors and transformers, the power factor can be considerably less than 1. For example, in a highly inductive motor circuit, if the voltage is 120V, the current is 12A, and the current lags the voltage by 60°, the power factor is 0.5 or 50% (cosine of 60° = 0.5). The apparent power is 1,440VA (120V × 12A), but the true power is only 720W [120V × (0.5 × 12A) = 720W]. This is a very inefficient circuit. What would you do to this circuit in order to achieve a circuit having a power factor as close to 1 as possible?

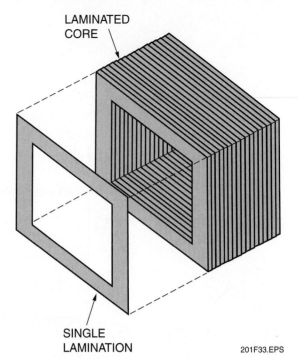

*Figure 33*   Steel laminated core.

*Figure 32*   Transformer action.

### 11.1.1 Core Characteristics

Commonly used core materials are air, soft iron, and steel. Each of these materials is suitable for particular applications and unsuitable for others. Generally, air-core transformers are used when the voltage source has a high frequency (above 20kHz). Iron-core transformers are usually used when the source frequency is low (below 20kHz). A soft-iron transformer is very useful where the transformer must be physically small yet efficient. The iron-core transformer provides better power transfer than the air-core transformer. Laminated sheets of steel are often used in a transformer to reduce one type of power loss known as eddy currents. These are undesirable currents, induced into the core, which circulate around the core. Laminating the core reduces these currents to smaller levels. These steel laminations are insulated with a nonconducting material, such as varnish, and then formed into a core as shown in *Figure 33*. It

takes about 50 such laminations to make a core one inch thick. The most efficient transformer core is one that offers the best path for the most lines of flux, with the least loss in magnetic and electrical energy.

### 11.1.2 Transformer Windings

A transformer consists of two coils called windings, which are wrapped around a core. The transformer operates when a source of AC voltage is connected to one of the windings and a load device is connected to the other. The winding that is connected to the source is called the primary winding. The winding that is connected to the load is called the secondary winding. *Figure 34* shows a cutaway view of a typical transformer.

The wire is coated with varnish so that each turn of the winding is insulated from every other turn. In a transformer designed for high-voltage applications, sheets of insulating material such as paper are placed between the layers of windings to provide additional insulation.

When the primary winding is completely wound, it is wrapped in insulating paper or cloth. The secondary winding is then wound on top of the primary winding. After the secondary winding is complete, it too is covered with insulating paper. Next, the iron core is inserted into and around the windings as shown.

Sometimes, terminals may be provided on the enclosure for connections to the windings.

*Figure 34* Cutaway view of a transformer core.

LAMINATED CORE

LEADS

PAPER INSULATION

PRIMARY WINDING

SECONDARY WINDING

201F34.EPS

# Transformers

Transformers are essential to all electrical systems and all types of electronic equipment. They are especially essential to the operation of AC high-voltage power distribution systems. Transformers are used to both step up voltage and step down voltage throughout the distribution process. For example, a typical power generation plant might generate AC power at 13,800V, step it up to 230,000V for distribution over long transmission lines, step it down to 13,800V again at substations located at different points for local distribution, and finally step it down again to 240V and 120V for lighting and local power use.

The figure shows four leads, two from the primary and two from the secondary. These leads are to be connected to the source and load, respectively.

## 11.2.0 Operating Characteristics

The operating characteristics of transformers are determined by the applied voltage and the winding design. These affect both the exciting current (no-load condition) and the phase relationship between the windings during transformer operation.

### 11.2.1 Energized with No Load

A no-load condition is said to exist when a voltage is applied to the primary, but no load is connected to the secondary. Assume the output of the secondary is connected to a load by a switch that is open. Because of the open switch, there is no current flowing in the secondary winding. With the switch open and an AC voltage applied to the primary, there is, however, a very small amount of current, called exciting current, flowing in the primary. Essentially, what this current does is excite the coil of the primary to create a magnetic field. The amount of exciting current is determined by three factors: the amount of voltage applied ($E_A$); the resistance (R) of the primary coil's wire and core losses; and the $X_L$, which is dependent on the frequency of the exciting current. These factors are all controlled by transformer design.

This very small amount of exciting current serves two functions:

- Most of the exciting energy is used to support the magnetic field of the primary.
- A small amount of energy is used to overcome the resistance of the wire and core. This is dissipated in the form of heat (power loss).

Exciting current will flow in the primary winding at all times to maintain this magnetic field, but no transfer of energy will take place as long as the secondary circuit is open.

### 11.2.2 Phase Relationship

The secondary voltage of a simple transformer may be either in phase or out of phase with the primary voltage. This depends on the direction in which the windings are wound and the arrangement of the connection to the external circuit (load). Simply, this means that the two voltages may rise and fall together, or one may rise while the other is falling. Transformers in which the secondary voltage is in phase with the primary are referred to as like-wound transformers, while those in which the voltages are 180° out of phase are called unlike-wound transformers.

Dots are used to indicate points on a transformer schematic symbol that have the same instantaneous polarity (points that are in phase). The use of phase-indicating dots is illustrated in *Figure 35*. In the first part of the figure, both the primary and secondary windings are wound from top to bottom in a clockwise direction, as viewed from above the windings. When constructed in this manner, the top lead of the primary and the top

**Figure 35** Transformer winding polarity.

201F35.EPS

lead of the secondary have the same polarity. This is indicated by the dots on the transformer symbol.

The second part of the figure illustrates a transformer in which the primary and secondary are wound in opposite directions. As viewed from above the windings, the primary is wound in a clockwise direction from top to bottom, while the secondary is wound in a counterclockwise direction. Notice that the top leads of the primary and secondary have opposite polarities. This is indicated by the dots being placed on opposite ends of the transformer symbol. Thus, the polarity of voltage at the terminals of the transformer secondary depends on the direction in which the secondary is wound with respect to the primary.

### 11.3.0 Turns and Voltage Ratios

To understand how a transformer can be used to step up or step down voltage, the term *turns ratio* must be understood. The total voltage induced into the secondary winding of a transformer is determined mainly by the ratio of the number of turns in the primary to the number of turns in the secondary, and by the amount of voltage applied to the primary. Therefore, to set up a formula:

$$\text{Turns ratio} = \frac{\text{Number of turns in the primary}}{\text{Number of turns in the secondary}}$$

The first transformer in *Figure 36* shows a transformer whose primary consists of 10 turns of wire, and whose secondary consists of a single turn of wire. As lines of flux generated by the primary expand and collapse, they cut both the 10 turns of the primary and the single turn of the secondary. Since the length of the wire in the secondary is approximately the same as the length of the wire in each turn of the primary, the EMF induced into the secondary will be the same as the EMF induced into each turn of the primary.

This means that if the voltage applied to the primary winding is 10 volts, the CEMF in the primary is almost 10 volts. Thus, each turn in the primary will have an induced CEMF of approximately $\frac{1}{10}$ of the total applied voltage, or one volt. Since the same flux lines cut the turns in both the secondary and the primary, each turn will have an EMF of one volt induced into it. The first transformer in *Figure 36* has only one turn in the secondary, thus, the EMF across the secondary is one volt.

The second transformer represented in *Figure 36* has a 10-turn primary and a two-turn secondary. Since the flux induces one volt per turn, the total

*Figure 36*  Transformer turns ratio.

voltage across the secondary is two volts. Notice that the volts per turn are the same for both primary and secondary windings. Since the CEMF in the primary is equal (or almost) to the applied voltage, a proportion may be set up to express the value of the voltage induced in terms of the voltage applied to the primary and the number of turns in each winding. This proportion also shows the relationship between the number of turns in each winding and the voltage across each winding, and is expressed by the equation:

$$\frac{E_S}{E_P} = \frac{N_S}{N_P}$$

*Where:*

$N_P$ = number of turns in the primary
$E_P$ = voltage applied to the primary
$E_S$ = voltage induced in the secondary
$N_S$ = number of turns in the secondary

The equation shows that the ratio of secondary voltage to primary voltage is equal to the ratio of secondary turns to primary turns. The equation can be written as:

$$E_P N_S = E_S N_P$$

For example, a transformer has 100 turns in the primary, 50 turns in the secondary, and 120VAC applied to the primary ($E_P$). What is the voltage across the secondary ($E_S$)?

---

## Think About It

### Turns and Voltage Ratios

What is the magnitude of the voltage and current supplied by the secondary of the transformer in the circuit shown below?

201SA05.EPS

---

$$N_P = 100 \text{ turns} \quad E_P = 120VAC$$

$$N_S = 50 \text{ turns}$$

$$\frac{E_S}{E_P} = \frac{N_S}{N_P} \text{ or } E_S = \frac{E_P N_S}{N_P}$$

$$E_S = \frac{120V \times 50 \text{ turns}}{100 \text{ turns}} = 60VAC$$

The transformers in *Figure 36* have fewer turns in the secondary than in the primary. As a result, there is less voltage across the secondary than across the primary. A transformer in which the voltage across the secondary is less than the voltage across the primary is called a step-down transformer. The ratio of a 10-to-1 step-down transformer is written as 10:1.

A transformer that has fewer turns in the primary than in the secondary will produce a greater voltage across the secondary than the voltage applied to the primary. A transformer in which the voltage across the secondary is greater than the voltage applied to the primary is called a step-up transformer. The ratio of a 1-to-4 step-up transformer should be written 1:4. Notice in the two ratios that the value of the primary winding is always stated first.

## 11.4.0 Types of Transformers

Transformers are widely used to permit the use of trip coils and instruments of moderate current and voltage capacities and to measure the characteristics of high-voltage and high-current circuits. Since secondary voltage and current are directly related to primary voltage and current,

measurements can be made under the low-voltage or low-current conditions of the secondary circuit and still determine primary characteristics. Tripping transformers and instrument transformers are examples of this use of transformers.

The primary or secondary coils of a transformer can be tapped to permit multiple input and output voltages. *Figure 37* shows several tapped transformers. The center-tapped transformer is particularly important because it can be used in conjunction with other components to convert an AC input to a DC output.

### 11.4.1 Isolation Transformer

Isolation transformers are wound so that their primary and secondary voltages are equal. Their purpose is to electrically isolate a piece of electrical equipment from the power distribution system.

Many pieces of electronic equipment use the metal chassis on which the components are mounted as part of the circuit (*Figure 38*). Personnel working with this equipment may accidentally come in contact with the chassis, completing the circuit to ground, and receive a shock as shown in *Figure 38(A)*. If the resistances of their body and the ground path are low, the shock can be fatal. Placing an isolation transformer in the circuit as shown in *Figure 38(B)* breaks the ground current path that includes the worker. Current can no longer flow from the power supply through the chassis and worker to ground; however, the equipment is still supplied with the normal operating voltage and current.

*Figure 37* Tapped transformers.

201F37.EPS

**(A) CURRENT PATH COMPLETE**

**(B) CURRENT PATH ISOLATED**

201F38.EPS

*Figure 38*  Importance of an isolation transformer.

### 11.4.2 Autotransformer

In a transformer, it is not necessary for the primary and secondary to be separate and distinct windings. *Figure 39* is a schematic diagram of what is known as an autotransformer. Note that a single coil of wire is tapped to produce what is electrically both a primary and a secondary winding.

The voltage across the secondary winding has the same relationship to the voltage across the pri-

mary that it would have if they were two distinct windings. The movable tap in the secondary is used to select a value of output voltage either higher or lower than $E_P$, within the range of the transformer. When the tap is at Point A, $E_S$ is less than $E_P$; when the tap is at Point B, $E_S$ is greater than $E_P$.

Autotransformers rely on self-induction to induce their secondary voltage. The term *autotransformer* can be broken down into two words: auto, meaning self; and transformer, meaning to change potential. The autotransformer is made of one winding that acts as both a primary and a secondary winding. It may be used as either a step-up or step-down transformer. Some common uses of autotransformers are as variable AC voltage supplies and fluorescent light ballast transformers, and to reduce the line voltage for various types of low-voltage motor starters.

### 11.4.3 Current Transformer

A current transformer differs from other transformers in that the primary is a conductor to the load and the secondary is a coil wrapped around the wire to the load. Just as any ammeter is

*Figure 39*  Autotransformer schematic diagram.

201F39.EPS

connected in line with a circuit, the current transformer is connected in series with the current to be measured. *Figure 40* is a diagram of a current transformer.

Since current transformers are series transformers, the usual voltage and current relationships do not apply. Current transformers vary considerably in rated primary current, but are usually designed with ampere-turn ratios such that the secondary delivers five amperes at full primary load.

Current transformers are generally constructed with only a few turns or no turns in the primary. The voltage in the secondary is induced by the changing magnetic field that exists around a single conductor. The secondary is wound on a

circular core, and the large conductor that makes up the primary passes through the hole in its center. Because the primary has few or no turns, the secondary must have many turns (providing a high turns ratio) in order to produce a usable voltage. The advantage of this is that you get an output off the secondary proportional to the current flowing through the primary, without an appreciable voltage drop across the primary. This is because the primary voltage equals the current times the impedance. The impedance is kept near zero by using no or very few primary turns. The disadvantage is that you cannot open the secondary circuit with the primary energized. To do so would cause the secondary current to drop rapidly to zero. This would cause the magnetic field generated by the secondary current to collapse rapidly. The rapid collapse of the secondary

LOAD

TO AMMETER,
RELAY, OR
CONTROL

201F40.EPS

*Figure 40*  Current transformer schematic diagram.

field through the many turns of the secondary winding would induce a dangerously high voltage in the secondary, creating an equipment and personnel hazard.

Because the output of current transformers is proportional to the current in the primary, they are most often used to power current-sensing meters and relays. This allows the instruments to respond to primary current without having to handle extreme magnitudes of current.

### 11.4.4 Potential Transformer

The primary of a potential transformer is connected across or in parallel with the voltage to be measured, just as a voltmeter is connected across a circuit. *Figure 41* shows the schematic diagram for a potential transformer.

Potential transformers are basically the same as any other single-phase transformer. Although primary voltage ratings vary widely according to the specific application, secondary voltage ratings are usually 120V, a convenient voltage for meters and relays.

Because the output of potential transformers is proportional to the phase-to-phase voltage of the primary, they are often used to power

voltage-sensing meters and relays. This allows the instruments to respond to primary voltage while having to handle only 120V. Also, potential transformers are essentially single-phase step-down transformers. Therefore, power to operate low-voltage auxiliary equipment associated with high-voltage switchgear can be supplied off the high-voltage lines that the equipment serves via potential transformers.

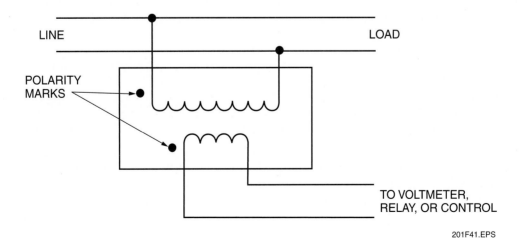

201F41.EPS

*Figure 41*  Potential transformer.

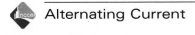

# SUMMARY

The process by which current is produced electro-magnetically is called induction. As the conductor moves across the magnetic field, it cuts the lines of force, and electrons within the conductor flow, creating an electromotive force (EMF). EMF is also known as voltage. There are three conditions that must exist before a current can be produced in this way:

- There must be a magnetic field through which the conductor can pass.
- There must be a conductor in which the voltage will be produced, and the conductor should be perpendicular to the field.
- There must be motion. Either the magnetic field or the conductor must move.

Three factors control the magnitude of the induced current. Voltage will be increased if:

- The speed with which the conductor cuts through the magnetic field is increased (the faster the conductor cuts through the field, the greater the current pulse)
- The strength of the field is increased (the stronger the field, the greater the current pulse)
- The conductor is wound to form a coil (the voltage increases directly with the number of turns of the coil)

A decrease in voltage occurs as the conductor intersects the magnetic field at an angle less than 90°. The greatest current is produced when the conductor intersects the magnetic field at right angles (perpendicular) to the flux lines.

It should be emphasized that a current may be induced by using the magnetic field of a permanent magnet or the magnetic field of another current-carrying conductor (electromagnet).

The magnetic field is among the reasons why phases of current-carrying conductors should not be separated in a raceway; all phase conductors (including the neutral) should be contained in the same raceway. For example, if phase A is separated from phase B and phase C by a metal enclosure, conduit wall, etc., the magnetic field around the conductors will cut across the conduit, causing the conduit to heat up.

# Review Questions

1. An electric current always produces _____.
   a. mutual inductance
   b. a magnetic field
   c. capacitive reactance
   d. high voltage

2. The number of cycles an alternating electric current undergoes per second is known as _____.
   a. amperage
   b. frequency
   c. voltage
   d. resistance

3. What is the peak voltage in a circuit with an rms voltage of 120VAC?
   a. 117 volts
   b. 120 volts
   c. 150 volts
   d. 170 volts

4. Which of the following conditions exist in a circuit of pure resistance?
   a. The voltage and current are in phase.
   b. The voltage and current are 90° out of phase.
   c. The voltage and current are 120° out of phase.
   d. The voltage and current are 180° out of phase.

5. In a purely resistive AC circuit where 240V is applied across a 10-ohm resistor, the amperage is _____.
   a. 10A
   b. 24A
   c. 60A
   d. 120A

6. When the current increases in an AC circuit, what role does inductance play?
   a. It increases the current.
   b. It plays no role at all.
   c. It causes the overcurrent protection to open.
   d. It reduces the current.

7. All of the following factors increase the inductance of a coil, *except* a _____.
   a. greater number of turns
   b. longer core
   c. wider coil
   d. magnetic core

8. Which of the following conditions exist in a circuit of pure inductance?
   a. The voltage and current are in phase.
   b. The voltage and current are 90° out of phase.
   c. The voltage and current are 120° out of phase.
   d. The voltage and current are 180° out of phase.

9. True or False? Reactance increases with an increase in inductance.
   a. True
   b. False

10. Capacitance is measured in _____.
    a. farads
    b. joules
    c. henrys
    d. amps

11. The total capacitance of two 15µF capacitors connected in parallel is _____.
    a. 5µF
    b. 10µF
    c. 15µF
    d. 30µF

12. The opposition to current flow offered by the capacitance of a circuit is known as _____.
    a. mutual inductance
    b. pure resistance
    c. inductive reactance
    d. capacitive reactance

13. The total opposition to current flow in an AC circuit is known as _____.
    a. resistance
    b. capacitive reactance
    c. inductive reactance
    d. impedance

14. A power factor is not an angular measure, but a numerical ratio with a value between 0 and 1, equal to the _____.
    a. sine of the phase angle
    b. tangent of the phase angle
    c. cosine of the phase angle
    d. cotangent of the phase angle

15. The two windings of a conventional transformer are known as the _____.
    a. mutual and inductive windings
    b. high and low voltage windings
    c. primary and secondary windings
    d. step-up and step-down windings

# Module 26201-11
## Supplemental Exercises

1. In a circuit containing resistance only, the voltage and current are _____ with one another.

2. The power consumed by _____ is called true power.

3. The power factor is the ratio of _____ power to _____ power.

4. LC circuits consist of a(n) _____ and a(n) _____ connected in series or in parallel with a voltage source.

5. As in all series circuits, the _____ in a series LC circuit is the same at all points.

6. On a transformer with 100 volts applied to the primary, 200 turns in the primary, and a secondary with 100 turns, the secondary voltage is _____ volts.

7. A(n) _____ can be used to electrically isolate a piece of equipment from the power distribution system.

8. Reactive power is the portion of apparent power that is caused by _____ and _____ in the circuit.

9. The primary winding in a transformer is connected to the _____.

10. Apparent power is the product of the source _____ and the _____.

11. The secondary winding in a transformer is connected to the _____.

12. True power is measured in _____.

13. VAR is the unit of measurement for _____ power.

14. The average value is calculated from all the values in a sine wave for only half a cycle.

   The average value formula is AVG = _____ × peak value.

15. In purely resistive circuits, the current and voltage are related by Ohm's law, which is E = IR. Define the variables.

   E = _____

   I = _____

   R = _____

# Trade Terms Introduced in This Module

**Capacitance:** The storage of electricity in a capacitor; capacitance produces an opposition to voltage change. The unit of measurement for capacitance is the farad (F) or microfarad (µF).

**Frequency:** The number of cycles an alternating electric current, sound wave, or vibrating object undergoes per second.

**Hertz (Hz):** A unit of frequency; one hertz equals one cycle per second.

**Impedance:** The opposition to current flow in an AC circuit; impedance includes resistance (R), capacitive reactance ($X_C$), and inductive reactance ($X_L$). Impedance is measured in ohms (Ω).

**Inductance:** The creation of a voltage due to a time-varying current; also, the opposition to current change, causing current changes to lag behind voltage changes. The unit of measure for inductance is the henry (H).

**Micro (µ):** Prefix designating one-millionth of a unit. For example, one microfarad is one-millionth of a farad.

**Peak voltage:** The peak value of a sinusoidally varying (cyclical) voltage or current is equal to the root-mean-square (rms) value multiplied by the square root of two (1.414). AC voltages are usually expressed as rms values; that is, 120 volts, 208 volts, 240 volts, 277 volts, 480 volts, etc., are all rms values. The peak voltage, however, differs. For example, the peak value of 120 volts (rms) is actually $120 \times 1.414 = 169.71$ volts.

**Radian:** An angle at the center of a circle, subtending (opposite to) an arc of the circle that is equal in length to the radius.

**Reactance:** The opposition to alternating current (AC) due to capacitance ($X_C$) and/or inductance ($X_L$).

**Root-mean-square (rms):** The square root of the average of the square of the function taken throughout the period. The rms value of a sinusoidally varying voltage or current is the effective value of the voltage or current.

**Self-inductance:** A magnetic field induced in the conductor carrying the current.

# Additional Resources

This module presents thorough resources for task training. The following resource material is suggested for further study.

*Introduction to Electric Circuits*, Latest Edition. New York: Prentice Hall.

*Principles of Electric Circuits*, Latest Edition. New York: Prentice Hall.

# Figure Credits

Associated Builders and Contractors, Inc., Module opener

Topaz Publications, Inc., Figure 13C, SA04

# CONTREN® LEARNING SERIES — USER UPDATE

NCCER makes every effort to keep its textbooks up-to-date and free of technical errors. We appreciate your help in this process. If you find an error, a typographical mistake, or an inaccuracy in NCCER's Contren® materials, please fill out this form (or a photocopy), or complete the online form at www.nccer.org/olf. Be sure to include the exact module number, page number, a detailed description, and your recommended correction. Your input will be brought to the attention of the Authoring Team. Thank you for your assistance.

*Instructors* – If you have an idea for improving this textbook, or have found that additional materials were necessary to teach this module effectively, please let us know so that we may present your suggestions to the Authoring Team.

### NCCER Product Development and Revision

3600 NW 43rd Street, Building G, Gainesville, FL 32606

**Fax:** 352-334-0932
**Email:** curriculum@nccer.org
**Online:** www.nccer.org/olf

☐ Trainee Guide    ☐ AIG    ☐ Exam    ☐ PowerPoints    Other _____

Craft / Level: _____ Copyright Date: _____

Module Number / Title: _____

Section Number(s): _____

Description: _____

_____

_____

_____

Recommended Correction: _____

_____

_____

_____

Your Name: _____

Address: _____

_____

Email: _____ Phone: _____

# Motors: Theory and Application

**Procter & Gamble Olean Facility**

This refinery produces the P&G patented no-fat Olean cooking oil in Cincinnati, Ohio. It won an ABC National Award of Excellence and involved the installation of over 300,000 feet of conduit, 10,000 feet of cable tray, 2,000,000 feet of power cable and wire, 100,000 feet of heat trace, and 2,700 light fixtures.

## 26202-11

Trainees with successful module completions may be eligible for credentialing through NCCER's National Registry. To learn more, go to **www.nccer.org** or contact us at **1.888.622.3720.** Our website has information on the latest product releases and training, as well as online versions of our *Cornerstone* newsletter and Pearson's Contren® product catalog.

Your feedback is welcome. You may email your comments to **curriculum@nccer.org,** send general comments and inquiries to **info@nccer.org,** or use the User Update form at the back of this module.

 V.1 7/11

## Objectives

Upon completion of this module, you will be able to do the following:

1. Define the following terms:
   - Controller
   - Duty cycle
   - Full-load amps
   - Interrupting rating
   - Thermal protection
   - NEMA design letter
   - Overcurrent
   - Overload
   - Power factor
   - Rated full-load speed
   - Rated horsepower
   - Service factor
2. Describe the various types of motor enclosures.
3. Explain the relationships among speed, frequency, and the number of poles in a three-phase induction motor.
4. Define *percent slip* and *speed regulation*.
5. Explain how the direction of a three-phase motor is changed.
6. Describe the component parts and operating characteristics of a three-phase wound-rotor induction motor.
7. Describe the component parts and operating characteristics of a three-phase synchronous motor.
8. Describe the design and operating characteristics of various DC motors.
9. Describe the methods for determining various motor connections.
10. Describe general motor protection requirements as delineated in the *National Electrical Code®* (*NEC®*).
11. Define the braking requirements for AC and DC motors.

## Performance Tasks

Under the supervision of the instructor, you should be able to do the following:

1. Collect data from a motor nameplate.
2. Identify various types of motors and their application(s).
3. Connect the terminals for a dual-voltage motor.

## Trade Terms

Armature
Branch circuit
Brush
Circuit breaker
Commutator
Continuous duty
Controller

Duty
Equipment
Field poles
Horsepower
Hours
Intermittent duty
Overcurrent

Overload
Periodic duty
Revolutions per minute (rpm)
Rotation
Synchronous speed
Thermal protector
Varying duty

## Required Trainee Materials

1. Pencil and paper
2. Appropriate personal protective equipment
3. Copy of the latest edition of the *National Electrical Code®*

Note: *NFPA 70®*, *National Electrical Code®*, and *NEC®* are registered trademarks of the National Fire Protection Association, Inc., Quincy, MA 02269. All *National Electrical Code®* and *NEC®* references in this module refer to the 2011 edition of the *National Electrical Code®*.

## Contents

*Topics to be presented in this module include:*

## Contents (*continued*) ——————————————

## Figures and Tables ——————————————

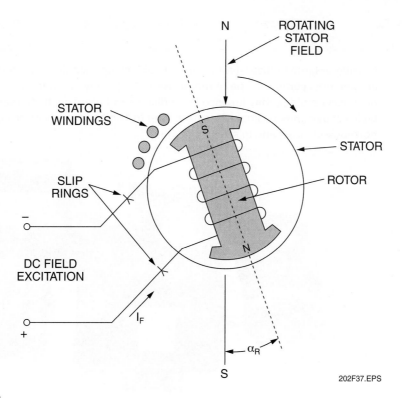

*Figure 37* Torque angle.

$N_R, S_R$ = ROTOR FIELD
$N_S, S_S$ = STATOR FIELD

motor. Its theory of operation is similar to that of the three-phase induction motor; therefore, it runs at a speed slightly lower than synchronous speed. Second, there is the single-phase synchronous motor.

### 4.4.1 Single-Phase Induction Motors

Single-phase AC induction motors are extremely popular. Unlike polyphase induction motors, the stator field in the single-phase motor does not rotate. Instead, it simply alternates polarity between poles as the AC voltage changes polarity.

Voltage is induced in the rotor, and a magnetic field is produced around the rotor. This field will always be in opposition to the stator field. However, the interaction between the rotor and stator fields will not produce rotation (see *Figure 38*). Because this force is across the rotor and through the pole pieces, there is no rotary motion, just a push and/or pull along this line.

If the rotor is rotated by some outside force (a twist of your hand, for example), the push-pull along the line is disturbed. Look at the fields shown in *Figure 38* as the motor begins to rotate. At this instant, the south pole on the rotor is being attracted to the left-hand pole. The north rotor pole is being attracted to the right-hand pole. All of this is a result of the rotor being rotated 90° by the outside force.

STATIONARY

ROTATING

*Figure 38* AC induction motor.

 Motors: Theory and Application

# Single-Phase Induction Motors

Outside of large industrial and commercial facilities, single-phase induction motors are the most common type of motor used. While they are initially less expensive than polyphase motors, they are also less efficient and more costly to maintain. They are typically available in small sizes from ⅛hp to 1hp (previously referred to as fractional horsepower sizes) and in sizes up to 10hp.

202SA02.EPS

The pull that now exists between the two fields becomes a rotary force, turning the rotor toward magnetic correspondence with the stator. Because the two fields continuously alternate, they will never actually line up and the rotor will continue to turn once started.

Since a single-phase rotor will rotate if it has a rotating magnetic field present, all that remains is to find a means of generating a rotating field at the start. There are a number of practical means for generating a rotating field. All the methods used for single-phase induction motors involve the simulation of a second phase for a starting circuit.

### 4.4.2 Split-Phase Induction Motor

The split-phase motor, shown schematically in *Figure 39*, has a stator composed of slotted laminations that contain an auxiliary (starting) winding and a running (main) winding. The axes of these two windings are displaced by an angle of 90 electrical degrees. The starting winding has fewer turns and smaller wire than the running winding and, therefore, has different electrical characteristics. The main winding occupies the lower half of the slots and the starting winding occupies the upper half. The two windings are connected in parallel across the single-phase line supplying the motor. The motor derives its name from the action of the stator during the starting period.

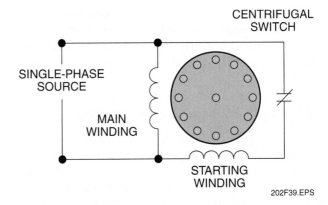

*Figure 39* Split-phase motor.

When energized with single-phase AC, the two windings are physically different enough in position and construction to produce a magnetic revolving field that rotates around the stator air gap at synchronous speed. As the rotating field moves around the air gap, it cuts across the rotor conductors and induces a voltage in them. The interaction between the rotor and stator causes the rotor to accelerate in the direction in which the stator field is rotating.

When the rotor has come up to about 75% of synchronous speed, a centrifugally operated switch disconnects the starting winding from the line supply, and the motor continues to run on

# Modern Split-Phase Induction Motor

Photo (A) shows a centrifugally-actuated start winding switch that is closed when the motor is at rest. The start winding switch is opened and closed by the movement of an actuator disk against a contact lever. The actuator disk is moved back and forth on the rotor shaft by a centrifugal weight assembly. When the motor is at rest, springs retract the weights and cause the actuator disk to move toward the bearing at the end of the shaft. This pushes on the contact lever, closing the switch contacts. After the motor starts and reaches about 75% of its rated speed, the weights swing out against the spring tension and retract the contact disk from the contact lever. This opens the switch contacts, removing the starting winding from the circuit. If the switch does not open after starting, the motor will operate at a reduced speed until it overheats the starting winding and activates the thermal relay. Photo (B) shows the starting winding (green) and the running winding (copper).

START WINDING SWITCH

ACTUATOR DISK CONTACT LEVER

CENTRIFUGAL WEIGHT ASSEMBLY

MOVABLE ACTUATOR DISK

202SA03.EPS

**(A)**

STARTING WINDING

RUNNING WINDING

202SA04.EPS

**(B)**

the main winding alone. As the motor ages, the centrifugal switch contacts pit and corrode. When this happens, they may get stuck in the closed position. To safeguard against the winding burning up, a thermal relay is also used. If the motor draws the high starting current for more than 5 or 10 seconds, the relay will de-energize.

In a split-phase motor, the starting torque is 150% to 200% of the full-load torque, and the starting current is six to eight times the full-load current. Fractional-horsepower split-phase motors are used in a variety of devices such as washers, oil burners, and ventilating fans. The direction of rotation of the split-phase motor can be reversed by interchanging the starting winding leads.

### 4.4.3 Capacitor-Type Induction Motor

The capacitor-type motor is a modified form of split-phase motor. A typical capacitor-type motor is shown in *Figure 40*. The capacitor is located on top of the motor.

To develop a larger starting torque than that available with a standard split-phase motor, a capacitor is placed in series with the auxiliary winding of a split-phase motor, as shown in *Figure 41*. This is called a capacitor-start motor. The capacitor tends to create a greater electrical phase separation of the two windings. Also, because the reactance of a capacitor is 180° out of phase with the inductive reactance of the motor

windings when they are combined, they yield a lower total impedance. This allows a larger current to produce a greater magnetic field.

The net effect of the capacitor is to give its motor a starting torque of about four times its rated torque. The split-phase motor, on the other hand, produces a starting torque of about one to two times its rated torque. Once the capacitor motor has come up to speed and the starting winding has been disconnected, it will have the same running characteristics as the split-phase motor.

To reverse the direction of rotation of the capacitor-start motor and split-phase motor, the connection of either winding would have to be reversed. Since the starting winding is disconnected at a high speed, this reversal can be accomplished only at standstill or at low speeds when the centrifugal switch is still closed.

The capacitor-start motor is made in sizes from ¼ to 10hp (150W to 7.5kW). The starting capacitor is the dry-type electrolytic capacitor made for AC use. Typical values are from 200 to 600 microfarads (µF). *Figure 42* shows a comparison of torque slip curves for a split-phase and capacitor-start motor. It also shows the typical effect of the starting capacitor.

A variation of the capacitor-start motor is one in which the capacitor and auxiliary winding are not disconnected. The centrifugal switch in *Figure 41* is eliminated, and the auxiliary

202F40.EPS

*Figure 40*  Capacitor motor.

202F41.EPS

*Figure 41*  Capacitor-start motor schematic.

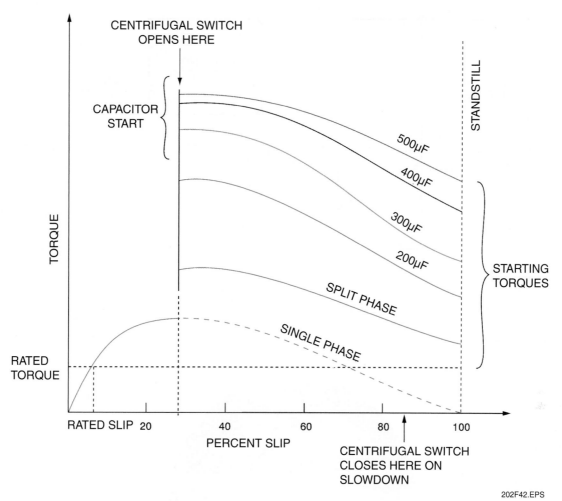

*Figure 42*   Torque-slip curves.

winding is left in all the time. This motor is called a capacitor-run motor.

The capacity used for running under load is not the same as that needed for starting. Furthermore, the capacitors used for starting cannot be used for continuous operation. Since the capacitor used in this motor is in all the time, it must be of a different type; that is, one capable of operating continuously. The net result is that the motor has improved running characteristics; however, it does not provide a starting torque as large as that of the capacitor-start motor.

Among the improvements are higher efficiency and power factor at rated load, lower line current, and very quiet operation. It should also be pointed out that the start winding must be designed for continuous operation. This makes the motor somewhat more costly.

Another variation is the capacitor-start, capacitor-run motor. This motor combines the useful features of the capacitor-start and the capacitor-run motors by using two different capacitors, as shown in *Figure 43*.

### 4.4.4 Shaded-Pole Induction Motor

The shaded-pole motor employs a salient-pole stator and a cage rotor. The projecting poles on the stator resemble those of DC machines, except that the entire magnetic circuit is laminated and a portion of each pole is split to accommodate a short circuited copper strap called a shading coil. This motor is generally manufactured in very small sizes and runs up to ¹⁄₂₀hp. A four-pole motor of this type is illustrated in *Figure 44*.

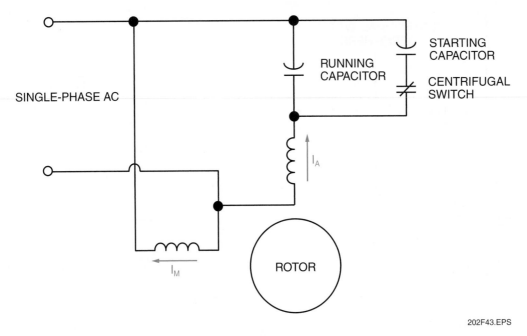

202F43.EPS

**Figure 43** Capacitor-start, capacitor-run motor schematic.

202F44.EPS

**Figure 44** Four-pole shaded-pole motor.

The shading coils are placed around the leading pole tip, and the main pole winding is concentrated and wound around the entire pole. The four coils that make up the main winding are connected in series across the motor terminals. An inexpensive type of two-pole motor that uses shading coils is illustrated in *Figure 45*.

*Figure 45* shows that during part of the cycle when the main pole flux ($\phi_1$) is increasing, the shading coil is cut by the flux, and the resulting induced EMF and current in the shading coil tend to prevent the flux from rising readily through it. Thus, the greater portion of the flux rises in the portion of the pole that is not in the vicinity of the shading coil ($\phi_1 > \phi_2$). When the flux reaches its maximum value, the rate of change of flux is zero, and the voltage and current in the shading coil are also at zero. At this time, the flux is distributed more uniformly over the entire pole face ($\phi_1 = \phi_2$).

**Figure 45** Two-pole shaded-pole motor.

---

# Capacitor-Start, Capacitor-Run Motors

These motors run quietly and smoothly and have a high starting torque. Their construction is similar to a split-phase motor in that they use a centrifugal switch to remove only the start capacitor from the starting winding while leaving the run capacitor connected to the winding. These motors have a higher power factor than ordinary split-phase motors.

---

As the main flux decreases toward zero, the induced voltage and current in the shading coil reverse their polarity, and the resulting force tends to prevent the flux from collapsing through the iron in the region of the shading coil ($\phi_2 > \phi_1$). The result is that the main flux rises first in the unshaded portion of the pole and later in the shaded portion. This action is equivalent to a sweeping movement of the field across the pole face in the direction of the shaded pole. The cage rotor conductors are cut by this moving field, and the force exerted on them causes the rotor to turn in the direction of the sweeping field.

Most shaded-pole motors have only one edge of the pole split, and therefore, the direction of rotation is not reversible. However, some shaded-pole motors have both leading and trailing pole tips split to accommodate shading coils. The leading pole tip shading coils form one series group, and the trailing pole tip shading coils form another series group. Only the shading coils in one group are simultaneously active, while those in the other group are on an open circuit.

The shaded-pole motor is similar in operating characteristics to the split-phase motor. It has the advantages of simple construction and low cost. It has no sliding electrical contacts and is reliable in operation. However, it has low starting torque, low efficiency, and a high noise level. It is normally used to operate small fans. The shading coil and split pole are also used in timers to make them self-starting.

### 4.4.5 Single-Phase Synchronous Motor

The single-phase synchronous motor, as its name implies, runs at synchronous speed. It finds use where a constant speed is needed, such as in tape recorders and clocks. It is started in the same way as any of the single-phase induction motors and therefore has a rotating field. By having a modified rotor, the motor pulls into synchronism and runs at synchronous speed.

## 5.0.0 MULTIPLE-SPEED INDUCTION MOTORS

The speed of an induction motor depends on the power supply frequency and the number of pairs of poles used in the motor. Obviously, to alter motor speed it is merely necessary to change one of these two factors. By far the most common method used involves changing the number of poles, generally at some type of external controller.

There are two types of multiple-speed squirrel cage induction motors in common use: the multiple-winding motor and the consequent-pole motor. Both feature poles that may be changed, as required, by shifting key external connections, and in this way they provide for operating the motor at a limited number of different speeds.

### 5.1.0 Multiple-Winding Motor

In the multiple-winding motor, two or more separate windings are placed in the stator core slots, one over the other, as shown in *Figure 46*. For example, a four-pole winding can be positioned in the core slots and have a two-pole winding placed on top of it. The windings are insulated from each other and arranged so that only one winding at a time can be energized. Switching speeds is normally accomplished by switching contacts that are in the motor controller external to the motor itself.

### 5.2.0 Consequent-Pole Motor

In the consequent-pole motor, there are two speeds. The motor is constructed to have a certain number of poles for high-speed operation and then, by a switching action, double this number of poles to give low-speed operation.

*Figure 46*  Two-winding, two-speed motor.

202F46.EPS

# Hysteresis Synchronous Motor

This unusual synchronous motor has an external rotor. These motors have low noise levels, high efficiency, and constant speed, and are used for applications such as tape recorders and clocks.

202SA06.EPS

The switching action is illustrated by the use of the two-phase motor in *Figure 47*. If you trace the wiring in *Figure 47*, you can see how the system is phased so that both magnetic north and south poles are produced at the winding projections. With two-phase power applied to the two-pole motor (two poles per phase), a rotating magnetic field of 3,600 rpm is produced.

In *Figure 48*, the connections are changed so that the system is phased to produce four magnetic north poles at the winding projections. Since every north pole must have a south pole, consequent south poles are produced between the projecting north poles as a consequence of having formed north poles. Accordingly, in *Figure 48*, there are twice as many pole groups as in *Figure 47*. Therefore, a four-pole rotating magnetic field of 1,800 rpm is produced.

*Figure 49* shows the short-jumpering arrangement of the consequent-pole motor. In this, all

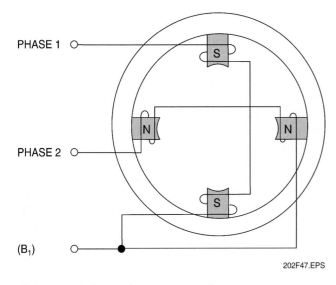

*Figure 47*   High-speed consequent-pole motor.

202F47.EPS

*Figure 48*   Low-speed consequent-pole motor.

202F48.EPS

SHORT JUMPER
SERIES CONNECTION

AC
SOURCE

LONG JUMPER
PARALLEL CONNECTION

AC
SOURCE

202F49.EPS

*Figure 49*  Single-phase consequent-pole motor.

windings are in series, and alternate north and south poles are produced. To produce consequent poles, the series connection is replaced with a parallel connection accomplished by the long-jumpering arrangement. By connecting the motor in this manner, four salient monopoles are produced and, as a result, create four opposite consequent poles. In the practical consequent-pole motor, all necessary internal connection rearrangements are accomplished at an external control panel.

Consequent-pole motor characteristics depend on the intended application. In a constant-horsepower motor, torque varies inversely with speed. It is used for driving machine tools.

In a constant-torque motor, horsepower varies directly with speed. It is used to drive pumps and air compressors, as well as in constant-pressure blowers. In a variable-torque, variable-horsepower motor, both torque and horsepower change with changes in speed. This is the type of motor found in household fans and air conditioners.

## 6.0.0 VARIABLE-SPEED DRIVES

The use of adjustable speed in industrial equipment is increasing due to the need for better equipment control and for energy savings where partial power is required. AC drives compete with DC drives, eddy current drives, and mechanical and hydraulic systems as methods to control speed. Reliability, cost, and control capabilities are the major factors in system selection.

A drive system includes both the drive controller and the motor being driven. This module focuses specifically on the electronic drive components and covers various types of control for both DC and AC drives. This section provides a basic review of some fundamental principles that are important to understand when starting up, operating, or troubleshooting a variable-speed drive system.

### 6.1.0 Types of Adjustable Speed Loads

Most drive controllers can be adjusted or modified to optimize performance and provide the most efficient and cost-effective drive, depending on the load characteristics of the application.

It is important to understand the speed and torque characteristics as well as the maximum horsepower requirements for the type of load to be considered. Based on this, either a constant-torque controller or a variable-torque controller is selected. The most common types of loads are shown in *Figure 50*. Note that a load requires the same amount of torque at low speed as at high speed.

For a constant-torque load, the torque remains constant throughout the speed range, and the horsepower increases and decreases in direct proportion to the speed. This applies to applications such as conveyors, as well as applications in which shock loads, overloads, or high inertia loads are encountered.

A variable-torque load requires much lower torque at low speeds than at high speeds. Horsepower varies approximately as the cube of the speed, and the torque varies approximately as the square of the speed. This applies to applications such as centrifugal fans, pumps, and blowers.

A constant-horsepower load requires high torque at low speeds, low torque at high speeds, and thus constant horsepower at any speed. It

Figure 50    Types of adjustable speed loads.

applies to applications such as lathes requiring slow speeds for deep cuts and high speeds for finishing. Usually, very high starting torques are required.

### 6.2.0 Motor Considerations

For industrial applications, motors are required to function at varying torques and speeds, and in forward and reverse directions. Besides operating as a motor, the machine may also function as a brake or a generator for short periods.

The various operating modes for industrial drives are shown in *Figure 51*. Positive and negative speed (rotation) are plotted on the horizontal axis and the torques are plotted on the vertical axis. The four quadrants of operation are labeled 1, 2, 3, and 4.

A machine operating in quadrant 1 has positive torque and speed, which means that they both act in the same direction (in this case, clockwise). A machine in this quadrant is functioning as a motor. It delivers mechanical power to a load. The machine will also act as a motor in quadrant 3, but torque and speed are reversed from quadrant 1 (counterclockwise).

While operating in quadrant 2, a machine will develop a positive torque and a negative speed. The torque is acting clockwise, and the speed is counterclockwise. In this quadrant, the machine is absorbing mechanical power from the load and functions as a generator. This mechanical power is converted into electric power and is generally transmitted back into the line. The electric power may also be dissipated in an external resistor, which is known as dynamic braking.

Depending on its connections, a machine may also be used as a brake while operating in quadrant 2. Absorbed mechanical power is converted to electric power, then converted into heat. If the machine absorbs electric line power as it is converting mechanical power into electric power, it functions as a brake. Both power inputs are dissipated as heat. Large power drives seldom use the brake mode of operation as it is very inefficient. The circuitry is generally chosen so that the machine will function as a generator when it is operating in quadrant 2. Quadrant 4 operation is identical to quadrant 2 except that speed and torque are reversed.

Figure 51    Electric drive operation in four quadrants.

### 6.2.1 Typical Torque-Speed Curves

A three-phase motor has a torque-speed curve that is a good example of an electrical machine's behavior as a generator brake. The solid curve in *Figure 52* is the torque-speed curve for a machine acting as a motor in quadrant 1, a brake in quadrant 2, and a generator in quadrant 4.

If the stator leads are reversed, the torque-speed curve is shown by the dotted curve. Now the motor operates as a motor in quadrant 3, a generator in quadrant 2, and a brake in quadrant 4. The machine functions as a brake or a generator in quadrants 2 and 4, but it always runs as a motor in quadrants 1 and 3.

*Figure 53* shows the torque-speed curve of a DC shunt motor. Motor, generator, and brake modes are apparent. The dotted curve represents reversed armature leads.

Variable-speed electric drives are designed to vary speed and torque in a smooth and continuous manner so as to satisfy load requirements. Typically, this is accomplished by shifting the torque-speed characteristic back and forth along the horizontal axis. The torque-speed characteristic of the motor is shifted by varying the armature voltage. Also, the curve of an induction motor can be shifted by varying the voltage and frequency applied to the stator.

In describing the various methods of motor control, only the behavior of power circuits will be discussed. The many ways of shaping and controlling triggering pulses will not be covered. They constitute a complex subject that involves sophisticated electronics, logic circuits, integrated circuits, and microprocessors.

### 6.2.2 Motor Heating

Since a variable-speed drive system includes both the drive controller and the motor, the design engineer should always consider the capabilities of the motor to perform acceptably under the desired operating conditions. One of the factors that you should be aware of is motor heating. When operating a motor at reduced speeds, the ability to dissipate heat is also reduced due to the slower cooling fan speed. This factor should be considered when maintaining the motor, modifying its enclosure or surrounding area, or troubleshooting the drive system.

## 6.3.0 Motor Speed Control

It is important to understand how DC or AC motor speed can be varied in order to understand how a drive controller accomplishes that task. This section reviews the fundamentals of DC and AC motor speed control.

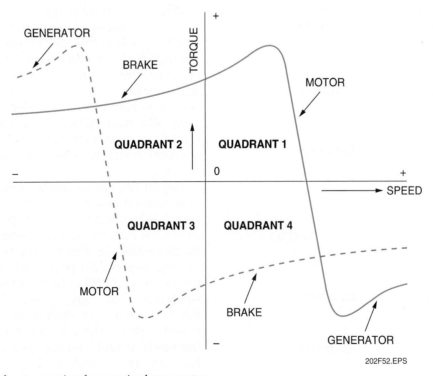

**Figure 52**   Four-quadrant operation for a squirrel cage motor.

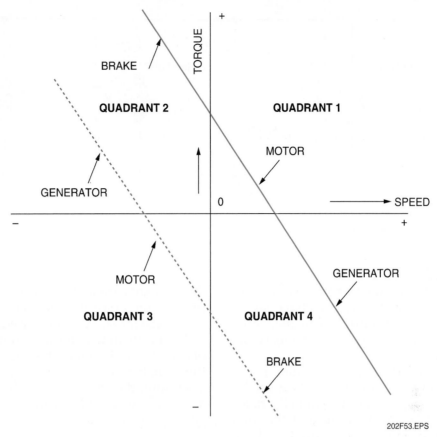

***Figure 53*** Four-quadrant operation for a DC motor.

### 6.3.1  *Varying the Speed of a DC Shunt Motor*

A DC shunt motor is shown in *Figure 54(A)*. Basically, there are two ways of varying the running speed of a DC shunt motor:

- Adjusting the voltage (and current) applied to the field winding. As the field voltage is increased, the motor slows down. This method is suggested by *Figure 54(B)*.
- Adjusting the voltage (and current) applied to the armature. As the armature voltage is increased, the motor speeds up. This method is suggested by *Figure 54(C)*.

### 6.3.2  *Field Control*

Here is how method 1, adjusting the field voltage, works. As the field voltage is increased, by reducing $R_V$ in *Figure 54(B)*, for example, the field current is increased. This results in a stronger magnetic field, which induces a greater CEMF in the armature winding. The greater CEMF tends to oppose the applied DC voltage and reduces the armature current, $I_A$. Therefore, an increased field current

causes the motor to slow down until the induced CEMF has returned to near its normal value.

Going in the other direction, if the field current is reduced, the magnetic field gets weaker. This causes a reduction in CEMF created by the rotating armature winding. The armature current

---

**GOING GREEN**

## Using a Motor as Both a Generator and a Brake

DC motors that power subway cars are also used for regenerative braking. When driven by the train's momentum, such as when slowing down upon approaching a station, the motor acts as a generator and puts current back into the system. While the train's motors are creating a significant amount of power from the train's movement, the train makes use of the resistive torque created by the power generation to slow down the train. The same theory applies to many hybrid cars.

---

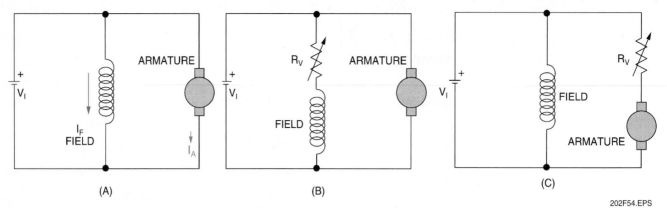

*Figure 54*    DC shunt motor schematic.

increases, forcing the motor to spin faster, until the CEMF is once again approximately equal to what it was before. The reduction in magnetic field strength is compensated for by an increase in armature speed.

This method of speed control has certain positive features. It can be accomplished by a small, inexpensive rheostat, since the current in the field winding is fairly low. Also, because of the low value of the field current, $I_F$, the rheostat $R_V$ does not dissipate very much energy. Therefore, this method is energy efficient.

However, there is one major drawback to speed control from the field winding: to increase the speed, you must reduce $I_F$ and weaken the magnetic field, thereby lessening the motor's torque-producing ability. The ability of a motor to create torque depends on two things: the current in the armature conductors and the strength of the magnetic field. As $I_F$ is reduced, the magnetic field is weakened, and the motor's torque-producing ability declines. Unfortunately, it is at this point that the motor needs all the torque-producing ability it can get, since it probably requires greater torque to drive the load at a faster speed.

### 6.3.3 Armature Control

From the torque-producing point of view, method 2, armature control, is much better. As the armature voltage and current are increased by reducing $R_V$, the motor starts running faster, which normally requires more torque. The reason for the rise in speed is that the increased armature voltage demands an increased CEMF to limit the increase in armature current to a reasonable amount. The only way the CEMF can increase is for the armature winding to spin faster, since the magnetic field strength is fixed. In this instance, the elements are all present for increased torque production, since the magnetic field strength is kept constant and $I_A$ is increased.

The problem with the armature control method of *Figure 54(C)* is that $R_V$, the rheostat, must handle the armature current, which is relatively large. Therefore, the rheostat must be physically large and expensive, and it will waste a considerable amount of energy.

### 6.3.4 Varying the Speed of an AC Motor

The principle of speed control for adjustable-frequency drives is based on the following fundamental formula for a standard AC motor:

$$N_s = \frac{120f}{P}$$

*Where:*

$N_s$ = synchronous speed (rpm)
$f$ = frequency
$P$ = number of poles
$120$ = constant

The number of poles of a particular motor is set in its design and manufacture.

The adjustable-frequency system controls the frequency (f) applied to the motor. The speed ($N_s$) of the motor is then proportional to this applied frequency. Control frequency is adjusted by means of a potentiometer or external signal, depending on the application.

The frequency output of the controller is adjustable over its design speed range. Therefore, the speed of the motor is adjustable over this same range. Because an electronic means of generating variable frequencies is being used, the speed range often exceeds the 60 hertz (Hz) rated speed of the motor.

When variable-frequency speed control is employed, the motor supply voltage cannot be allowed to remain at a steady value. The magnitude of the motor voltage must be increased or decreased in proportion to the frequency. That is, the voltage-to-frequency ratio, V/f, must remain approximately constant.

For instance, if the motor has a nameplate rating of 240V at 60Hz, the voltage-to-frequency ratio is 4 (240 ÷ 60 = 4). If the motor is speeded up by adjusting its variable-frequency inverter to 90Hz, the voltage magnitude must be increased to 360V, since 4 × 90 = 360. If the motor is slowed down by adjusting the inverter frequency to 45Hz, the voltage magnitude must be decreased to 180V, since 4 × 45 = 180.

The stator's magnetic field strength must remain constant under all operating conditions. If the stator field strength should happen to rise much above the design value, the motor's core material would go into magnetic saturation. This would effectively lower the core's permeability, thereby inhibiting proper induction of voltage and current in the rotor loops (or bars), and detracting from the torque-producing capability of the motor. On the other hand, if the stator field strength should happen to fall much below the design value, the weakened magnetic field would simply induce lower values of voltage and current in the rotor loops. This would also detract from the torque-producing ability of the motor.

Therefore, the magnetic field produced by the stator windings must hold a constant rms value, regardless of frequency. The magnetizing current of an induction motor is the current that flows through the stator winding when the rotor is spinning at steady-state speed with no torque load. The magnetizing current for an induction motor is given by Ohm's law:

$$I_{mag} = \frac{V}{X_L}$$

*Where:*

V = rms value of the applied stator voltage

$X_L$ = inductive reactance of the stator winding

In the equation, $X_L$ does not remain constant as the supply frequency is adjusted; it varies in proportion to the frequency ($X_L = 2\pi fL$). Therefore V must also be varied in proportion to the frequency, so that the Ohm's law division operation yields an unvarying value of magnetizing current.

Alternatively, using $X_L = 2\pi fL$, the equation can be rewritten as follows:

$$I_{mag} = \frac{V}{X_L}$$

$$I_{mag} = \frac{V}{2\pi fL}$$

$$I_{mag} = \frac{1}{2\pi L} \times \frac{V}{f}$$

Since $1 \div 2\pi L$ is a constant determined by the motor's construction, the magnetizing current is kept constant by maintaining the V/f ratio.

The controller can automatically maintain the required volts/cycle (V/Hz) ratio to the motor at any speed. This provides maximum motor capability throughout the speed range.

The V/Hz setting is typically preset at the factory. However, on many controllers it can be adjusted or changed to fine-tune controller operation.

## 7.0.0 MOTOR ENCLOSURES

Motors are usually designed with covers over the moving parts. These covers, called enclosures, are classified by NEMA (National Electrical Manufacturers Association) according to the degree of environmental protection provided and the method of cooling. If the cover has openings, the motor is classified as an open motor; if the enclosure is complete, the motor is classified as an enclosed motor. Each of these types of motors has many modifications. *Table 1* lists the various types for both open and totally enclosed motors.

The different types of open motors as defined by NEMA are as follows:

• *General purpose* – Has ventilating openings which permit the passage of external cooling air over and around the windings of the machine.

**Table 1** Motor Enclosure Types

| Open | Totally Enclosed |
| --- | --- |
| General purpose | Nonventilated |
| Drip-proof | Fan-cooled |
| Splash-proof | Fan-cooled guarded |
| Guarded | Explosion-proof |
| Semi-guarded | Dust- and ignition-proof |
| Drip-proof guarded | Pipe-ventilated |
| Externally ventilated | Water-cooled |
| Pipe-ventilated | Water-to-air-cooled |
| Weather-protected (Type I & Type II) | |
| Encapsulated windings | |
| Sealed windings | |

- *Drip-proof* – Ventilating openings are so constructed that successful operation is not interfered with when drops of liquid or solid particles strike or enter the enclosure at any angle from 0° to 15° downward from the vertical.
- *Splash-proof* – Ventilating openings are so constructed that successful operation is not interfered with when drops of liquid or solid particles strike or enter the enclosure at any angle not greater than 100° downward from the vertical.
- *Guarded* – Openings giving direct access to live metal or rotating parts (except smooth surfaces) are limited in size by the structural parts or by screens, baffles, grills, expanded metal, or other means to prevent accidental contact with hazardous parts.
- *Semi-guarded* – Some of the ventilating openings, usually in the top half, are guarded as in the case of a guarded machine, but the others are left open.
- *Drip-proof guarded* – This type of machine has ventilating openings as in a guarded machine.
- *Externally ventilated* – Designating a machine that is ventilated by a separate motor-driven blower mounted on the machine enclosure. Mechanical protection may be as defined above. This machine is sometimes known as a blower-ventilated or force-ventilated machine.
- *Pipe-ventilated* – Openings for the admission of ventilating air are so arranged that inlet ducts or pipes can be connected to them.
- *Weather-protected* – Type I: Ventilation passages are so designed as to minimize the entrance of rain, snow, and airborne particles to the electrical parts. Type II: In addition to the enclosure described for a Type I machine, ventilating passages at both intake and discharge are so arranged that high-velocity air and airborne particles blown into the machine by storms or high winds can be discharged without entering the internal ventilating passages leading directly to the electric parts.
- *Encapsulated windings* – An AC squirrel cage machine having random windings filled with an insulating resin, which also forms a protective coating.
- *Sealed windings* – An AC squirrel cage machine making use of form-wound coils and an insulation system that, through the use of materials, processes, or a combination of materials and processes, results in a sealing of the windings and connections against contaminants.

The different types of totally enclosed motors as defined by NEMA are as follows:

- *Nonventilated* – Not equipped for cooling by means external to the enclosing parts.

- *Fan-cooled* – Equipped for exterior cooling by means of a fan or fans that are integral with the machine but external to the enclosing parts.
- *Fan-cooled guarded* – All openings giving direct access to the fan are limited in size by design of the structural parts or by screens, grills, expanded metal, etc., to prevent accidental contact with the fan.
- *Explosion-proof* – Designed and constructed to withstand an explosion of a specified gas or vapor that may occur within it and to prevent the ignition of the specified gas or vapor surrounding the machine by sparks, flashes, or explosions of the specified gas or vapor that may occur within the machine casing.
- *Dust- and ignition-proof* – Designed and constructed in a manner that will exclude ignitable amounts of dust or amounts which might affect performance or rating, and that will not permit arcs, sparks, or heat otherwise generated or liberated inside the enclosure to cause ignition of exterior accumulations or atmospheric suspensions of a specific dust on or in the vicinity of the enclosure.
- *Pipe-ventilated* – Openings are so arranged that when inlet and outlet ducts or pipes are connected to them there is no free exchange of the internal air and the air outside the case.
- *Water-cooled* – Cooled by circulating water, with the water or water conductors coming in direct contact with the machine parts.
- *Water-to-air-cooled* – Cooled by circulating air, which in turn is cooled by circulating water.

### 7.1.0 Open Motor

The most common type of motor is the open motor. It has ventilating openings that permit the passage of external cooling air over and around its windings. If these are limited in size and shape, the motor is called a protected motor, since it is protected from any large pieces of material that may somehow enter the motor, thus damaging its internal parts. A protected motor also prevents a

### 9.1.2 Full-Load Amps (FLA)

The FLA rating appearing on the nameplate indicates the current the motor will draw at nameplate horsepower, frequency, and voltage. Most manufacturers test to determine this value on a periodic basis during production, ensuring reasonable accuracy. The *NEC®* requires that the rated full-load current be the basis for determining the proper sizing of cable, overload protective devices, and other overcurrent protection in the motor circuit. Since many motors can be connected for one of two voltage ratings, they have two FLA ratings.

The FLA ratings are guaranteed if the induction motor is operating at full-load conditions and the applied voltage and frequency are the same as stated on the nameplate. When voltage or frequency are not the same, however, the current drawn by the motor at full-load conditions will be different from the nameplate indication (see *Table 4*). It is possible to damage a motor operated below its rated voltage or frequency, since the current the motor draws at full-load conditions increases in both cases. If the overload protective device is not sized properly, motor life may be shortened by this overcurrent condition.

### 9.1.3 Rated Full-Load Speed

The rated full-load speed is the value indicated in rpm on the nameplate. It is the speed at which the shaft will turn at the nameplate horsepower when supplied with power at the nameplate voltage and frequency. If the driven load is less than the nameplate horsepower, the shaft will turn faster than full-load speed.

If the motor is operating unloaded, the shaft will turn very close to synchronous speed. For example, with a full-load speed of 1,750 rpm, it can be inferred that the motor's synchronous speed is 1,800 rpm. The machine will operate from close to 1,800 rpm down to 1,750 rpm, from no-load to full-load conditions.

Common synchronous speeds are 3,600, 1,800, 1,200, 900, and 600 rpm. Synchronous speed is rarely found on the motor nameplate unless the machine has been retrofitted and has not yet been tested for new full-load speed.

**Table 4** Motor Operation

| Mode | Full-Load Current |
|------|-------------------|
| 110% of rated volts | 7% decrease |
| 90% of rated volts | 11% increase |
| 105% of rated frequency | 5%–6% decrease |
| 95% of rated frequency | 5%–6% increase |

### 9.1.4 Rated Horsepower

An induction motor is really a torque generator. It delivers a needed torque to a driven machine at a certain speed. Thus:

$$\text{Horsepower} = \frac{\text{load torque in ft-lbs} \times \text{rpm}}{5{,}250}$$

For induction motors that are built to NEMA standards, the ratings will range from ½hp to 400hp, with 24 categories in all. If horsepower requirements fall between any two ratings, the larger motor size should be selected.

Remember, an induction motor will try to deliver any amount of horsepower the load requires. If properly sized, most motors operate at something less than the motor nameplate horsepower. Standard motors are designed to operate at nameplate values from sea level up to an altitude of 3,300 feet if the ambient temperature does not exceed 104°F (40°C). Above this altitude, the nameplate horsepower no longer applies.

NEMA standards provide a method for determining the proper temperature rise, or the new maximum ambient temperature, at higher elevations. However, the standards do not provide a direct method for deriving the horsepower. Several methods are available to estimate true motor horsepower output.

### 9.1.5 Duty or Time Rating

All polyphase induction motors have either a **duty** or a time rating, which is the elapsed time the motor can operate at nameplate horsepower without shortening its life. The time rating of a motor is determined by operating the machine at full-load conditions and measuring the time it takes for the windings to heat up to the temperature rating of the insulation.

Standard time ratings are 5, 15, 30, and 60 minutes, and continuous or 24 **hours**. A motor with a time (or duty) rating other than continuous is a smaller motor that is given a higher horsepower

---

### Think About It

## Motor Horsepower and Speed

Can you replace a fan motor rated at a specific horsepower and speed with a motor rated at the same horsepower but a higher speed to increase airflow? If not, why not?

---

rating for a shorter period of time, thus reducing size and cost. In power plant uses, most motor ratings are continuous at 104°F (40°C).

### 9.1.6 NEMA Design Letters

The NEMA design letter defines the starting torque characteristics of an induction motor. It is one of the most important pieces of information on the nameplate; unfortunately, when a motor is replaced, the NEMA design letter is usually ignored, often leading to misapplication of the new machine. For fans or centrifugal pumps, starting torque requirements increase with the square of the change in speed. For mixers or loaded conveyor belts, however, starting torque requirements change very little with speed.

To account for these differences, NEMA has formulated design letters A, B, C, D, and F. The difference between motors with these letters is mainly in the design of the rotor, although there are also a few external differences. Design A and B motors are intended to drive conventional loads such as fans, blowers, and centrifugal pumps. About 80% of industrial motors are NEMA Design B.

Generally, the starting current is about five to seven times the rated full-load current. From *Table 5*, it can be seen that for larger motors, the starting current can be very significant, and across-the-line starting of larger motors could result in objectionable line-voltage dips. These voltage dips could result in other control equipment dropping out on low voltage and could even cause lights to dim.

### 9.1.7 Insulation Class

The electrical insulation system in a motor determines the machine's ultimate life span more than any other component. By some estimates, over 60% of all motors brought to repair shops are

**Table 5** Typical Currents for 220V, 60-Cycle Squirrel Cage Motors

| HP | Rated Full-Load Current | Starting (Maximum) Current | |
|---|---|---|---|
| | | Classes B, C, D | Class F |
| ½ | 2.0 | 12 | — |
| 1 | 3.5 | 24 | — |
| 1½ | 5.0 | 35 | — |
| 2 | 6.5 | 45 | — |
| 3 | 9 | 60 | — |
| 5 | 15 | 90 | — |
| 7½ | 22 | 120 | — |
| 10 | 27 | 150 | — |
| 15 | 40 | 220 | — |
| 20 | 52 | 290 | — |
| 25 | 64 | 365 | — |
| 30 | 78 | 435 | 270 |
| 40 | 104 | 580 | 360 |
| 50 | 125 | 725 | 450 |
| 60 | 150 | 870 | 540 |
| 75 | 185 | 1,085 | 675 |
| 100 | 246 | 1,450 | 900 |
| 125 | 310 | 1,815 | 1,125 |
| 150 | 360 | 2,170 | 1,350 |
| 200 | 480 | 2,900 | 1,800 |

there because of premature failure of the insulation system.

The insulation class is a NEMA designation that identifies the class of material used to insulate the windings. Four letters designate the four classifications. They are A, B, F, and H. The insulation class defines the temperature that the insulation can be subjected to without suffering damage. The insulation class is shown on the nameplate in *Figure 59*.

Class A is now obsolete insofar as industrial motors are concerned. Class A was once the most common classification for motor insulation, especially for small motors. Class A comprises materials or combinations of materials such as

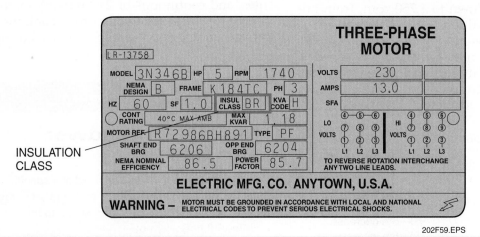

*Figure 59* Nameplate showing insulation class.

cotton or paper, when suitably impregnated or coated, or other materials capable of operation at the temperature rise assigned for Class A insulation for the particular machine.

Class B is the predominant class of insulation used in motor manufacturing and rewinding today. This class is the basic standard of the industry. It includes materials such as mica, glass fiber, polyester, and aramid laminates, etc., with suitable bonding substances, or other materials, not necessarily inorganic, capable of operation at the temperature rise assigned for Class B insulation for the particular machine. (The insulation class may be designated more specifically by the use of additional letters, such as the BR shown in *Figure 59*.)

Class F incorporates materials that are similar to those in Class B but are capable of operation at the temperature rise assigned for Class F for the particular machine.

Class H insulation systems comprise materials or combinations of materials such as silicone elastomer, mica, glass fiber, polyester, and aramid laminates, etc., with suitable bonding substances such as silicone resins, or other materials capable of operation at the temperature rise assigned for Class H insulation for the particular machine.

When replacing motors, ensure that the insulation class is equal to or better than that of the motor removed from service.

### 9.1.8 Nominal Rated Voltage

Nominal rated voltage is defined as the voltage rating at which the motor is designed to operate.

### 9.1.9 Minimum Starting Voltage

Minimum starting voltage may be defined as the lowest voltage at which a motor will start without drawing an excessive/trip current.

### 9.1.10 Frequency

Frequency is given for AC motors in hertz or cycles per second. Standard frequencies for AC motors are 50Hz and 60Hz. Alternating current in the U.S. is 60Hz.

### 9.1.11 Service Factor

The service factor is a multiplier for the nameplate horsepower rating that determines the amount of overload the motor can withstand. This extra horsepower is available if the motor is already operating at rated voltage and frequency and is in an environment that does not exceed the ambient temperature rating. The most common service factor appearing on a motor nameplate is 1.15.

### 9.1.12 NEMA Code Letters

The high current draw of the motor during the first moments of startup is called the in-rush or locked-rotor current. It can be derived from the kVA code letter on the motor nameplate. The letter corresponds to the kilovolt-amperes per rated horsepower (kVA/hp) required during the first moments of motor startup. *Table 6* provides the kVA/hp value for each kVA code—a letter from A to V, excluding I, O, and Q. The locked-rotor current is required when sizing fuses or determining a circuit breaker setting in an induction motor circuit.

### 9.1.13 Bearings

Polyphase induction motors require either anti-friction or sleeve bearings. Anti-friction bearings are standard in medium (integral) horsepower motor sizes through 125hp/1,800 rpm. They are optional in 150 to 600hp/1,800 rpm sizes. Sleeve bearings are standard in 500hp/3,600 rpm and larger sizes.

Since radial loads are higher at the drive end of the motor, the drive-end bearing has a higher load rating than the bearing at the opposite end.

**Table 6**  Locked-Rotor Code Letters
[Data from *NEC Table 430.7(B)*]

| Code Letter | kVA Per Horsepower with Locked Rotor |
|---|---|
| A | 0–3.14 |
| B | 3.15–3.54 |
| C | 3.55–3.99 |
| D | 4.0–4.49 |
| E | 4.5–4.99 |
| F | 5.0–5.59 |
| G | 5.6–6.29 |
| H | 6.3–7.09 |
| J | 7.1–7.99 |
| K | 8.0–8.99 |
| L | 9.0–9.99 |
| M | 10.0–11.19 |
| N | 11.2–12.49 |
| P | 12.5–13.99 |
| R | 14.0–15.99 |
| S | 16.0–17.99 |
| T | 18.0–19.99 |
| U | 20.0–22.39 |
| V | 22.4–AND UP |

A typical nameplate (*Figure 60*) might depict both bearing duties as:

- Shaft end brg: 6,206
- Opp end brg: 6,204

Bearing internal clearances are: C1 and C2 (smaller-than-normal clearance); standard clearance (normal); and C3, C4, and C5 (larger-than-normal clearance). Electric motors usually require a C3 internal clearance. Some bearing manufacturers have a different designation for motor bearings that have a larger-than-normal internal clearance.

### 9.1.14 Rated Amperage

Rated amperage may be defined as the full-load current required to produce full-rated horsepower at the motor's rated voltage and frequency. *Figure 60* shows the amperage for a typical motor.

### 9.1.15 Rated Horsepower

Horsepower is a rating used to specify the capacity of an electric motor to produce mechanical power to drive a specific piece of equipment (see *Figure 60*).

### 9.1.16 Locked-Rotor Current

The locked-rotor current is the steady-state current of a motor with the rotor locked and with rated voltage applied at rated frequency. NEMA has designated a set of code letters [*NEC Table 430.7(B)*] to define locked-rotor kilovolt-amperes (kVA) per horsepower (*Table 6*). This code letter appears on the nameplate of all AC squirrel cage induction motors. The kVA rating is an indication of the current draw and, indirectly, the impedance of the locked rotor.

The current drawn by the motor under stall conditions can be calculated using the values given in *Table 6*. The current drawn by the motor under stalled conditions must be considered when selecting the motor protection and starting package and in coordination with the power system protective devices.

### 9.1.17 Starting Current

The total instantaneous starting current comprises the locked-rotor current plus the transient in-rush that flows until the motor magnetic circuit stabilizes.

### 9.1.18 Temperature Rise

The temperature rise may be defined as the measure of the heat produced by the operation of the motor. Several conditions contribute to temperature rise. Examples include running current, hysteresis losses, and friction of rotating parts.

### 9.1.19 Power Factor

The power factor (pf) is the ratio of active power of an alternating or pulsating current (when measured with a wattmeter) to the apparent power indicated by an ammeter and voltmeter. It is also referred to as the phase factor. The power factor is the measure of the system or equipment efficiency.

## 9.2.0 Motor Protection

Fuses are normally used for motor overload protection. When used, fuses must be provided in each ungrounded conductor and also the grounded conductor of a three-wire, three-phase AC system with one conductor grounded. When non-fuse overload protective devices are used (*NEC*

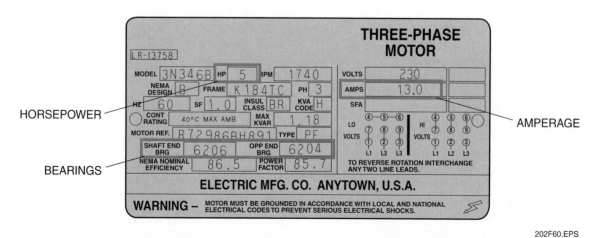

*Figure 60*  Nameplate showing bearings, horsepower, and amperage.

*Section 430.37*), follow the guidelines as stated in *Table 7*. Note that each motor winding must be individually protected against short circuits and ground faults.

In general, when providing overcurrent protection for motor circuits against overcurrents due to grounds and short circuits (*NEC Section 430.52*), follow the guidance listed in *Table 8*. *Table 8* provides time delay and instantaneous trip values for various types of motors as a percentage of full-load current.

When sizing the overload protection for continuous duty motors larger than 1hp, refer to *NEC Section 430.32(A)*. Motors that have a service factor of 1.15 and/or a maximum temperature rise of 40°C shall be provided with overload protection limited to 125% of the full-load current of the motor (this also applies to the secondary circuit of a wound rotor motor). All other motors shall be limited to 115% of the full-load current. Motors that are rated at less than 1hp have separate rules found in *NEC Section 430.32(B)*.

If the desired overload ratings are not available when sizing overload protection for the motor, then use the next highest available overload rating. This is allowed provided that 140% of full-load current is not exceeded by motors that have a service factor of 1.15 and/or a maximum temperature rise of 40°C. For all other motors, this maximum value would be 130% rather than 115% as stated earlier.

Additionally, in certain situations, motors with installed overloads rated as discussed previously may not start or be able to carry system load. In these instances, it is permissible to increase the overload settings to the respective 140%/130% values. When motor starting is still a problem, the overload protective device may be shorted out during the equipment startup sequence provided that the circuit breaker or fuse protecting the branch is not set at greater than 400% of the full-load current value and the motor does not have an automatic starter.

Overload protection for adjustable speed drives is based on the rated input to the power conversion equipment. If overload protection is supplied with the equipment, then no further overload protection is required. The rating of this disconnecting means shall be no less than 115% of the power conversion equipment rated input current and it shall be physically located in the incoming line. Overload protection, if not shunted, should allow a sufficient time delay for the motor to start and accelerate.

### 9.2.1 Thermal Protectors

Thermal protectors that are integral with the motor are often used to protect the motor from overloads and starting failures. All motors with a voltage rating greater than 600V must have a thermal protector and its overload must not have an

**Table 7**  Minimum Number of Overload Units (Data from *NEC Table 430.37*)

| Type of Motor | Supply System | Number and Location of Overload Units (such as trip coils or relays) |
|---|---|---|
| Single-phase AC or DC | Two-wire, single-phase AC or DC, ungrounded | One in either conductor |
| Single-phase AC or DC | Two-wire, single-phase AC or DC, one grounded conductor | One in ungrounded conductor |
| Single-phase AC or DC | Three-wire, single-phase AC or DC, grounded neutral | One in either ungrounded conductor |
| Single-phase AC | Any three-phase supply | One in ungrounded conductor |
| Two-phase AC | Three-wire, two-phase AC, ungrounded | Two, one in each phase |
| Two-phase AC | Three-wire, two-phase AC, one grounded conductor | Two, one in each ungrounded conductor |
| Two-phase AC | Four-wire, two-phase AC, grounded or ungrounded | Two, one per phase in ungrounded conductors |
| Two-phase AC | Five-wire, two-phase AC, grounded neutral or ungrounded | Two, one per phase in any ungrounded phase wire |
| Three-phase AC | Any three-phase supply | Three, one in each phase* |

*Exception: An overload unit in each phase shall not be required where overload protection is provided by other approved means.

**Table 8**  Motor Protection (Data from *NEC Table 430.52*)

| Type of Motor | Percent of Full-Load Current | | | |
| --- | --- | --- | --- | --- |
| | Nontime Delay Fuse | Dual Element (Time Delay) Fuse** | Instantaneous Trip Breaker | Inverse Time Breaker* |
| Single-phase motors | 300 | 175 | 800 | 250 |
| AC polyphase motors other than wound rotor: | | | | |
|   Squirrel cage: | | | | |
|     Other than Design B, energy efficient | 300 | 175 | 800 | 250 |
|     Design B, energy efficient | 300 | 175 | 1,100 | 250 |
|   Synchronous† | 300 | 175 | 800 | 250 |
| Wound rotor | 150 | 150 | 800 | 150 |
| Direct current (constant voltage) | 150 | 150 | 250 | 150 |

For certain exceptions to the values specified, see *NEC Section 430.54*.

*The values given in the last column also cover the ratings of nonadjustable inverse time types of circuit breakers that may be modified per *NEC Section 430.52*.

**The values in the Nontime Delay Fuse column apply to time delay Class CC fuses.

†Synchronous motors of the low-torque, low-speed type (usually 450 rpm or lower), such as are used to drive reciprocating compressors, pumps, etc., that start unloaded, do not require a fuse rating or circuit breaker setting in excess of 200% of the full-load current.

Reprinted with permission from *NFPA 70®*, the *National Electrical Code®*. Copyright © 2010, National Fire Protection Association, Quincy, MA 02269. This reprinted material is not the complete and official position of the National Fire Protection Association on the referenced subject, which is represented only by the standard in its entirety.

automatic reset feature. They shall trip no higher than the following percentage of full-load current:

- Motor full-load current not exceeding 9 amps—170%
- Motor full-load current between 9.1 and 20 amps—156%
- Motor full-load current greater than 20.1 amps—140%

This requirement is based on the maximum full-load motor current as listed in the tables provided in *NEC Article 430*.

Motors over 1,500hp include a device that is set to de-energize the motor once the actual temperature rise of the motor equals the rated temperature rise of the motor insulation. Thermal protectors are usually sized and installed by the motor manufacturer.

### 9.2.2 Branch Considerations

According to *NEC Section 430.22*, when a single motor used in a continuous duty application is supplied from a branch circuit, the ampacity of the branch circuit must be not less than 125% of the motor full-load current as determined by *NEC Section 430.6(A)(1)*, or not less than required by *NEC Section 430.22*. If a multiple-speed motor is used, then the ampacity shall be based on the highest of the full-load current ratings on the motor nameplate. Where motors have unusual duty cycle requirements, use the requirements listed in *Table 9*, as referenced in *NEC Section 430.22(E)*.

Per *NEC Section 430.24*, when sizing conductors supplying several motors, the capacity shall not be less than 125% of the largest motor plus the sum of the full-load current ratings of all other motors in the group, as determined by *NEC Section 430.6(A)*. Several motors or loads are permitted to be provided for on one branch circuit if:

- The system voltage is less than 600 volts.
- The branch protective device protects the smallest installed motor.
- All motors on the circuit must not be over 1hp and less than 20A (15A) on 120V (600V) circuits where each motor draws less than 6A, overloads must be installed on the motors, and short circuit current and ground fault current must not exceed the branch circuit rating.
- It is part of a factory-listed assembly.

In instances where taps are used, short circuit current and ground fault current protection may not be required for the taps used. This is true provided that the tap used has the same ampacity as the branch circuit it is connected to. Additionally, the tap cannot be longer than 25' and it must also be physically protected from damage.

**Table 9** Duty Cycle Service [Data from *NEC Table 430.22(E)*]

| Classification of Service | Percentages of Nameplate Current Rating | | | |
|---|---|---|---|---|
| | 5-Minute Rated Motor | 15-Minute Rated Motor | 30- and 60-Minute Rated Motor | Continuous Rated Motor |
| Short-Time Duty | | | | |
|   Operating valves, raising or lowering rolls, etc. | 110 | 120 | 150 | — |
| Intermittent Duty | | | | |
|   Freight and passenger elevators, tool heads, pumps, drawbridges, turntables, etc. | | | | |
|   For arc welders, see *NEC Section 630.11* | 85 | 85 | 90 | 140 |
| Periodic Duty | | | | |
|   Rolls, ore- and coal-handling machines, etc. | 85 | 90 | 95 | 140 |
| Varying Duty | 110 | 120 | 150 | 200 |

Any motor application shall be considered as continuous duty unless the nature of the apparatus it drives is such that the motor will not operate continuously with load under any condition of use.

## 10.0.0 Connections and Terminal Markings for AC Motors

The markings on the external leads of an induction motor are sometimes missing or illegible, and proper identification must be made before the motor can be connected to the line. This section describes the procedures for identifying leads in either a wye-connected or delta-connected, three-phase, nine-lead motor.

The required materials for this procedure are:

- Appropriate personal protective equipment
- 12V battery such as an automotive battery
- Analog meter with a large scale and low range (digital meters may not clearly capture the voltage kick)
- Test leads and jumpers
- Momentary contact, normally open (N.O.) pushbutton switch
- Labels to mark leads as they are identified
- Three-phase, nine-lead induction motor

Before starting, identify whether the motor to be tagged is wye-connected or delta-connected. Compare the two connection diagrams in *Figure 61* and *Figure 62*. You will see that both types of motors have nine leads and six coils. In a wye-connected motor, three coils are connected together and three coils are isolated. A delta-connected motor has three sets of two coils connected together. Using an ohmmeter, insulate all motor leads from one another and check for continuity between each lead. Start by placing one probe on one lead and check through the remaining leads. As you identify

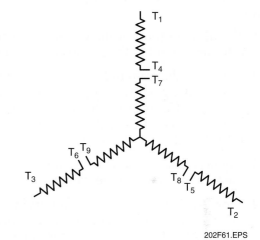

202F61.EPS

*Figure 61* Dual-voltage, three-phase wye connection.

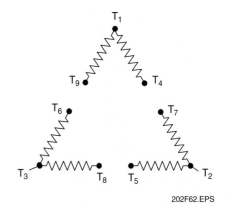

202F62.EPS

*Figure 62* Dual-voltage, three-phase delta connection.

which leads show continuity, group them together. Continue this procedure with each lead until all leads are grouped. When you have completed this procedure, you should have either three wires in one group and three sets of two wires grouped together, which would indicate a wye-connected motor, or three sets of three wires grouped together, indicating a delta-connected motor.

> **NOTE**
>
> If the leads are partially tied together, mark or identify them in a way that will allow you to reconnect them after the testing procedure is completed.

### 10.1.0 Identifying the Terminals of Wye-Connected Motors

*Figure 63* shows the coil arrangement for a three-phase, wye-connected motor. To identify the terminals of a wye-connected motor, proceed as follows:

*Step 1* Taking the group of three common leads, arbitrarily identify them as leads 7, 8, and 9.

*Step 2* Using the diagram in *Figure 64* as a guide, connect the positive lead from the battery to lead 7. Connect the lead from the battery negative through the N.O. switch to leads 8 and 9 simultaneously.

*Step 3* Connect one of the three remaining lead pairs to the voltmeter terminals.

*Step 4* Close the N.O. switch while observing the DC voltmeter. Use the lowest scale practical without over-ranging the meter. If the meter deflection is upward, note the voltage reading. Observe that the reading occurs only on the initial energization of

*Figure 64*  Battery hookup for wye-connected motor lead identification.

the windings and then decays. Note the peak reading only and ignore the deflection in the opposite direction that occurs when the switch is opened. If the meter initially deflects downward, reverse the test lead connections.

*Step 5* Continue with the remaining two lead pairs. The pair with the highest voltage reading is the winding associated with lead 7. The lead with positive polarity is identified as lead 4 and the lead with negative polarity is lead 1.

*Step 6* Repeat Step 3, but apply the positive lead of the battery to lead 8 and the negative lead of the battery to leads 7 and 9. The positive lead of the pair with the highest voltage is identified as lead 5 and the lead with negative polarity is lead 2.

*Step 7* Repeat Step 3, but apply the positive lead of the battery to lead 9 and the negative lead of the battery to leads 7 and 8. The positive lead of the pair with the highest voltage is marked lead 6 and the negative lead is lead 3.

*Step 8* To confirm that all leads are correctly identified, connect the motor to the circuit. Be sure to observe proper connection procedures for the applied voltage. Once connected to the circuit, start the motor and take current readings on all three lines. If the motor starts correctly and the current readings are approximately equal, the procedure was a success.

*Figure 63*  Coil arrangement in a wye-connected motor.

## 10.2.0 Identifying the Terminals of Delta-Connected Motors

A delta-connected motor has three sets of three leads. *Figure 65* shows how the coils are arranged in a delta-connected, three-phase motor. In this figure, the coils that are side by side are actually wound on the same poles on the motor. As discussed previously, this will allow some transformer interaction between adjacent coils. To identify the terminals of a delta-connected motor, proceed as follows:

*Step 1*  Using an ohmmeter on a low scale, measure the resistance between each of the three leads in one group. When performing this measurement, you should see that the resistance between two of the leads is about twice that between either of those two and the third. The lead that shows the least resistance to the other two will be lead 1. Refer to all three wires in this set as set 1.

*Step 2*  Repeat Step 1 with the second set to identify the lead with the least resistance as lead 2. Refer to all three wires in this set as set 2.

*Step 3*  Repeat Step 1 with the final set of leads to identify the lead with the least resistance as lead 3. Refer to all three wires in this set as set 3.

*Step 4*  Using the diagram in *Figure 66* as a guide, connect lead 1 to the positive terminal of a DC voltage source and one of the remaining leads in that set (set 1) to the negative terminal. Attach the red lead of the voltmeter to lead 2 and the black lead to one of the two unknown leads in set 2. Press the pushbutton and observe the meter needle. If the correct leads have been selected, a voltage will be induced into this coil. If not, connect the second lead to the voltmeter and

repeat the test. If there is still no induced voltage, disconnect the unknown lead in set 1 from the DC voltage and connect the remaining unknown lead to the negative source. Repeat the test until the leads with an induced voltage have been identified. Once these leads are located, identify the lead connected to the negative terminal of the DC voltage source as lead 4 and the lead connected to the negative voltmeter probe as lead 7. Identify the other lead in set 2 as lead 5.

*Step 5*  The remaining lead in set 1 will be lead 9. Leaving lead 1 on the positive DC terminal, connect the negative terminal to lead 9. Attach the red lead of the voltmeter to lead 3 and the black lead to one of the two unknown leads in set 3. Press the pushbutton and observe the meter needle. If the correct leads have been selected, a voltage will be induced into this coil. If not, connect the second lead to the voltmeter and repeat the test. Once these leads are located, identify the lead connected to the negative voltmeter probe as lead 6.

*Step 6*  The remaining lead in set 3 is lead 8. To verify this, connect lead 3 to the positive terminal of the DC voltage source and lead 8 to the negative terminal. Attach the red lead of the voltmeter to lead 2 and the black lead to lead 5. Press the pushbutton and observe the meter needle. If the results are correct, a voltage will be induced into this coil, resulting in meter needle deflection.

*Figure 66*  Battery hookup for delta-connected motor lead identification.

*Figure 65*  Coil arrangement in a delta-connected motor.

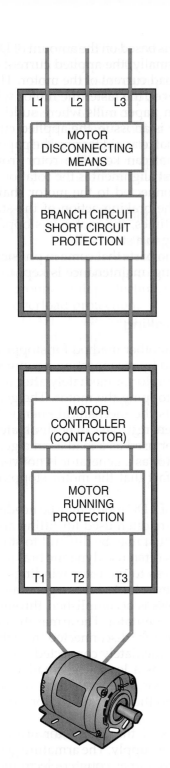

*NEC Article 430, Part IX*
*Sections 430.101*
*through 430.113*

**Disconnects motor and controllers from circuit.**
1. Continuous rating of 115% or more of motor FLC. Also see *NEC Article 430, Part II*.
2. All disconnecting means in the motor circuit must be per *NEC Sections 430.108, 430.109, and 430.110*.
3. Must be located in sight of motor location and driven machinery. The controller disconnecting means can serve as the disconnecting means if the controller disconnect is located in sight of the motor location and driven machinery.

*NEC Article 430, Part IV*
*Sections 430.51*
*through 430.58*

**Protects branch circuit from short circuits or grounds.**
1. Must carry starting current of motor.
2. Rating must not exceed values in *NEC Table 430.52* unless not sufficient to carry starting current of motor.
3. Values of branch circuit protective devices shall in no case exceed exceptions listed in *NEC Section 430.52*.

*NEC Article 430, Part VII*
*Sections 430.81*
*through 430.90*

**Used to start and stop motors.**
1. Must have current rating of 100% or more of motor FLC.
2. Must be able to interrupt LRC.
3. Must be rated as specified in *NEC Section 430.83*.

*NEC Article 430, Part III*
*Sections 430.31*
*through 430.44*

**Protects motor and controller against excessive heat due to motor overload.**
1. Must trip at following percent or less of motor FLC for continuous motors rated more than one horsepower.
   a) 125% FLC for motors with a marked service factor of not less than 1.15 or a marked temperature rise of not over 40°C.
   b) 115% FLC for all others. (See the *NEC*® for other types of protection.)
2. Three thermal units required for any three-phase AC motor.
3. Must allow motor to start.
4. Select size from FLC on motor nameplate.

*NEC Article 430, Part II*
*Sections 430.21*
*through 430.29*

**Specifies the sizes of conductors capable of carrying the motor current without overheating.**
1. To determine the ampacity of conductors, switches, branch circuit overcurrent devices, etc., the full-load current values given in *NEC Tables 430.247 through 430.250* shall be used instead of the actual current rating marked on the motor nameplate. *(See NEC Section 430.6.)*
2. According to *NEC Section 430.22*, branch circuit conductors supplying a single motor used in a continuous duty application shall have an ampacity of not less than 125% of motor FLC, as determined by *NEC Section 430.6(A)(1)*.

202F67.EPS

*Figure 67* Summary of requirements for motors, motor circuits, and controllers.

stop using dynamic braking. Dynamic braking also cannot hold a load once it is stopped because without the generated CEMF there is no more braking action.

For this reason, electromechanical friction brakes are sometimes used along with dynamic braking in applications that require the load to be held, or in applications where a large, heavy load is to be stopped. This is similar to using a parachute to slow a racecar before applying the brakes.

## 13.0.0 MOTOR INSTALLATION

The best motors on the market will operate improperly if they are installed incorrectly. Therefore, all personnel involved with the installation of electric motors should understand the procedures for installing the various types of motors that will be used.

> **WARNING!**
>
> When a motor is received at the job site, always refer to the manufacturer's instructions and follow them to the letter. Failure to do so could result in serious injury or death. Install and ground according to *NEC*® requirements and good practices. Consult qualified personnel with any questions or problems.

Keep the following in mind when installing new motors:

- *Uncrating* – Once the motor has been carefully uncrated, check to see if any damage has occurred during handling. Be sure that the motor shaft and armature turn freely. This is also a good time to check to determine if the motor has been exposed to dirt, grease, grit, or excessive moisture during shipment or storage. Motors in storage should have shafts turned over once each month to redistribute grease in the bearings. The measure of insulation resistance is a good dampness test. Clean the motor of any dirt or grit.

> **WARNING!**
>
> Never start a motor that has been wet until it has been completely dried and thoroughly tested.

## Installing Motors

The shaft of this air conditioner motor must be aligned precisely with the shaft of the driven device. Note the micrometer attached to the motor and the load to achieve exact alignment.

202SA07.EPS

- *Lifting* – Eyebolts or lifting lugs on motors are intended only for lifting the motor and factory motor-mounted standard accessories. These lifting provisions should never be used when lifting or handling the motor when the motor is attached to other equipment as a single unit. The eyebolt lifting capacity rating is based on its alignment with the eyebolt centerline. The capacity decreases if it is not aligned properly.
- *Guards* – Rotating parts such as pulleys, couplings, external fans, and shaft extensions must be permanently guarded against accidental contact with clothing or body extremities.
- *Requirements* – All motors must be installed, protected, and fused in accordance with *NEC Article 430*. For general information on grounding, refer to *NEC Article 250* and *NEC Article 430, Part XIII*.

- *Thermal protector information* – The motor nameplate may or may not be stamped to indicate thermal protection.

---

### Think About It

## Putting It All Together

Count the motors in your home. The typical home may easily have over 30 motors (including electronic equipment and tools). A century ago, a typical home might have had none. Determine what types of motors you have. Are they AC or DC? What identifying information can you determine by examining each motor nameplate?

---

## SUMMARY

This module discussed AC and DC motor theory, construction, and various motor types and applications. This discussion included torque, speed, and speed regulations as well as the fundamental concepts associated with variable speed drive systems. Motor enclosures were described, including open and totally enclosed motors and motor frame designations. Discussions of horsepower and calculation of load under various conditions were included. *NEC*® requirements and installation considerations were also discussed.

# Review Questions

1. The name of the motor part that rotates during operation is the _____.
   a. stator
   b. shunt/capacitor
   c. brushes/commutator
   d. armature/rotor

2. The electrical energy required to produce one horsepower of mechanical energy is _____.
   a. 476W
   b. 647W
   c. 746W
   d. 864W

3. When there is a wide variation in both torque and speed requirements, such as a blower or a hoist, use a _____.
   a. brushless DC motor
   b. compound DC motor
   c. series DC motor
   d. shunt DC motor

4. The principal reason for developing the brushless DC motor was to _____.
   a. eliminate commutator problems
   b. improve efficiency
   c. increase horsepower
   d. improve airplanes

5. The name of the stationary motor part that produces the magnetic field during operation is the _____.
   a. stator
   b. shunt/capacitor
   c. brushes/commutator
   d. armature/rotor

6. The most popular rotor in use for polyphase AC motors is the _____.
   a. wound rotor
   b. squirrel cage rotor
   c. multiphase rotor
   d. induction rotor

7. The starting torque characteristics of an induction motor are defined by _____.
   a. rotor resistance
   b. starting current
   c. load
   d. stator resistance

8. A three-phase synchronous motor is started with a(n) _____.
   a. squirrel cage bar
   b. moveable actuator disc
   c. centrifugal weight assembly
   d. amortisseur winding

9. The speed of an induction motor depends on the power supply frequency and the _____.
   a. current
   b. voltage
   c. size
   d. number of poles

10. The speed on a multiple-winding motor is normally changed by _____.
   a. switching contacts in the motor controller
   b. reversing any two of the incoming leads
   c. an external wye-connected resistor
   d. timing relays

11. The two primary types of motor enclosures classified by NEMA are _____.
   a. general purpose and guarded
   b. open and totally enclosed
   c. guarded and semi-guarded
   d. nonventilated and externally ventilated

12. An open AC squirrel cage motor that makes use of form-wound coils and an insulation system is known as a(n) _____.
   a. semi-guarded type
   b. sealed type
   c. non-ventilated type
   d. explosion-proof type

13. The NEMA frame designation for gasoline pump motors is _____.
   a. B
   b. C
   c. G
   d. H

14. The letters FLA on a motor nameplate stand for _____.
   a. fused last application
   b. fast lower arm
   c. full-load amps
   d. full-load armature

15. To determine the horsepower rating of a motor that is already installed and in place, _____.

    a. multiply the rated voltage times the FLA
    b. divide the nominal efficiency by the power factor
    c. check the nameplate data
    d. use a torque wrench and horsepower data

16. The multiplier to the nameplate horsepower rating is called the _____.

    a. maximum kVAR
    b. service factor
    c. power factor
    d. NEMA nominal efficiency

17. To protect the motor from overloads and starting failures, use a _____.

    a. fuse
    b. circuit breaker
    c. centrifugal switch
    d. thermal protector

18. The code requirements for motor disconnects are covered in _____.

    a. *NEC Article 430, Part VIII*
    b. *NEC Article 430, Part IX*
    c. *NEC Article 430, Part X*
    d. *NEC Article 430, Part XI*

19. The method of braking in which DC is applied to the motor windings after AC is removed is known as _____.

    a. injection braking
    b. friction braking
    c. dynamic braking
    d. regenerative braking

20. Reconnecting a running motor to act as a generator is known as _____.

    a. injection braking
    b. friction braking
    c. regenerative braking
    d. dynamic braking

1. The field coils of a series motor are connected in series with the _____.

2. A(n) _____ is used to provide the switching action required for continuous DC motor rotation.

3. The most common class of insulation used in motor manufacturing and rewinding is NEMA _____.

4. All motors with a voltage rating greater than _____ must have a thermal protector.

5. What type of motor requires brushes, slip rings, and an amortisseur winding? _____ _____

6. The two basic types of DC motor connections are the _____ and _____.

7. Explain slip and why it is present. _____

_____

_____

_____

_____

_____

_____

8. Most industrial motors are NEMA Design _____.
   a. A
   b. B
   c. D
   d. F

9. What type of rotor is most commonly used in an induction motor? _____

_____

10. An insulation resistance test can be used to determine _____.
    a. hp rating of a motor
    b. FLA rating of a motor
    c. motor dampness
    d. field strength of a motor

11. When a single motor is supplied from a branch circuit, the ampacity of the branch circuit must be _____ of the motor full-load current as determined by *NEC Section 430.6(A)(1)*.

12. A consequent-pole motor has _____ speeds.

13. How is the speed normally changed in a multiple-speed induction motor?

_____

14. Name three conditions that can cause a synchronous motor to become out of step.

_____

_____

_____

15. The rpm of a 60Hz motor with two poles is _____.

**Armature:** The rotating windings of a DC motor.

**Branch circuit:** The circuit conductors between the final overcurrent device protecting the circuit and the outlet(s).

**Brush:** A conductor between the stationary and rotating parts of a machine. It is usually made of carbon.

**Circuit breaker:** A device designed to open and close a circuit by nonautomatic means and to open the circuit automatically on a predetermined overcurrent without injury to itself when properly applied within its rating.

**Commutator:** A device used on electric motors or generators to maintain a unidirectional current.

**Continuous duty:** Operation at a substantially constant load for an indefinitely long time.

**Controller:** A device that serves to govern, in some predetermined manner, the electric power delivered to the apparatus to which it is connected.

**Duty:** Describes the length of operation. There are four designations for circuit duty: continuous, periodic, intermittent, and varying.

**Equipment:** A general term including material, fittings, devices, appliances, fixtures, apparatus, and the like used as a part of, or in connection with, an electrical installation.

**Field poles:** The stationary portion of a DC motor that produces the magnetic field.

**Horsepower:** The rated output capacity of the motor. It is based on breakdown torque, which is the maximum torque a motor will develop without an abrupt drop in speed.

**Hours:** The duty cycle of a motor. Most fractional horsepower motors are marked continuous for around-the-clock operation at the nameplate rating in the rated ambient conditions. Motors marked one-half are for ½-hour ratings, and those marked one are for 1-hour ratings.

**Intermittent duty:** Operation for alternate intervals of (1) load and no load; or (2) load and rest; or (3) load, no load, and rest.

**Overcurrent:** Any current in excess of the rated current of equipment or the ampacity of a conductor. It may result from an overload, short circuit, or ground fault.

**Overload:** Operation of equipment in excess of the normal, full-load rating, or of a conductor in excess of rated ampacity, which, after a sufficient length of time, will cause damage or dangerous overheating. A fault, such as a short circuit or ground fault, is not an overload.

**Periodic duty:** Intermittent operation at a substantially constant load for a short and definitely specified time.

**Revolutions per minute (rpm):** The approximate full-load speed at the rated power line frequency. The speed of a motor is determined by the number of poles in the winding. A four-pole, 60Hz motor runs at an approximate speed of 1,725 rpm. A six-pole, 60Hz motor runs at an approximate speed of 1,140 rpm.

**Rotation:** For single-phase motors, the standard rotation, unless otherwise noted, is counterclockwise facing the lead or opposite shaft end. All motors can be reconnected at the terminal board for opposite rotation unless otherwise indicated.

**Synchronous speed:** The speed of the revolving field of the stator, which is dependent on the supply frequency and number of poles in a stator winding.

**Thermal protector:** A protective device for assembly as an integral part of a motor or motor compressor that, when properly applied, protects the motor against dangerous overheating due to overload or failure to start.

**Varying duty:** Operation at varying loads and/or intervals of time.

# Additional Resources

This module presents thorough resources for task training. The following resource material is suggested for further study.

*National Electrical Code® Handbook*, Latest Edition. Quincy, MA: National Fire Protection Association.

# Figure Credits

Timothy A. Ely, Module opener

Topaz Publications, Inc., Figure 25, SA02, Figure 40, SA03, SA04, Figure 45B, SA05, SA06, Figure 55B, and SA07

Reprinted with the permission of the National Electrical Manufacturers Association, Tables 2 and 5

National Fire Protection Association, Tables 6–9

Reprinted with permission from *NFPA 70®* -2011, *National Electrical Code®*, Copyright © 2010, National Fire Protection Association, Quincy, MA. This reprinted material is not the complete and official position of the National Fire Protection Association on the referenced subject, which is represented only by the standard in its entirety. *NFPA 70®*, *National Electrical Code®*, and *NEC®* are registered trademarks of the National Fire Protection Association, Inc., Quincy, MA.

# CONTREN® LEARNING SERIES — USER UPDATE

NCCER makes every effort to keep its textbooks up-to-date and free of technical errors. We appreciate your help in this process. If you find an error, a typographical mistake, or an inaccuracy in NCCER's Contren® materials, please fill out this form (or a photocopy), or complete the online form at www.nccer.org/olf. Be sure to include the exact module number, page number, a detailed description, and your recommended correction. Your input will be brought to the attention of the Authoring Team. Thank you for your assistance.

*Instructors* – If you have an idea for improving this textbook, or have found that additional materials were necessary to teach this module effectively, please let us know so that we may present your suggestions to the Authoring Team.

**NCCER Product Development and Revision**
3600 NW 43rd Street, Building G, Gainesville, FL 32606

**Fax:** 352-334-0932
**Email:** curriculum@nccer.org
**Online:** www.nccer.org/olf

☐ Trainee Guide ☐ AIG ☐ Exam ☐ PowerPoints Other _____

Craft / Level: _____ Copyright Date: _____

Module Number / Title: _____

Section Number(s): _____

Description: _____

_____

_____

_____

Recommended Correction: _____

_____

_____

_____

Your Name: _____

Address: _____

_____

Email: _____ Phone: _____

# Electric Lighting

**Walt Disney World's Wilderness Lodge**

The Wilderness Lodge is modeled after the Old Faithful Inn that was built in Yellowstone National Park in 1902; it even includes a functional reproduction of Old Faithful Geyser. Installation of the electrical work involved highly detailed coordination to ensure that the various lighting and power systems would be concealed and not detract from the period feel of the building.

## 26203-11

Trainees with successful module completions may be eligible for credentialing through NCCER's National Registry. To learn more, go to **www.nccer.org** or contact us at **1.888.622.3720.** Our website has information on the latest product releases and training, as well as online versions of our *Cornerstone* newsletter and Pearson's Contren® product catalog.

Your feedback is welcome. You may email your comments to **curriculum@nccer.org,** send general comments and inquiries to **info@nccer.org,** or use the User Update form at the back of this module.

 V.1 6/11

## Objectives

When you have completed this module, you will be able to do the following:

1. Describe the characteristics of light.
2. Recognize the different kinds of lamps and explain the advantages and disadvantages of each type:
   - Incandescent
   - Halogen
   - Fluorescent
   - High-intensity discharge (HID)
3. Properly select and install various lamps into lighting fixtures.
4. Recognize and describe the installation requirements for various types of lighting fixtures:
   - Surface-mounted
   - Recessed
   - Suspended
   - Track-mounted
5. Recognize ballasts and describe their use in fluorescent and HID lighting fixtures.
6. Explain the relationship of Kelvin temperature to the color of light produced by a lamp.
7. Recognize basic occupancy sensors, photoelectric sensors, and timers used to control lighting circuits and describe how each device operates.

## Performance Tasks

Under the supervision of the instructor, you should be able to do the following:

1. Read and interpret information given in lamp manufacturers' catalogs for one or more selected lamps.
2. Properly select and install lamps into lighting fixtures.
3. Install one or more of the following lighting fixtures and their associated lamps:
   - Surface-mounted
   - Recessed
   - Suspended
   - Track-mounted

## Trade Terms

| | | |
|---|---|---|
| Ballast | Lumen (lm) | Reflected |
| Color rendering index (CRI) | Lumen maintenance | Reflected light |
| Dip tolerance | Lumens per watt (LPW) | Refracted/refraction |
| Efficacy | Luminaire (fixture) | Starter |
| Incandescence | Luminance | Troffer |
| Incident light | | |

## Required Trainee Materials

1. Pencil and paper
2. Appropriate personal protective equipment
3. Copy of the latest edition of the *National Electrical Code®*

Note: *NFPA 70®*, *National Electrical Code®*, and *NEC®* are registered trademarks of the National Fire Protection Association, Inc., Quincy, MA 02269. All *National Electrical Code®* and *NEC®* references in this module refer to the 2011 edition of the *National Electrical Code®*.

# Contents

*Topics to be presented in this module include:*

## Figures and Tables

## Figures and Tables (*continued*)

# 1.0.0 INTRODUCTION

Electric lighting is used extensively throughout residential structures, commercial businesses, industrial plants, and outdoor sites. Its use in residential structures can improve their appearance and that of the objects in them. In stores, hotels, and office buildings, electric lighting is used on a large scale to improve the efficiency of the employees and aid in selling merchandise. In industry, electric lighting helps to increase production, reduce errors, and increase safety. Regardless of the application, correct electric lighting serves to provide illumination for the performance of visual tasks with a maximum of comfort and a minimum of eyestrain and fatigue, allowing individuals to perform their daily living and work-related tasks more easily.

This module is the first of two modules that cover the subject of electric lighting. It introduces the basic principles of human vision and seeing and describes the characteristics of light. It also covers different kinds of lamps, lighting fixtures, and related components. The remainder of this module focuses on basic guidelines and procedural information for the receiving, storing, handling, and installation of electric lamps and lighting fixtures. It also provides an overview of common lighting circuit control devices and energy management systems. The second module, presented later in your training, will cover the practical applications of lighting fixtures.

# 2.0.0 HUMAN VISION

Human vision is a process that occurs partly in the eye and partly in the brain. Light reflected from objects stimulates the eye. This stimulation is conveyed to the brain, where it is registered as a conscious sensation.

The structure of the eye is similar in many ways to a camera, which consists of a lens system, a variable diaphragm, and film. The variable diaphragm is the iris of the eye (*Figure 1*), and the film is the retina. Light enters the eye through a transparent layer called the cornea. The amount of light that is allowed to strike the lens is controlled by the contraction and expansion of the iris. During low-light levels, the iris expands, and during high-light levels, it contracts.

Light passes through the pupil, which is the opening of the iris, and then through the lens, which is directly behind the iris. There, the light is focused to form an image that is passed through a transparent, jelly-like substance, called the vitreous humor, to the back wall or retina of the eye. The light on the retina

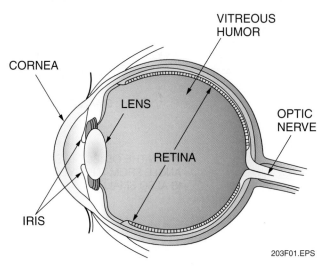

*Figure 1*   Structure of the human eye.

stimulates nerve terminals called rods and cones. These nerve terminals are connected to the brain by the optic nerve, which furnishes the path by which the light impulses are transferred from the eye to the brain.

The field of vision covers a wide angle of about 200° horizontally and 120° vertically. In the central region of the field of vision, the eye is responsive to color and detail, whereas in the outer region, it is chiefly sensitive to motion. The limits of vision are determined mainly by the following factors:

- *Intensity threshold* – The lowest brightness level that can stimulate the eye.
- *Contrast* – The difference in the degree of luminance (brightness).
- *Visual angle* – When an object is made smaller or is placed at a greater distance from the eye, the angle formed by the light rays from the extremities of the object to the eye becomes smaller (*Figure 2*), and vice versa.
- *Time threshold* – The minimum time during which a light stimulus must be present in order to be effective. If the interval is too short, the rods and cones of the eye do not have time to respond to an image on the retina. The time threshold is also dependent on the size, brightness, and color of the object being observed.

Other characteristics of vision are:

- High contrast, sharp edges, and motion increase the sensitivity to detail.
- Straight lines are easier to see than curved lines. Horizontal or vertical lines are more easily seen than diagonal lines.

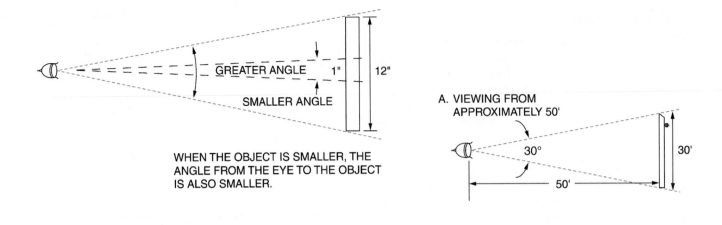

WHEN THE OBJECT IS SMALLER, THE
ANGLE FROM THE EYE TO THE OBJECT
IS ALSO SMALLER.

A. VIEWING FROM
APPROXIMATELY 50'

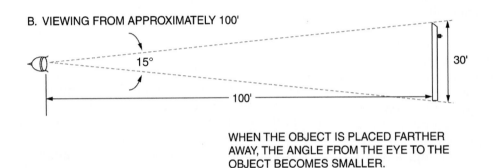

WHEN THE OBJECT IS PLACED FARTHER
AWAY, THE ANGLE FROM THE EYE TO THE
OBJECT BECOMES SMALLER.

203F02.EPS

*Figure 2*   Visual angle.

- The background surrounding an object changes its appearance. For example, a gray object appears lighter when it is placed upon a black background, but it appears darker when placed on a white background.

## 3.0.0 LIGHT CHARACTERISTICS

When discussing light, most people think of natural light from the sun or the light emitted from a light source, such as an electric light bulb. This type of light is called incident light. Another type of light is reflected light, which is light given off by an object when incident light strikes it. The difference between the two is that reflected light is dependent on the incident light. When light is not shining on an object, light will not be given off unless the object contains self-luminating properties.

### 3.1.0 Absorption, Reflection, and Refraction of Light

Three things can happen when light strikes objects or surfaces. Some of the light can be absorbed, some can be transmitted through the object media and refracted or diffused, and some can be reflected back from the surface (*Figure 3*). Absorption occurs when light rays pass through a transparent or translucent medium or meet a dense body, such as an opaque reflector surface. The amount of energy absorbed depends on the object's molecular construction, the wavelength of the incident light striking it, and the angle at which the light strikes it. Light energy absorbed by an object is dissipated in the form of heat. This explains why walls, floors, metals, etc., feel warm

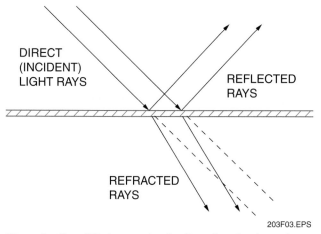

*Figure 3* Simplified example of reflected and refracted light rays.

**Table 1** Percentages of Light Reflected by Common Surface Materials

| Surface Material | Percentage of Light Reflected |
| --- | --- |
| White plaster | 90% to 92% |
| Mirrored glass | 80% to 90% |
| White paint | 75% to 90% |
| Metalized plastic | 75% to 85% |
| Polished aluminum | 75% to 80% |
| Stainless steel | 55% to 65% |
| Limestone | 35% to 65% |
| Marble (white) | 45% |
| Concrete | 40% |
| Dark red–glazed bricks | 30% |

or hot when exposed to direct sunlight or light from a strong artificial light source.

The energy not absorbed by an object when light strikes it can be transmitted through the object or reflected back from its surface, or both. If light rays strike the object perpendicular (at 90°) to its surface, the rays will be transmitted through it in a straight line and/or be reflected back from its surface in the same direction from which it came. If the light strikes the object's surface at an angle other than 90°, as shown in *Figure 3*, then the light transmitted through the object is bent in a different direction. This is called refraction. The light may also be reflected back from the object in a different direction from that in which it came. The resultant directions taken by reflected and/or refracted light waves can be very complex. This is because the redirection imparted to the waves depends on the type and density of the object's material, as well as whether its surface is smooth or rough, or if the material contains particles beneath its surface.

The greater the amount of light that is reflected by an object, the brighter the object will appear to the eye. For example, if all the direct light striking an object's surface is reflected from the surface, the object appears light; if it is entirely absorbed, the object appears dark. In addition, the more intense the light source, the brighter the object will become. For example, if you cast a shadow on a portion of an object and look at the difference in brightness between the two areas, the portion without the shadow will be brighter.

Some surfaces and materials reflect light better than others. Light-colored or highly-polished surfaces reflect more light than darker-colored or dull surfaces. This is because white and light-colored materials absorb less of the light energy than do dark or dull surfaces, thus leaving more energy available to be reflected. The color of the walls, ceilings, and floors and their reflecting ability are major considerations in interior lighting design. Typically, darker areas require the use of more artificial light. *Table 1* shows some typical examples of the percentages of light reflected by common surface materials.

### 3.2.0 Light Colors

Any energy that travels by wave motion is considered radiant energy. Light is one of the many forms of radiant energy. Radio waves and X-rays are also forms of radiant energy. Visible light occupies only a small portion of the radiant energy spectrum. Light is made up of that portion of the spectrum encompassing wavelengths between 380 and 780 nanometers (*Figure 4*). A nanometer is one billionth of a meter.

When all the wavelengths of the light spectrum are presented to the eye in nearly equal proportions, white light is seen. This white light is made up of the wavelengths that create the different colors. This composition can be demonstrated by passing light through a prism. As shown in *Figure 5*, the light spectrum is broken up into its component wavelengths, with each representing a different color. The ability of the prism to separate the different wavelengths, and therefore colors, is because the white light is dispersed into its component wavelengths by refraction as it passes through the prism. The spectrum ranges from violet on the lower end to red on the upper end. In between fall blue, green, yellow, and orange. A total of six colors are visible when passing white light through a prism. Since the colors of the spectrum pass gradually from one to the other, the theoretical number of colors becomes infinite. It has been determined that about 125 colors can be identified over the visible spectrum.

Figure 4 Visible light spectrum.

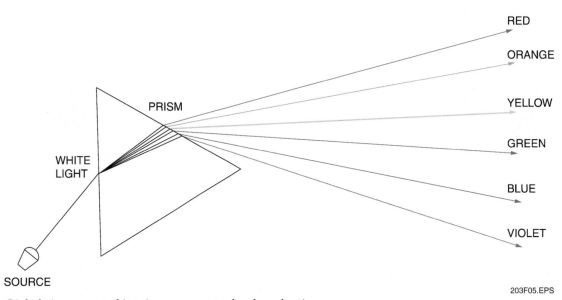

Figure 5 Light being separated into its component colors by refraction.

As previously described, the amount of light energy absorbed by an object depends on the wavelengths of the incident light striking the object. All objects absorb light of different wavelengths in different proportions. This is called selective absorption. Practically all colored objects owe their color to selective absorption in some part of the visible light spectrum, with resulting reflection and transmission in other selected parts of the spectrum. It is this characteristic that gives us our color distinction when viewing different objects, surfaces, etc. An object's appearance results from the way it reflects the particular light that is falling on it. For example, under white light an apple appears red because it tends to reflect light in the red portion of the spectrum and absorb light of other wavelengths (colors).

Now that you understand the basic concepts of light and how it is perceived, you can better choose the various electric lamps and **luminaires (fixtures)** that are used to provide light in residential, commercial, and industrial applications.

---

### Think About It

## Visible Light Spectrum

Can you think of one example that sometimes occurs in nature that confirms the fact that white light is actually made up of different colors?

## 4.0.0 LAMPS

There are three main categories of lamps used for lighting: incandescent, fluorescent, and HID. Each category of lamp is made in a wide variety of shapes, sizes, finishes, and mounting bases.

---

## 4.1.0 Standard Incandescent Lamps

Incandescent lamps, also called filament lamps, are used for general lighting and typically provide a warm and natural light. Incandescent lamps were invented over a century ago. With some refinements, the basic construction of a standard incandescent lamp remains the same today. Incandescent lamps consist of a thin coiled or shaped tungsten-wire filament (*Figure 6*) supported inside an evacuated glass envelope (bulb) filled with an inert gas, typically a mix of argon and nitrogen. The inert gas helps to prevent the filament from combining with oxygen and burning out. The envelopes of most lamps are made of regular lead or soda lime (soft) glass. Envelopes of lamps that must withstand higher temperatures are typically made of borosilicate heat-resistant (hard) glass. The lamp's base supports the lamp envelope and filament and provides the electrical connection between the lamp and its power source.

In an incandescent lamp, light is generated by passing an electric current through the filament, and its resistance causes it to heat to **incandescence.** The hotter the filament gets, the more efficient it is in converting electricity to light output. Tungsten has a positive resistance characteristic that makes its resistance at operating temperature much greater than its cold resistance (typically 12 to 16 times greater). It should be pointed out that when a filament operates hotter, its life is shortened. This makes the design of each type of

lamp a trade-off between efficiency and lamp life. This is why lamps of equal wattage may have different **lumen (lm)** and life ratings.

Of all the lamp types, standard incandescent lamps are the most inefficient. Because they produce light by heating the filament until it glows,

*Figure 6* Components of an incandescent lamp.

most of the energy they consume is given off as heat, resulting in a low efficiency (**efficacy**) of typically 5 to 22 **lumens per watt (LPW)**. Incandescent lamps also have the shortest life expectancy of all lamp types, typically between 750 and 2,000 average hours, depending on the type. This is because tungsten from the filament evaporates over time and is deposited on the walls of the bulb, thus reducing the light output. Also, the filament gets thinner and thinner with use and eventually breaks, causing the lamp to fail.

Incandescent lamps are made in numerous sizes and shapes, with different filament form arrangements and mounting bases. *Figure 7* shows some

examples of typical lamp sizes and shapes. Lamps are identified by a letter referring to their shape and a number that indicates the maximum diameter stated in eighths of an inch. For example, A-40 identifies a lamp with an A-shape that is ⁴⁰⁄₈" (5") in diameter. Lamps with standard, tubular, and similar envelope shapes provide lighting in all directions (omni-directional). Those with shapes designated as R, ER, and PAR are all reflector-type lamps. They direct their light out in front by reflecting it from their cone-shaped inside walls.

*Figure 8* shows some examples of incandescent lamp filament forms used by one lamp manufacturer. The form of a filament is determined mainly

203F07.EPS

***Figure 7*** Incandescent lamp shapes. *(Courtesy of GE Consumer Products.)*

NCCER — *Contren® Learning Series* 26203-11

C-9    C-6     C-17    C-22    C-13    C-13D    CC-8    C-8     C-5     C-7A    C-2V
       CC-6                                             CC-8                    CC-2V

203F08.EPS

*Figure 8*   Examples of incandescent lamp filament forms. *(Courtesy of GE Consumer Products.)*

by service requirements. Filament forms are identified by a letter or letters followed by an arbitrary number. Commonly used letters are C (coiled), indicating that the filament wire is wound into a helical coil; CC (coiled coil), indicating the coil itself is wound into a helical coil; and S (straight), indicating that the filament wire is uncoiled. The numbers shown in *Figure 8* indicate the arrangement of the filament on the supports.

The lamp base supports the lamp and provides the connection between the lamp and the power source. *Figure 9* shows some examples of common incandescent lamp bases.

Incandescent lamps and other types of lamps are available in many different voltage and wattage ratings. When installing incandescent lamps, it is important to select the correct voltage rating. This is because a small difference between the rating and the actual supply voltage has a great effect on the lamp life and light output. The wattage rating is an indication of the consumption of electrical energy used by a lamp to produce its rated light output. It is not a measure of the lamp's light output. For example, a standard 100W lamp may produce 1,600 lumens, while a typical 32W fluorescent lamp may produce about 2,600 lumens. *Figure 10*

CAND SCREW E12    INTERMEDIATE E17    3 CONTACT MED E26    MED SCREW E26    MOG SCREW E39    3 CONTACT MOGUL E39D    MOG PF P40s

MED SKIRT E26/50×39    CAND DC BAY    CAND SC BAY B15    2-LUG SLEEVE B22d    3-LUG SLEEVE B22-3    MEDPF P28s    MOGUL BIPOST G38

PAR    MED BIPOST    MEDIUM SIDE PRONG    MOGUL END PRONG

203F09.EPS

*Figure 9*   Examples of incandescent lamp bases. *(Courtesy of GE Consumer Products.)*

shows the relationships between watts, lumens, and lamp life for a typical lamp when operated at different percentages of its rated voltage. For example, reducing the supply voltage for a 120V lamp to approximately 94% (113V) will increase its life by 220%, but reduce the light output to 80% and the wattage to about 90%.

### 4.1.1 Tungsten Halogen Incandescent Lamps

Tungsten halogen incandescent lamps (halogen lamps) are a refinement over the standard incandescent lamp. Like standard incandescent lamps, halogen lamps are made in many sizes, shapes, and wattages. When compared to the standard incandescent lamp, they provide greater efficacy (12 to 36 LPW), longer service life (2,000 to 5,000 average hours), and improved light quality. Their light output contains more blue and less yellow than standard incandescent lamps, making their light appear whiter and brighter.

203F10.EPS

*Figure 10*  Relationship of rated lamp voltage to watts, lumens, and lamp life.

## Efficiency

Different lamp types have varying abilities to convert electrical power into visible light. The quantity of light emitted (measured in lumens) is divided by the input power (in watts) to determine a lamp's efficiency (efficacy). This number is expressed in lumens per watt (LPW) and is a measure of energy efficiency. For example, a 60W lamp that produces 800 lumens of light has an efficacy of 13.333 LPW.

Halogen lamps (*Figure 11*) typically have a short, thick tungsten filament encased in a capsule filled with halogen gases, such as iodine or bromine, that allow the filaments to operate at higher temperatures than a standard incandescent lamp. This increases their efficacy (LPW) by more than 20%. The use of halogen gas in the lamp accounts for the longer life and

● Halogen Atoms
● Tungsten Atoms

1. Tungsten atoms evaporate from filament.
2. Tungsten atoms combine with halogen atoms.
3. Gaseous compound returns to hot filament, redepositing tungsten atoms.
4. Halogen atoms are released to combine with additional tungsten atoms.

203F11.EPS

*Figure 11*  Basic tungsten halogen lamp.

---

NCCER — *Contren® Learning Series*  26203-11

excellent **lumen maintenance**. During operation, tungsten atoms evaporated from the filament combine with halogen atoms to form a gaseous compound that circulates inside the lamp, causing the tungsten atoms to be redeposited on the hot filament, rather than on the inside surface of the lamp envelope. The halogen atoms are then released, allowing them to combine with additional tungsten atoms, thus repeating the process. This action slows down any deterioration of the filament, thereby improving lumen maintenance and extending the lamp's life.

Because of the higher filament operating temperatures, there is more ultraviolet (UV) radiation generated from a halogen lamp than from standard incandescent lamps. The amount of UV radiation emitted is determined by the lamp envelope material. Fused quartz and high-silica glass transmit most of the UV radiation emitted by the filament; special high-silica glass and alumino-silicate glasses absorb UV radiation. For general lighting applications, it is recommended that lighting fixtures for halogen lamps have a lens or cover glass that will, in addition to providing the required safety protection, filter out most of the UV radiation.

> **CAUTION**
>
> Operating halogen lamps at voltages above and below the manufacturer's recommendations can have adverse effects on the internal chemical process because the temperature will differ from the design value. Also, it is important to follow the manufacturer's instructions as to burning position, lamp handling, and lighting fixture temperatures.

## 4.2.0 Fluorescent Lamps

Fluorescent lamps are low-pressure mercury discharge lamps that are very energy efficient (75 to 100 LPW) and have a long service life (12,000 to more than 24,000 average hours). Each requires a **ballast** to effectively start the lamp and regulate its operation. Light is produced by passing an electric arc between two tungsten cathodes at opposite ends of a glass tube filled with a low-pressure mercury vapor and other gases (*Figure 12*). The arc excites the atoms of mercury. This generates UV radiation, which causes the phosphor coating on the inside of the tube to fluoresce and produce visible light. By using different phosphor coatings, the spectral light output of a fluorescent lamp can be made to produce warm, intermediate, or cool color temperatures. The color temperatures of lamps are covered later in this module.

*Figure 12*  Basic fluorescent lamp.

Fluorescent bulbs are made in straight, U-bent, circular, and compact varieties, as shown in *Figure 13*. Not only do they come in a wide variety of wattages, sizes, and bases, but they are also available in several color temperatures and color rendition capabilities. Fluorescent lamps are designated by the letter T followed by the diameter of the lamp tube expressed in eighths of an inch. They vary in diameter from T-5 (⅝") to PG (power groove)-17 (2⅛"). In overall length, straight fluorescent lamps range from 6" to 96". Higher wattages go with longer tubes. For example, a 20W straight T-12 tube is shorter than a 40W T-12 tube.

> **On Site**
>
> ## Bi-Pin Lamps
>
> Most bi-pin lamps, like the one shown here, have indents or index marks to aid in the proper alignment when inserting into a fixture. When properly inserted, the indents or marks will be perpendicular to the lamp holders.
>
>
>
> 203SA02.EPS

(A) CURVALUME®

(B) T8 ARRAY

(C) T12 ARRAY

203F13.EPS

*Figure 13* Typical fluorescent lamps.

Fluorescent lamps have two electrical requirements. To start the lamp, a high-voltage surge is needed to establish an arc in the mercury vapor. Once the lamp is started, the gas offers a decreasing amount of resistance, which means that current must be regulated to match this drop. Otherwise, the lamp would draw more and more power and rapidly burn itself out. This is why fluorescent lamps are operated in lighting fixture circuits containing a ballast that provides the required voltage surge at startup and then controls the subsequent flow of current to the lamp. Ballasts are covered in detail later in this module.

There are three electrical classes of fluorescent lamps and lighting fixtures: preheat, rapid start, and instant start. The term *preheat* refers to a lighting fixture circuit used with fluorescent lamps wherein the lamp electrodes are heated or warmed to a glow stage by a replaceable starter separate from the ballast. When power is applied to the lighting fixture, the starter functions to preheat the lamp's cathodes before the lamp is started. Note that preheat lamps and lighting fixtures are nearly obsolete and are not used in new construction. The term *rapid start* refers to a lighting fixture circuit designed to start the lamp by continuously heating or preheating the lamp electrodes by means of heater windings built into the ballast. Unlike the preheat circuit, the ballast does not require a separate starter, but the fixture must be properly grounded. During operation, standard rapid-start lamps draw about 430mA of current. The term *instant start* refers to a lighting fixture circuit used to start specially designed lamps without the aid of a starter. To strike the arc instantly, the circuit uses a higher open circuit voltage than is required for a preheat or rapid-start lamp of the same length, a voltage that is approximately three times the normal lamp operating voltage. Both preheat and rapid-start lamps have a bi-pin (2-pin) base at each end. Instant-start lamps have a single pin at each end of the lamp. Normally, lamps identified as preheat, rapid-start, or instant-start types should be used only with the corresponding type of ballast. Other terms commonly used to designate types of fluorescent lamps include:

- *Slimline lamps* – A group of instant-start lamps with single-pin bases.
- *High-output (HO) lamps* – A group of rapid-start lamps designed to operate at higher operating currents (800mA to 1,000mA) that produce higher levels of light output. Because of the higher operating currents involved, the lamps have a recessed double-contact base. HO lamps are typically used in industrial areas and retail stores with high ceilings.

**On Site**

## Fluorescent Lamps

The life of a fluorescent lamp is affected by the number of times the lamp is started. Frequent switching on and off of fluorescent lamps results in shorter lamp life, while continuous operation provides the longest lamp life. The typical fluorescent lamp ratings given by one major manufacturer of fluorescent lamps are based on three hours per start. For example, if a lamp is rated as having an average life of 12,000 hours, this is based on the lamp being turned on 4,000 times and being left on for a minimum of at least three hours.

As fluorescent lamps age, they tend to darken at the ends. This helps in identifying lamps for future replacement.

- *Very high-output (VHO) lamps* – A group of rapid-start lamps designed to operate at high currents (1,500mA) in order to produce high light output levels. Because of the higher operating currents involved, the lamps have a recessed double-contact base. VHO lamps are typically used in factories, warehouses, gymnasiums, and open areas.

**NOTE**

Although the bases on HO and VHO lamps are the same, each must be matched to the correct ballast.

- *Compact lamps* – Lamps made of ½" to ⅝" single or multiple U-shaped tubes that terminate in a plastic base (*Figure 14*). The base contains the cathodes and, in some versions, a magnetic or electronic ballast. Some have replaceable tubes. A few of the newer versions are dimmable, and others look much like a standard lamp in size and shape. They are used both to replace incandescent lamps and in lighting fixtures designed for their use. They can provide up to 75% energy cost savings when compared to incandescent lamps of comparable light output. They also have a lifespan of up to 13 times longer than standard incandescent lamps.
- *T-5 lamps* – Newer and more energy-efficient lamps than T-8 or T-12 lamps. They are the first type of linear lamp to use only electronic ballasts, and produce nearly twice the light output of a T-8 or T-12 system. In addition, they use an improved triphosphor coating that allows them to maintain higher light levels over the life of the lamp.

Lamp holders (*Figure 15*) are made in several variations for each lamp base style to allow for various spacings and mounting methods in fixtures. When fluorescent lamps are used in circuits providing an open circuit voltage in excess of 330V, or in circuits that may permit a lamp to ionize and conduct current with only one end inserted in the lamp holder, electrical codes require some automatic means for de-energizing the circuit when the lamp is removed. This is usually accomplished by the lamp holder so that upon removal, the ballast primary circuit is opened. Note that the use of recessed contact bases for HO and VHO lamps has eliminated the need for this disconnect feature in the lamp holders for these lamps.

### Cost Effectiveness of Compact Fluorescent Lamps

*GOING GREEN*

The 14W compact fluorescent lamp (CFL) shown here has approximately the same light output in lumens as a 60W incandescent lamp. However, the CFL has a much longer lifespan at one-fourth the operating cost and energy usage.

203SA03.EPS

(A) CIRCLINE FLUORESCENT

(B) CLASSIC BULB SHAPE FLUORESCENT

(C) TRIPLE COMPACT FLUORESCENT

203F14.EPS

*Figure 14* Compact fluorescent lamps.

(A) BI-PIN SLIDE-ON
AND SCREW MOUNT
LAMP HOLDER

(B) HIGH-OUTPUT
(HO) LAMP HOLDER

(C) SINGLE-PIN (SLIMLINE)
LAMP HOLDER

203F15.EPS

*Figure 15*  Typical fluorescent lamp holders.

## 4.3.0 High-Intensity Discharge (HID) Lamps

High-intensity discharge (HID) lamps provide long life and high efficiency (*Figure 16*). They are somewhat similar to fluorescent lamps in that they produce light when electricity excites specific gases in pressurized bulbs. An arc is established between two electrodes in a gas-filled tube, which causes mercury vapor to produce radiant energy. Unlike a fluorescent lamp, a combination of factors shifts the wavelength of much of the energy to within the visible range, so light is produced without phosphors. First, the electrodes are only a few inches apart at opposite ends of a sealed arc tube, and the gases in the tube are highly pressurized. This allows the arc to generate extremely high temperatures, causing metallic elements within the gas atmosphere to vaporize and release large amounts of visible radiant

203F16.EPS

*Figure 16*  Typical high-intensity discharge lamps.

energy. Like fluorescent lamps, HID lamps must be used in matched lighting fixtures with a ballast specifically designed for the lamp type and wattage. In addition, HID lamps require a warmup period to achieve full light output.

There are three types of HID lamps: mercury vapor, metal halide, and high-pressure sodium. The names refer to elements that are added to the gases in the lamp, which cause each type to have somewhat different color characteristics and efficiency. Mercury vapor lamps are the oldest HID technology. They are energy efficient (50 to 60 LPW) and have a long service life (12,000 to more than 24,000 average hours). These lamps produce light energy by radiation from excited mercury vapor in both the visible and ultraviolet range. They normally have specially formulated glass outer jackets to filter the UV energy. The phosphor coatings used in some mercury vapor lamp types add additional light and improve color rendering. Today, the use of mercury vapor lamps is limited mainly to the replacement of existing lamps and landscape lighting of evergreen trees. Other HID lamps that have better efficiency and color properties are being used for new construction.

Metal halide lamps are the most energy-efficient source of white light. They have high efficacy (80 to 115 LPW), excellent color rendition, long service life (10,000 to more than 20,000 average hours), and good lumen maintenance (longevity). The metal halide HID lamp combines mercury and metal halide atoms under high pressure (*Figure 17A*). In the arc stream, these atoms generate both UV radiation and visible light. A special glass bulb filters the UV radiation without affecting the visible light.

If the arc tube is ruptured in an open metal halide lamp, the outer bulb may explode, causing

(A) BASIC METAL HALIDE LAMP

203F17A.EPS

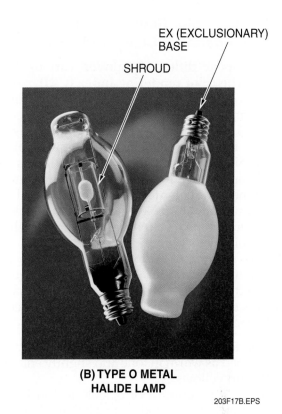

(B) TYPE O METAL HALIDE LAMP

203F17B.EPS

*Figure 17*   Basic metal halide lamp.

personal injury or a fire. Per *NEC Section 410.130(F)(5)*, all metal halide lamps except thick-glass PARs must either have a containment barrier (lens) or be Type O with a protected socket. Type O lamps have an internal shroud that protects the outer bulb from damage if the arc tube is ruptured (*Figure 17B*). The protected socket prevents the use of any lamp except a Type O lamp.

Low-pressure sodium lamps use sodium in a low-pressure arc stream and produce light that is limited to a single wavelength in the yellow portion of the spectrum. These lamps are the most efficient of any lamp type, but are used only where energy efficiency and long life are the only requirements. Technically speaking, low-pressure sodium lamps are not actually a type of HID lamp.

## Comparing Lamp Efficiency Ratings

The following listing shows how incandescent, halogen, fluorescent, and HID lamps compare in terms of efficiency and rated average life.

| Lamp Type | Typical LPW | Rated Avg. Life (Hrs.) |
|---|---|---|
| Incandescent | 5–22 | 750–2,000 |
| Halogen | 12–36 | 2,000–5,000 |
| Fluorescent (compact) | 27–80 | 9,000–10,000 |
| Fluorescent | 75–100 | 12,000–24,000+ |
| Mercury | 50–60 | 12,000–24,000+ |
| Metal halide | 80–115+ | 10,000–20,000+ |
| High-pressure sodium | 90–140 | 10,000–24,000+ |

Note that the typical LPW value given above is based on the initial lumen output produced by the lamp when it is new. Most lamps degrade in efficiency over time. For example, mercury vapor lamps have a long average life, but with a continual reduction in lumen output, especially after 10,000 hours.

## 4.4.0 Lamp Color Rendering and Color Temperature Characteristics

Colors appear differently under various light sources. The color rendering index (CRI), a scale from 0 to 100, is used by lamp manufacturers to indicate how normal and natural a specific lamp makes objects appear. Generally, the higher the CRI, the better it makes people and objects appear. Note that the CRI of different lamps can be compared only if the sources have approximately the same color temperature. Also, CRI differences among lamps are not usually visible to the eye unless the difference is greater than three to five points.

Lamps can create atmospheres that are warm or cool in appearance. The color temperature, expressed in kelvins (K), is one way lamp manufacturers describe the color tone (warmth or coolness) produced by a lamp. For example:

- Color temperatures of 3,000K and lower are described as warm in tone and slightly enhance reds and yellows.
- A color temperature of 3,500K is considered moderate in tone, producing a balance between warmth and coolness.
- Color temperatures of 4,100K and higher are considered cool in tone, slightly biased toward blues and greens.

Some typical color temperatures are 2,200K for high-pressure sodium lamps, 2,800K for incandescent lamps, 3,000K for halogen lamps, 3,500K for metal halide lamps, 4,100K for cool white fluorescent lamps, and 5,000K for daylight-simulating fluorescent lamps.

Fluorescent lamps have more options in terms of light quality than any other lamp type. This is because of the variations (formulations) available in the composition of the phosphor coating on the inside of the lamp tube. Early fluorescent lamps used a single halophosphor coating and could offer improved color quality with only a decrease in efficacy (LPW). With the newer lamps, triphosphor coatings are used that allow precise control over the generation of red, green, and blue (the primary colors of light). This enables the manufacture of high-LPW lamps in a variety of color temperatures that provide excellent color rendition.

*Table 2* provides a brief comparison of the major lamp types.

## 4.5.0 Guidelines for Installing Lamps

Listed below are some general guidelines to follow when installing lamps in their related lighting fixtures.

- Make sure that the lamp and lighting fixture are compatible. Install incandescent lamps in incandescent lighting fixtures, fluorescent lamps in fluorescent fixtures, etc.
- Make sure that the lamp or lamps being installed in a lighting fixture have the correct voltage rating for use with the fixture. This is especially important when installing incandescent lamps because a small difference between the rating and the actual voltage has a great effect on a lamp's life and light (lumen) output.
- Make sure that you do not install a lamp, or combination of lamps, in a lighting fixture that exceeds the wattage rating for the lighting fixture; otherwise, the fixture can overheat and may cause a fire. In recessed lighting fixtures with built-in thermal protectors, it will cause the lamps to cycle on and off as the fixture overheats, then cools down.
- When installing fluorescent and HID lamps, make sure that the lamp type and wattage are matched to the lighting fixture and ballast being used. Always install the lamps specified by the lighting fixture manufacturer.
- When installing bi-pin fluorescent lamps, make sure to rotate them ¼ turn to seat and connect them.
- Do not install lamps that have scratched or otherwise damaged envelopes.
- Use a soft cloth or gloves to handle an unencapsulated halogen tubular lamp. This prevents oil from your hands from contacting the bulb. This is important because its quartz glass walls can withstand high operating temperatures, but may crack if etched with oil. Also make sure that no combustible materials can come in contact with the lamp.
- Unlike standard lamps, halogen and metal halide lamps may continue to light even if the exterior lamp glass is broken. If this occurs, replace the lamp immediately to prevent creating a hazard or personal injury.
- In some applications, such as food processing plants, bakeries, dairies, schools, warehouses, etc., safety sleeves made to fit over fluorescent lamps should be used in order to retain broken glass and phosphors if the lamp ever breaks, falls, or bursts.

In general, energy legislation promotes replacing standard incandescent lamps with halogen and compact fluorescent lamps, replacing incandescent reflector lamps with halogen PARs, and using higher-efficiency fluorescent lamps, such as T-8s, instead of full wattage types. If it is necessary to relamp existing light fixtures designed for use with a discontinued or inefficient type of lamp,

**Table 2** Comparison of Major Lamp Types

| Type of Lamp | Advantages | Disadvantages |
|---|---|---|
| Incandescent | • Low initial cost<br>• Small size<br>• Excellent color rendering index (CRI)<br>• Variety of shapes<br>• Dimmable<br>• Wattage interchangeable | • Inefficient<br>• Excessive heat output<br>• High operating costs<br>• Short service life<br>• Glare potential |
| Halogen | • Small size<br>• Increased efficiency*<br>• Longer life*<br>• Excellent CRI<br>• Bright white light<br>• Dimmable | • Excessive heat output<br>• Glare potential |
| Fluorescent | • Highly efficient<br>• Long service life<br>• Choice of color temperatures and CRI<br>• Low operating costs<br>• Low heat output<br>• Diffuse light source | • High initial cost<br>• Temperature sensitive<br>• Limited optical control<br>• Requires matched fixture/ballast |
| Metal halide (HID) | • Highly efficient<br>• Long service life<br>• Low operating costs<br>• Good color rendering | • High initial cost<br>• Requires matched fixture/ballast<br>• Long startup/restrike period<br>• Glare potential |
| High-pressure sodium (HID) | • Long service life<br>• Exceptionally efficient<br>• Very low operating costs<br>• High lumen maintenance | • High initial cost<br>• Requires matched fixture/ballast<br>• Long startup/restrike period<br>• Poor color rendering<br>• Glare potential |

\* Compared to standard incandescent lamps

replace the lamp with an approved energy-efficient substitute. Lists of approved substitutes are available at all lighting and electrical distributors.

## 5.0.0 BALLASTS

Fluorescent and HID fixtures require the use of a ballast. The main function of a ballast is to provide the voltage needed to strike an arc between the electrodes.

### 5.1.0 Fluorescent Lighting Fixture Ballasts

In fluorescent lighting fixtures, the ballasts perform the following functions:

• Provide the proper voltage to establish an arc between two electrodes
• Regulate the electric current flowing through the lamp to stabilize the light output
• Supply the correct voltage required for proper lamp operation and compensate for voltage variations in the electrical current

• Provide continuous voltage to maintain heat in the lamp electrodes while the lamp operates (rapid-start circuits)

Fluorescent ballasts (*Figure 18*) are made to operate in the three basic fluorescent lighting fixture operating circuits: preheat, rapid start, and instant start.

### 5.1.1 Preheat Lamp and Ballast Operation

In preheat circuits, such as the one shown in *Figure 19(A)*, the lamp electrodes (cathodes) are heated before application of the high voltage across the lamp(s). The preheating requires a few seconds, and the necessary delay is provided by an automatic switch called a starter. When power is first applied to the lamp circuit, the starter places the lamp's electrodes in series across the ballast, causing current to flow through both electrodes, heating them. After the electrodes are sufficiently preheated, the switch opens and applies the voltage across the lamp. Because the switch opens under load, a transient voltage (inductive

Figure 18    Fluorescent ballast.

spike) is developed in the circuit, which aids in the ignition of the lamp. Note that the first fluorescent lamps developed were of the preheat type. This type of lamp is now obsolete and is seldom used except in smaller sizes, such as those used in desk lamps and similar luminaires.

### 5.1.2 Rapid-Start Lamp and Ballast Operation

This is probably the most common type of lamp and luminaire used today. Lamps designed for rapid-start operation, such as the one shown in *Figure 19(B)*, normally have low-resistance

electrodes. These remain energized by low voltage applied from the ballast while the lamps are in operation. They usually start in one second, the time required to bring the electrodes up to proper temperature. The standard rapid-start circuit operates with a typical lamp current of about 430mA. Rapid-start circuits used with high-output (HO) and very high-output (VHO) lamps draw currents of about 800mA and 1500mA, respectively. In some energy-saving circuits, the electrode voltage is reduced or disconnected after the starting of the lamps. Heating is accomplished through low-voltage windings built into the ballast or through separate low-voltage transformers designed for this purpose. Fluorescent lamps used with rapid-start ballasts are bi-pin lamps. Rapid-start lamps can be dimmed using special dimming ballasts. These are covered later in this module.

Another version of a rapid-start ballast is the trigger-start ballast. Trigger-start ballasts are used with preheat fluorescent lamps up to 32W without the need for a starter.

### 5.1.3 Instant-Start Lamp and Ballast Operation

The lamp electrodes in instant-start lamps are not preheated. The ballasts provide a high voltage (100V to 1,000V) across the electrodes that causes electrons to be emitted from the electrodes,

Figure 19    Basic fluorescent lighting fixture circuits.

as shown in *Figure 19(C)*. These electrons flow through the tube, ionizing the gas and initiating an arc discharge. Thereafter, the arc current provides electrode heating. Because no preheating of electrodes is required, instant-start lamps need only a single contact at each end. Thus, a single pin is used on most instant-start lamps, commonly called slimline lamps. However, some bi-pin lamps can be operated with either rapid-start or instant-start electronic ballasts. When used with an instant-start ballast, the terminals in the lamp holders must be connected together. New fixtures come with the lamp holders wired in this way.

### 5.1.4 Types of Fluorescent Ballasts

Each fluorescent lamp must be operated by a ballast that is specifically designed to provide the proper starting and operating voltage required by the particular lamp. Lamp and lighting fixture manufacturers make a wide variety of ballasts designed for use in lighting fixtures that operate in all three of the lamp starting modes previously described. Ballasts are made for single-lamp, two-lamp, three-lamp, and four-lamp operation. The names used to identify the different types of ballasts can vary by manufacturer. Some common categories of ballasts are:

- *Standard ballast* – Lowest priced, least efficient, and highest wattage ballast. Their use is obsolete.
- *High-efficiency ballast* – Lower wattage, better efficiency, and longer life than a standard ballast.
- *Hybrid ballast* – Lower wattage, higher efficiency, and longer life than a standard high-efficiency ballast.
- *Electronic rapid-start ballast* – Low wattage and longest life available with various ballast factors and wattage packages for various lamp types.
- *Electronic instant-start ballast* – Lowest system wattage and highest system efficiency. The lamp life is slightly shorter than with rapid-start ballasts.

High-efficiency ballasts can be of the magnetic or electronic type. The magnetic type usually contains coils, capacitors, transformers, and a thermal protector installed in a metal case. The coils and transformers are generally made with steel

laminations and copper wire. Once assembled, the ballast components are encapsulated in the ballast case with a potting compound to improve the heat dissipation and reduce ballast noise.

If the ballasts are of the electronic type, they may be made with discrete electronic components and/or with integrated circuits. Some of those with integrated circuits are compatible for use with dimming systems, occupancy sensors, and daylight sensors. Electronic ballasts are quieter, more efficient, and weigh less, but are also more expensive. With electronic ballasts, the power input is 50Hz to 60Hz, but the ballast operates the lamps at 20kHz to 50kHz, with resulting improvements in ballast and lamp efficacy. The operating frequency is selected so that it is high enough to increase the lamp's efficacy and to shift the ballast noise to the inaudible range, but not so high as to cause electromagnetic interference (EMI) problems. Note that many electronic ballasts do not come equipped with wire leads. These ballasts are connected into the circuit using wire harnesses specifically designed for that purpose.

*NEC Section 410.130(E)(1)* requires that all fluorescent fixture ballasts used indoors, including any replacement ballasts, have a thermal protection device integral within the ballast. All such ballasts are marked Class P. The exceptions to this requirement include simple reactance ballasts used with fluorescent fixtures using straight tubular lamps, ballasts used with exit fixtures and so identified, and egress lighting energized only during an emergency. Class P ballasts with thermal protection operate to open the circuit at a predetermined temperature in order to prevent abnormal heat buildup caused by a fault in one or more of the ballast components, or by some lamp holder or wiring fault.

High-efficiency, energy-saving ballasts have a high power factor rating. The power factor is the ratio of watts to volt-amperes. To be classified as a high power factor ballast, a ballast must have a power factor of at least 90%. Anything less is considered a normal or low power factor. A ballast's power factor rating is marked on the ballast nameplate. Energy-saving high power factor ballasts cost more than low power factor ballasts, but over time, the savings in energy consumption far exceed the higher initial cost. In any application where there are to be a large number of ballasts, it is best to install ballasts with a high power factor. When compared to magnetic ballasts, electronic ballasts are more energy efficient.

Ballasts can emit a hum, especially the magnetic types. This is caused by magnetic vibration in the ballast core. Ballast manufacturers give their ballasts a sound rating ranging from A to F, with A being the quietest. The need for quiet ballast operation is determined mainly by the desired ambient noise level of the location where the ballast is to be installed. For example, a ballast with an A rating might be used in a doctor's office, while one with an F rating might be suitable for a factory application.

When installing a replacement ballast, make sure to dispose of the old ballast in a proper manner. Unless you see a label stamped *No PCBs* on the failed ballast you are disposing of, you must assume that it contains toxic PCBs and must be disposed of in accordance with the prevailing EPA and local requirements. Failure to do so can expose you and your employer to potential liability for cleanup in the event of PCB leakage.

### 5.1.5 Fluorescent Dimming Ballasts

Special dimming ballasts and dimmer switches are required to dim fluorescent lamps. The dimmer control allows the dimming ballast to maintain a voltage to the lamp's electrodes that will maintain the electrodes' proper operating temperature. It also allows the dimming ballast to vary the current flowing in the arc. This in turn varies the intensity of light coming from the lamp. Dimming fluorescent lamps differs from dimming incandescent lamps in two main ways: first, fluorescent dimmers do not provide dimming to zero light as do incandescent dimmers; and second, when dimming fluorescent lamps, the color temperature does not vary much over the dimming range. This is unlike incandescent lamps, which tend to turn yellower when dimmed.

Most fluorescent dimming ballasts are of the electronic type. However, older autotransformer magnetic types are also available. Electronic dimming ballasts are normally more efficient and less bulky than magnetic ballasts. *Figure 20* shows a wiring diagram for an electronic dimmer used with a dimming ballast for rapid-start lamps. It is important to point out that the performance of a dimming system may not be satisfactory if the lamp is not correctly matched with the dimming ballast and the controller. Also, when connecting such dimmer circuits, always check the dimmer and ballast manufacturer's wiring diagrams to determine the proper connections.

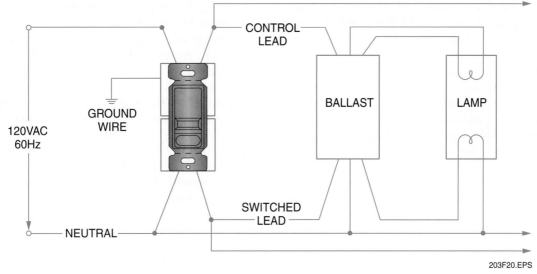

**Figure 20**  Dimmer circuit using dimming ballast for rapid-start lamp.

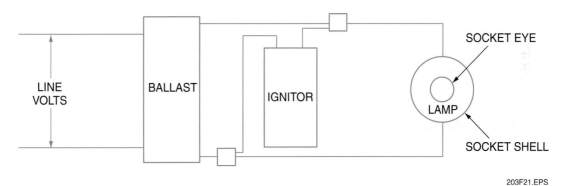

**Figure 21**  Simplified HID lamp ignitor circuit.

### 5.1.6 Emergency Lighting Ballasts

Special emergency ballasts with self-contained battery-operated power packs are made for use in some fluorescent lighting fixtures. Upon the loss of input power to the lighting fixture, these ballasts typically function to operate one 8' lamp at emergency lighting levels for about 90 minutes, or one 4' lamp for about 120 minutes.

### 5.2.0 HID Lighting Fixture Ballasts

HID lamps require the use of a ballast to provide enough voltage to strike the arc in the lamp. This function may be accomplished by the ballast itself or in conjunction with a separate electronic ignitor circuit. An ignitor (*Figure 21*) is an electronic device used in the circuitry for high-pressure sodium and some metal halide HID lamps. It provides a pulse of at least 2,500V peak root mean square (rms) to initiate the lamp arc.

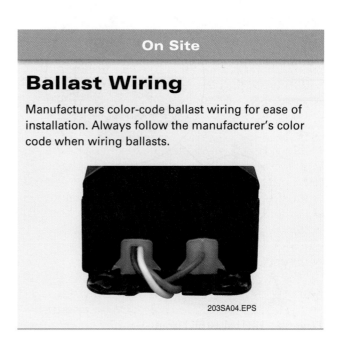

When the system is energized, the ignitor provides the required pulse until the lamp is completely lit and then automatically stops pulsing.

The HID lamp ballast also acts to control the arc wattage during warmup and normal operation. In addition, some ballasts may also provide a line voltage matching transformer function, enhance lamp wattage regulation with respect to changes in line voltage and/or lamp voltage, and dimming or other control interface functions.

Physically, there are numerous types and shapes of ballasts used with HID lamps (*Figure 22*). The same is true electrically. Some ballasts are made with primary leads that allow the ballast to be connected to different supply voltages, such as 120V, 208V, 240V, or 277V. Such ballasts are called multi-tap ballasts. It is extremely important that only the proper voltage lead be connected to the supply voltage. The types of ballasts used by a major HID lighting fixture manufacturer (Hubbell) are described here. The ballasts used by other manufacturers are similar. HID ballasts can be grouped into three basic categories:

- Linear, nonregulating circuit ballasts
- Constant wattage autotransformer ballasts
- Three-coil ballasts

### 5.2.1 Linear, Nonregulating Circuit Ballasts

Linear, nonregulating circuits include reactor ballasts and auto-lag ballasts. These provide for the basic operation of some mercury and high-pressure sodium HID lamps. With the exception

*Figure 22*   HID ballast and lamp.

of the self-ballasted lamp, the reactor ballast is the most basic form of ballast. See *Figure 23(A)*. It consists of a single coil wound on an iron core placed in series with the lamp. The only function performed by the reactor is to limit the current delivered to the lamp. This type of ballast can only be used when the line voltage applied to the lamp is within the required starting voltage range of the lamp. The power factor for a reactor ballast is typically in the 40% to 50% range. However, a capacitor is normally added to the circuit to improve the power factor to better than 90%. Because this type of ballast provides for no line voltage regulation, outages due to line dips and brownouts are typical. The auto-lag ballast, as shown in *Figure 23(B)*, is a reactor ballast combined with a step-up or step-down autotransformer, which provides for some input voltage regulation.

### 5.2.2 Constant-Wattage Autotransformer Ballasts

The constant-wattage autotransformer (CWA) ballast, shown in *Figure 23(C)*, is a ballast circuit that uses magnetic saturation to maintain better lamp wattage regulation and improved **dip tolerance.** CWA ballasts are used mainly with mercury and high-pressure sodium lamps. A variation of the CWA, called the peaked lead auto-regulator (PLA), is used with metal halide lamps. Another variation, called the constant-wattage isolated (CWI) ballast, shown in *Figure 23(D)*, is an isolated winding version of the PLA. It is typically used with mercury lamps.

The term *dip tolerance* relates to the dips in line voltage experienced by all power systems as loads are switched in and out, or as other transitory conditions occur. A well-regulated distribution system will seldom experience voltage dips of more than 10%, but on some circuits, dips of 20% or more may occur. If a ballast is not capable of riding through the voltage dip and sustaining the lamp, it will extinguish and have to cool down before reignition. Lamp dropout due to line voltage dips generally increases with lamp age. The use of ballasts with improved dip tolerance may delay the onset of such lamp dropout problems.

### 5.2.3 Three-Coil Ballasts

Three-coil ballasts are isolated winding ballasts in which the input and lamp windings are separated (isolated) by a third winding, which helps to eliminate drastic changes in the demands of the lamp on the supply system and maintains lamp stability during supply system variations. See *Figure 23(E)*. This type of circuit provides

(A) REACTOR BALLAST

(B) AUTO-LAG BALLAST

(C) CONSTANT-WATTAGE AUTOTRANSFORMER (CWA) BALLAST

(D) CONSTANT-WATTAGE ISOLATED (CWI) BALLAST

(E) THREE-COIL BALLAST

203F23.EPS

*Figure 23* Simplified HID ballast circuits.

the highest degree of lamp operating stability and waveform control, with the lowest harmonics and best performance consistency through the life of the lamp. A magnetic regulator version is used with high-pressure sodium lamps and an electro-regulator version is used with metal halide lamps.

### 5.2.4 Dimming HID Lamps

HID lamps can be controlled using equipment similar to that used for dimming fluorescent lamps. However, the long warmup and restrike times associated with HID lamps may limit their applications. Multi-level ballasts are made for HID lamps that allow their light to be reduced. This type of ballast is typically used in lighting fixtures for warehouses, parking garages, tunnels, and daytime lighting applications. Some equipment is made that allows HID lamps to be dimmed to less than 20% of their full light output; however, most lamp manufacturers will not guarantee full life expectancy if their lamps are operated below 50%. Also, color shifts in the light output of the lamps may limit their use in some applications.

## 6.0.0 LIGHTING FIXTURES

In the electrical trade, the terms *lighting fixture* and *luminaire* are used interchangeably (luminaire is an international term). A lighting fixture (luminaire) is defined as a complete lighting unit consisting of the lamp or lamps, together with the parts designed to distribute the light, position and protect the lamps, and connect the lamps to a power supply.

Lighting fixtures are classified in many ways, including: incandescent, fluorescent, and HID; indoor or outdoor; surface-mounted, recessed, or suspended; residential, commercial, or industrial. There are thousands of lighting fixtures manufactured in different styles for all of these applications. From an installation point of view, lighting fixtures can best be classified by the way in which they are mounted. These include:

- Surface-mounted
- Recessed
- Suspended
- Track-mounted

### 6.1.0 Surface-Mounted Lighting Fixtures

Surface-mounted lighting fixtures include those fixtures that are directly mounted on a ceiling or wall. They can be incandescent, fluorescent, or HID units. Some typical examples of surface-mounted indoor ceiling and wall lighting fixtures are shown in *Figures 24* and *25*, respectively. *Figure 26* shows an example of a wall-mounted outdoor lighting fixture.

### 6.2.0 Recessed Lighting Fixtures

As the name implies, recessed lighting fixtures are mounted into a ceiling so that the main body of the fixture is not visible. *Figure 27* shows examples of recessed lighting fixtures typical of those used with incandescent, metal halide HID, or compact fluorescent lamps. As shown, a typical unit for new work consists of a housing, junction box, and a mounting frame used to fasten it to the ceiling structure. These components of the lighting fixture are all located above the ceiling line. Only the related lens/diffuser and trim are viewable at the ceiling surface.

203F24.EPS

*Figure 24* Surface-mounted ceiling light fixtures.

Figure 25 Surface-mounted indoor wall lighting fixtures.

Figure 26 Surface-mounted outdoor wall fixture.

Other recessed fluorescent fixtures, called **troffers,** are square or rectangular units installed above the ceiling (*Figure 28*). These are typically 2' × 4' or 2' × 2' units, containing two or three T-8, T-10, or T-12 lamps. Some troffers may incorporate a mounting flange or are used with a mounting flange kit, allowing them to be installed in ceiling recesses over plaster, drywall, or concealed-frame ceiling systems. Other troffers, called grid-type troffers, are designed for installation in grid T-bar suspended ceilings (*Figure 29*).

> **CAUTION**
>
> Incandescent recessed lighting fixtures generate a considerable amount of heat within their enclosure and are a definite fire hazard if not installed properly. For this reason, several precautions and *NEC*® requirements must be observed when installing them. These precautions and requirements are covered later in the installation section of this module.

### 6.3.0 Suspended Lighting Fixtures

Suspended lighting fixtures are hung from a ceiling using a chain, aircraft wire, cord, or metal stem. Included in this group are chandeliers and pendant units (*Figure 30*).

Figure 27 Typical recessed lighting fixtures.

FOR USE IN INVERTED-T OR GRID CEILINGS

ADJUSTABLE MOUNTING BRACKETS

FOR USE IN DRYWALL OR PLASTER CEILINGS

203F28.EPS

*Figure 28*  Typical recessed fluorescent troffers.

203F29.EPS

*Figure 29*  Fluorescent troffer mounted in a suspended ceiling.

### 6.4.0 Track-Mounted Lighting Fixtures

Track-mounted lighting systems (*Figure 31*) include track, an electrical feed box, and two or more fixtures that can be positioned along the track. Track lighting wiring is in the metal track channels. It can be single-circuit or multiple-circuit wiring to allow two or more sets of lights to be independently controlled.

The tracks are normally mounted on or recessed in the ceiling, but can also be mounted horizontally or vertically on walls.

### 6.5.0 Storage, Handling, and Security of Lamps and Lighting Fixtures

Prior to installation, it is extremely important that lamps and lighting fixtures be handled and stored properly, both while in transit and at the job site. Failure to do so can result in broken and/or damaged lamps/lighting fixtures. This can result in expensive replacement costs and possible delays.

The proper care, handling, and storage of lamps/lighting fixtures begins by using some common sense. Manufacturers of lamps and

203F30.EPS

*Figure 30*  Suspended lighting fixtures.

*Figure 31* Track lighting.

203F31.EPS

lighting fixtures mark their shipping cartons, pallets, etc., with handling and storage instructions. The first thing you should always do is read and follow all such instructions. Obviously, lamps and lighting fixtures are fragile; therefore, pallets and/or cartons should not be dropped or otherwise handled roughly.

If you are responsible for receiving an order of lamps and/or lighting fixtures from a supplier or distributor, make sure that there is no obvious damage to the shipping cartons. Also make sure that the entire quantity that was ordered is delivered or that the delivery ticket shows what is missing.

If the lamps/lighting fixtures are not going to be installed immediately, store them in a dry place where they will be protected from loss or damage. If they must be stored in several locations, keep a record of what is stored at each of the locations.

## 7.0.0 LIGHTING FIXTURE INSTALLATION

Before starting the actual installation of lighting fixtures, it is extremely important to review the lighting floor plan and related lighting fixture schedules to thoroughly familiarize yourself with the different types of lighting fixtures to be installed and their locations. Given this information, you can distribute the different kinds of lighting fixtures to the areas within the building where they are to be installed. *Figure 32* shows an example of a typical lighting floor plan and lighting fixture schedule. *Appendix A* shows a summary of several symbols commonly used on floor and building plans to identify different kinds of lighting fixtures while *Appendix B* provides troubleshooting procedures for various fixtures. If available, shop drawings should be studied prior to installation, and if there are any questions, the fixture manufacturer or lighting designer should be consulted.

When opening lighting fixture shipping cartons, make sure to open the carton carefully so as not to damage the fixture. Do not rip or cut open the carton using knives with long blades, screwdrivers, or similar tools that can cut, scratch, gouge, or otherwise damage the fixture. When handling the fixture itself, wear gloves in order to prevent any grease, dirt, etc., on your hands from getting on its finished and/or decorative surfaces. Also, make sure not to lose or misplace the mounting hardware and insulation (if any) supplied with the fixture. Most lighting fixtures require a minimum amount of preparation before use. Depending on how they are purchased, some are shipped fully assembled, while others may require minor assembly, which should always be done according to the manufacturer's instructions.

**Figure 32** Typical lighting floor plan and lighting fixture schedule.

| TYPE | MANUFACTURING CATALOG NUMBER | MOUNTING | LAMP(S) | DESCRIPTION |
|------|------------------------------|----------|---------|-------------|
| FA | Hubbell RDL 84 | Recessed | 2 F32-T8, 3100K | 2 x 4 Troffer |
| FB | Hubbell WIM42WACB1 | Surface | 2 F32-T8, 3100K | Wrap around |
| FC | Hubbell D332-120V/194W | Recessed | GX24q3 | Downlight |
| HC | Hubbell LS716148 | Recessed | 175W MH supplied | Downlight |
| EX | Hubbell CWP-RB-L-9 | Surface | LED included | Exit sign |
| EM | Hubbell PE6-12-2-R61-A3 | Wall | Included | Emergency |

203F32.EPS

**Think About It**

## Tracking the Installation

When roughing-in lighting circuit wiring, how can you keep track of the work that has been accomplished?

Always read the labels on the lighting fixture and/or the installation instructions supplied with the fixture to familiarize yourself with its characteristics. Knowing this information can help prevent improper or inappropriate installation. *Figure 33* shows an example of one lighting manufacturer's installation instructions.

Typical information that can be found marked on lighting fixture labels and/or given in the installation instructions includes:

- Type(s) of lamps to use with the fixture and the maximum lamp wattage
- Approval for use in suspended ceilings, non-insulated ceilings (non-IC), or insulated ceilings (IC)
- Range of ambient temperatures for fixture operation

# INSTALLATION INSTRUCTIONS
# FOR MSV

See warnings on carton.

## WARNING

- Disconnect power before installation or servicing.
- Install, operate and maintain to meet all applicable codes.
- Labeled ballast voltage must match supply voltage.
- Protect all wiring connections with approved insulators (by others).
- Insure that **all** internal wiring, excluding lead wires from reflector, do not contact reflector assembly due to high lamp temperatures.
- Do not **overtighten** lens door hinge and retaining screws.

## PRE-INSTALLATION

1. Remove lens door and set aside **(See Figure 1)**. To remove lens door, **loosen** lens door retaining screws (2) and one hinge screw (1).
2. Remove the two reflector retaining screws and lift out reflector assembly.
3. Remove shipping bolt located in arm mounting hole.
4. When a multi-tap ballast is supplied, the fixture is factory wired to the highest voltage. Verify that fixture wiring matches the supply voltage.

## ARM INSTALLATION (See Figure 2)

**NOTE:** Guide electrical supply wires from fixture through the bearing plate, fixture, gasket, cover, positioning bracket, pole and bolt plate as each mounting step is completed.

1. Insert longer threaded end of hex rods (5) through lockwashers, positioning bracket (4) and pole.
2. Tighten securely to bolt plate (3).
3. Slide cover (6) over hex rods. Drain hole in cover must be on bottom.
4. Slide gasket (7) over threaded end of hex rods.
5. Place fixture onto hex rods.
6. Slide bearing plate (8) and lockwashers over end of hex rods on inside of fixture.
7. Install and tighten nuts securely.
8. Install supply wire (by others) thru the cord grip bracket in pole using appropriate supplied cord grip (9).
9. Make wiring connections.
10. Push wires and wirenuts into pole top and install cap (10).

## FINAL ASSEMBLY

1. Re-install and secure reflector assembly using the reflector retaining screws. The reflector assembly is fully rotatable 360° by 90° increments.
2. Install proper lamp.
3. Re-install lens door and insure that all fasteners are securely tightened.

**Hubbell Lighting**
® **A Division of Hubbell Lighting, Inc.**
**2000 Electric Way**
**Christiansburg, VA 24073-2500**
**(540) 382-6111**
**FAX (540) 382-1526**

**FIGURE 1**
**FIGURA 1**

3/8"
(9.5mm)

4-1/16"
(103.2mm)

**FIGURE 2**
**FIGURA 2**

| Fixture | |
|---|---|
| Black | Line (HOT) |
| White | Neutral |
| Ground | Green/Ground |

**WIRING DIAGRAM**

| Luminaire | |
|---|---|
| Noir | Alimentation (VIVANT) |
| Blanc | Neutre |
| Mise à la terre | Vert/mise à la terre |

**SCHÉMA DE CÂBLAGE**

| Artefacto | |
|---|---|
| Negro | Vivo |
| Blanco | Neutro |
| Tierra | Verde/tierra |

**DIAGRAMA DE CABLEADO**

**268-1072-9901**

203F33.EPS

*Figure 33* Example of lighting fixture installation instructions.

- Type of wire to use for fixture supply connections
- Thermal protection
- Suitability for wall or ceiling mount only
- Suitability for use in damp and/or wet locations
- Suitability for use as a raceway
- Suitability for installation in concrete
- Mechanical installation instructions
- Electrical connection instructions

*NEC Article 410* sets forth most of the requirements pertaining to the installation of lighting fixtures. *NEC Article 411* covers lighting systems of 30V or less. You must meet all *NEC®* and/or local code requirements, including those for mounting, supporting, grounding, live-parts exposure, insulation clearances, supply conductor types, maximum lamp wattages, etc. Whenever you have questions or need clarification about the method to use, always refer to the appropriate codes.

It is important to understand how different types of lighting fixtures are safely and securely attached to a structure or outlet box. Code requirements for the installation of lighting fixtures, lighting outlet boxes, and related methods of support are covered in *NEC Sections 314.23(F), 314.27(A) and (C)*, and *NEC Article 410, Parts III and IV*. The specific methods and hardware used for mounting and supporting lighting fixtures depend on several factors:

- Type of lighting fixture
- Location of lighting fixture
- Construction of mounting surface
- Weight of lighting fixture

### 7.1.0 Mechanical Installation of Surface-Mounted Lighting Fixtures

Surface-mounted lighting fixtures include those that are directly mounted on a ceiling or wall. They can be incandescent, fluorescent, or HID units. The general methods for mechanically installing ceiling and wall surface-mounted lighting fixtures are basically the same; only the location of the outlet box is different. Normally, boxes for indoor and outdoor wall lighting fixtures are installed at heights of 6'-0" and 6'-6" above the floor, respectively. A lighting fixture should be mounted to its outlet box using the mounting hardware supplied with the fixture and installed according to the manufacturer's instructions. If you are not thoroughly familiar with the lighting fixture to be installed, first read and be sure that you understand the related manufacturer's product literature and installation instructions supplied either with the fixture or printed on the shipping carton.

In buildings of all types, branch circuit wiring for surface-mounted ceiling and wall lighting fixture outlets normally terminates in octagonal or round boxes (*Figure 34*). Some may terminate in square or rectangular boxes. Both metal and non-metallic types of boxes may be used. These boxes are designed to be covered by the mounting base of a lighting fixture with a similar shape. For concealed wiring in wooden frame construction, the outlet boxes are fastened directly to studs, joists, or other framing members using nails or screws. Some ceiling outlet boxes are mounted on an adjustable bar hanger so that the box can be installed in the space between two ceiling joists. The box is attached to the bar hanger by a fixture stud or two bolts. When properly installed, the front face of all concealed ceiling and wall boxes should be flush with the finished ceiling or wall. For exposed conduit installed on concrete or other masonry surfaces, the outlet boxes should be fastened to the ceiling or wall with screws and expansion shields, anchors, powder-actuated fasteners, or similar appropriate fasteners. On steel work, they should be bolted or clamped to the steel. Outlet boxes embedded in concrete are held securely in position by the hardened concrete.

The size and weight of the fixture determine what type of outlet box should be used with a fixture. When mounting lighting fixtures that weigh more than 50 pounds in a ceiling, *NEC Section 314.27(A)(2)* requires that the fixture be supported independently of the box or mounted to a listed outlet box designed and marked to support a fixture weighing more than 50 pounds. Boxes used at luminaire or lampholder outlets in a wall must be designed for the purpose and marked to indicate the maximum weight of the fixture to be supported, if other than 50 pounds, per *NEC Section 314.27(A)(1)*. In addition to complying strictly with the code, good judgment must also be exercised. Remember, codes only give the minimum requirements for electrical installations, not necessarily those that make a quality installation.

---

### Think About It

## Installing Surface-Mounted Fixtures in Closets

What precautions should you take when installing surface-mounted ceiling or wall fixtures in a clothes closet?

---

ADJUSTABLE
BAR HANGER

PANCAKE BOX

203F34.EPS

*Figure 34*   Typical outlet boxes used with surface-mounted lighting fixtures.

### 7.1.1 Drum, Globe, Sconce, and Similar Lighting Fixtures

*Figure 35* shows some common methods used to mount drum, globe, sconce, and similar types of indoor lighting fixtures to an outlet box. Usually, these types of lighting fixtures can be directly mounted to the outlet box with no additional support. As shown in *Figure 35(A)*, the common porcelain-type lamp holder fixture screws directly to the ears of the outlet box. Other ceiling fixtures and wall sconces are fastened to the outlet box using a mounting bar that is attached to a threaded stud in the box, as shown in *Figure 35(B)*. The mounting bar is positioned on the stud and fastened in place with a locknut. After the fixture wiring has been completed, the mounting holes in the base of the fixture are slipped over the machine screws installed in the mounting bar. Cap nuts are then installed on the machine screws and tightened until the fixture is drawn flush with the surface. Fixtures can be mounted to an outlet box without a box stud in a similar manner by screwing a mounting bar to the ears of the box, as shown in *Figure 35(C)*. Another common way some fixtures are installed is by screwing a reducing nut and nipple onto a box stud, as

shown in *Figure 35(D)*. Following this, the hole in the center of the fixture base is placed over the nipple, then a cap nut is installed on the nipple and tightened until the fixture is drawn flush with

**EARS**

**STUD**

**MOUNTING BAR**

**MACHINE SCREWS**

**FIXTURE BASE**

**CAP NUT**

(A)

(B)

**REDUCING NUT**

**NIPPLE**

(C)

(D)

203F35.EPS

*Figure 35*  Common methods of installing surface-mounted lighting fixtures.

the surface. A similar method is to screw a nipple into the threaded center hole of a mounting bar attached to the ears of the outlet box.

### 7.1.2 Fluorescent Lighting Fixtures

Some lightweight, surface-mounted fluorescent fixtures can be supported directly from an outlet box. Such types typically use a stud, reducing nut, and nipple in the box, with a mounting bar inside the housing, as shown in *Figure 36*. The assembly is held with a locknut. Heavier fluorescent fixtures are supported directly from the ceiling using toggle bolts, wood screws in joists, anchors in masonry, or other appropriate fasteners. These fixtures have an access opening that, when the fixture is in place, must be positioned directly under the outlet box. Many surface-mounted fluorescent fixtures have bumps or ridges on their mounting surfaces that are designed to hold the fixtures

slightly off the ceiling in order to allow for the dissipation of heat.

### 7.1.3 Chandeliers and Pendants

*Figure 37* shows some methods used for attaching chandeliers to an outlet box. Chandeliers are normally heavy, requiring that they be mounted on a box with a stud. As shown, a hickey (or a hickey together with a threaded adapter) is installed over the fixture chain and wires, then screwed onto the stud. The fixture base is then fastened to the hickey or adapter with a locknut or cap nut. Some are secured to a mounting bar by means of a nipple. Pendant lighting fixtures are typically supported by chains, metal stems, thin cables, or retractable cords. Most pendants are designed to be wired directly into a ceiling outlet box in a similar manner as that described for chandeliers.

STUD    REDUCING NUT

FIXTURE CHANNEL

NIPPLE

MOUNTING BAR

LOCKNUT

203F36.EPS

*Figure 36*    Typical method for mounting a smaller fluorescent fixture to an outlet box.

STUD

HICKEY

FIXTURE BASE

LOCKNUT OR CAP NUT

STUD

HICKEY

THREADED ADAPTER

FIXTURE BASE

LOCKNUT OR CAP NUT

STUD

HICKEY
FIXTURE BASE

LOCKNUT OR CAP NUT

NIPPLE

MOUNTING BAR

FIXTURE BASE

203F37.EPS

*Figure 37*    Methods for attaching chandeliers to an outlet box.

*NEC Section 410.10(D)* prohibits pendants and other hanging-type fixtures from being installed within a zone measured 3' horizontally and 8' vertically from the top of a bathtub rim or shower stall threshold.

TYPICAL FAN JOIST HANGER

### 7.1.4 Ceiling Fans/Fixtures

*NEC Section 422.18* covers the support of ceiling-suspended (paddle) fans. It refers to *NEC Section 314.27(C)*, which states that for fans weighing 35 pounds or less, the outlet box may be used as the only support of the fan if it is listed for such use. The code also requires that fans weighing over 35 pounds either be supported independently of the outlet box or the box must be listed for such use and marked with the maximum weight to be supported. Extra support can be provided by mounting the box on a fan joist hanger (*Figure 38*).

The specific method for mounting ceiling fans varies among ceiling fan manufacturers; therefore, the fan should be mounted to the outlet box per the manufacturer's directions. *Figure 38* shows an example of one mounting method. *NEC Section 410.10(D)* also prohibits ceiling-suspended (paddle) fans from being installed within a zone measured 3' horizontally and 8' vertically from the top of a bathtub rim or shower stall threshold.

### 7.1.5 Outdoor Lighting Fixtures

The methods for attaching outdoor surface-mounted lighting fixtures to outlet boxes are basically the same as those described for indoor surface-mounted fixtures. However, the installation must be done so that the final assembly is watertight.

All fixtures installed in wet locations must be marked as suitable for wet locations. All fixtures installed in damp locations must be marked as either suitable for wet locations or suitable for damp locations. When mounting a lighting fixture to an outlet box that is fully exposed to the weather, the outlet box must be a metal weatherproof-type box.

All washers, gaskets, etc., required to make a watertight assembly must be properly installed as directed by the manufacturer. *Figure 39* shows an example of a common outdoor spotlight fixture. With this fixture, waterproofing fiber washers must be inserted into the fixture sockets and rubber gaskets installed over the end of each socket.

203F38.EPS

*Figure 38*   Typical ceiling fan/light mounting method.

In addition, a gasket must be installed between the fixture base and the face of the outlet box.

### 7.2.0 Mechanical Installation of Recessed Lighting Fixtures

Recessed lighting fixtures are mounted into ceiling recesses so that the main body of the fixture is not visible.

*Figure 40* lists some of the code requirements for recessed fixtures.

Recessed lighting fixtures generate a considerable amount of heat within their enclosures and are a definite fire hazard if not installed properly. For this reason, the code requirements given in *NEC Article 410* must be strictly observed when installing these fixtures.

GASKET

FIBER WASHER

203F39.EPS

*Figure 39* Installing washers and gaskets in outdoor lighting fixtures.

### 7.2.1 Incandescent, Compact Fluorescent, and HID Recessed Fixtures

Incandescent, compact fluorescent, and HID recessed lighting fixtures are all mounted in a ceiling in basically the same way. The rough-in for these types of fixtures must be completed before the ceiling material is installed. As shown in *Figure 41*, they typically include captive adjustable bar hangers for attachment to wooden joists or T-bar grid ceiling members. Recessed fixtures that are identified as acceptable for through wiring are equipped with feed-through junction boxes, which allow a series of them to be wired from one to another or branch circuits to be fed through the housing junction box. Recessed fixtures are typically wired with conduit, nonmetallic-sheathed cable, or armored cable, depending on the application.

A general installation procedure involves the following steps:

*Step 1* Extend the fixture mounting bars to reach the framing members.

*Step 2* Position the lighting fixture on the mounting bars to locate it properly.

*Step 3* Align the bottom edges of the mounting bars with the bottom face of the framing members, then nail, screw, or otherwise attach the bars to the framing members.

*Step 4* Remove the junction box cover and open one knockout for each cable entering the box.

*Step 5* If required, install cable clamps for each open knockout and tighten the locknut.

*Step 6* Run the branch circuit wiring through the cable clamps and into the fixture connection box.

*Step 7* After the electrical connections are made and with the finished ceiling in place, install the lens/diffuser and/or trim on the fixture. The fixture trim combinations are marked on the label inside the fixture.

**CAUTION**

The use of a trim type that is not listed on the fixture can result in overheating and possible on-off cycling of the fixture's thermal protective device. Mismatching fixtures and trim is a violation of *NEC Section 110.3(B)*, which states that "Listed or labeled equipment shall be installed and used in accordance with any instructions included in the listing or labeling."

---

**On Site**

## Installing Chandeliers and Pendants

It is important to make sure that a chandelier or pendant fixture is installed at the proper height. For example, you should allow for door clearances when installing the fixture in a foyer or entrance hallway. Such a fixture should also be hung at a height of 7' or more so that it allows safe passage for tall people.

If the fixture is being hung over a table, its width should be at least 12" less than the width of the table (6" from each edge) so that people do not hit their heads on the fixture when sitting down at or getting up from the table. It should be hung at a height of about 30" to 36" above the table, unless the ceiling is higher than 8'. If the ceiling height is greater than 8', the fixture may look better raised even higher than 36". To provide light for demanding tasks, pendants should be hung about 15" above the work surface.

---

Branch circuit conductors with insulation suitable for the temperature requirements marked on the listed prewired recessed fixture may be run directly to the junction box on the fixture.
*NEC Section 410.117(B)*

Junction box must be at least one foot from the fixture.
*NEC Section 410.117(C)*

This box must be accessible.
*NEC Section 314.29*

Fixture whip is not required to be supported.
*NEC Section 348.30(A), Exception No. 3*

Fixture whip must be at least 18 inches but not more than 6 feet of suitable raceway, with conductors having insulation suitable for the temperatures encountered.
*NEC Section 410.117(C)*

May be ⅜".
*NEC Section 348.20(A)(2)*

**LISTED**
prewired fixture with junction box

**LISTED**
fixture without junction box

At least ½" clearance from combustible material, except at point of support (non-IC fixture).
*NEC Section 410.116(A)(1)*

All incandescent recessed fixtures must have thermal protection and must be so marked.
*NEC Section 410.115(C)*

Keep insulation at least 3" from fixture unless it is identified for direct contact with insulation.
*NEC Section 410.116(B)*

**NON-IC FIXTURE**
(NOT SUITABLE FOR DIRECT CONTACT)

**TYPE IC FIXTURE**
(SUITABLE FOR DIRECT CONTACT)

203F40.EPS

*Figure 40* Summary of *NEC®* requirements for recessed fixtures.

**Figure 41** Typical mounting of a recessed lighting fixture.

### 7.2.2 Fluorescent Troffers

The code requirements for the installation of recessed fluorescent troffers are basically the same as those for other types of recessed fixtures. Recessed fluorescent troffers are inserted into the rough-in opening in the drywall, plaster, or other ceiling types and fastened in place using various devices. Some have mounting brackets used to fasten the fixture to the ceiling structure that are adjustable from within the installed fixture. Other types of fixtures are held in place by plaster frame kits or flange kits used to trim out an opening

## What's wrong with this picture?

203SA05.EPS

in the ceiling (*Figure 42*). Note that flange kits are used to convert grid troffers to flange troffer applications. Other fixtures with a flange on the housing are supported on the ceiling structure by swing-out supports adjusted in position by a screwdriver from below the ceiling. When in position, the adjacent ceiling structure is clamped between the supports and the housing flange holding the fixture in place. The screwdriver access holes are sealed after installation using the plugs supplied with the fixture.

If a suspended ceiling supports recessed fluorescent troffers or other lighting fixtures, *NEC Section 410.36(B)* requires that all the ceiling framing members be securely fastened together and to the building structure. This job is normally done by the installer of the ceiling grid. The proper support of suspended fixtures is a critical safety issue for people that occupy the building and for any emergency personnel that may be called into the building in the event of an earthquake, fire, etc. For this reason all lighting fixtures, including any lay-in fixtures such as fluorescent troffers, must also be securely

---

**Think About It**

## Installing Recessed Fixtures

When installing recessed fixtures, is it permissible to wire a series of lights from one to another or to run branch circuit wiring through a fixture junction box? If so, when?

---

SIDE RAIL

POP RIVET

#6 SCREW

FLANGE KIT

#8 SCREWS

END RAIL ASSEMBLY

SIDE RAIL

PLASTER FRAME KIT

203F42.EPS

*Figure 42*   Fluorescent fixture plaster kit and flange kit.

fastened to the main ceiling framing members by mechanical means such as bolts, screws, rivets, or listed clips identified for use with the type of ceiling framing members and fixtures involved. These are installed by the electrician. Most manufacturers now provide locking clips in the end plates of their troffers. These clips are simply bent perpendicular to the end plate so that as the troffer is lowered into the ceiling grid, the clip locks over the top of the T. *Figure 43* shows some types of listed clips commonly used to fasten lay-in troffers to suspended grid systems.

Wiring for lighting fixtures in a suspended ceiling is installed above the suspended ceiling using electrical metallic tubing (EMT), armored cable (AC), nonmetallic-sheathed cable (NM), or other acceptable methods as listed in the *NEC®*. Then, from properly supported outlet boxes placed above the ceiling and near the intended locations of the lay-in fluorescent fixtures, a flexible

FIXTURE WHIP

203F43.EPS

*Figure 43*   Typical clips used to fasten lay-in troffers to suspended ceiling grid systems.

## Installing Lighting Fixtures

A group of electricians and apprentices were wiring fluorescent fixtures in a suspended ceiling in the new wing of a hospital. Emergency fixtures on a separate circuit were to be wired first, then the existing temporary lights were to be de-energized. The crew decided that this process was too slow and, contrary to supervisory instructions, disconnected the temporary lights and began to wire the remaining fixtures while some circuits were still energized. Co-workers warned the crew to test the circuits to see if they were energized.

An apprentice was standing on a wooden ladder with his body extended above the ceiling grid while wiring a fluorescent fixture. He was wet with sweat and was leaning against a ceiling grid as he worked. Co-workers heard a noise and upon investigation, found the apprentice dangling unconscious from the ceiling grid. The victim was pulled free, then given CPR, but he had received a fatal shock.

**The Bottom Line:** This accident could have been prevented if standard safety measures had been taken:

- The crew had followed the supervisor's instructions.
- The branch circuit feeding the lighting fixture had been de-energized and the proper lockout/tagout procedures followed.
- The apprentice had been working under the supervision of a qualified electrician who had first checked the branch circuit wires feeding the fixture with a voltage tester to make sure that the wires were not energized.

---

connection, commonly called a fixture whip, is made between the outlet box and the fixture (*Figure 43*). As appropriate, the fixture whip is:

- An 18" (minimum) to 6' (maximum) length of ⅜" flexible metal conduit using conductors suitable for the temperature requirement stated on the fixture label
- An 18" to 6' length of armored cable containing conductors suitable for the temperature as stated on the fixture label
- An 18" to 6' length of ⅜" flexible nonmetallic conduit with conductors and grounding conductor suitable for at least 90°C

### 7.3.0 Mechanical Installation of Suspended Lighting Fixtures

Suspended lighting fixtures used in commercial and industrial buildings can be supported in numerous ways, depending on the construction of the building and the particular lighting application. Because the methods of support used can vary so widely, it would be impossible to describe them all here. *Figure 44* shows some types of hangers commonly used to attach supporting rods, cables, chains, etc., to the building ceiling or beam structure.

*Figure 45* shows some examples of the different kinds of supports used with suspended fluorescent fixtures.

*Figure 46* also shows fixture support methods typical of those found in commercial and industrial sites. Note that the method of presentation used in this figure is representative of what you will see on electrical installation drawings and blueprints.

### 7.4.0 Mechanical Installation of Track Lighting Fixtures

Track-mounted lighting systems include the track (rail), an electrical feed box, two or more fixtures that can be positioned along the track, and various types of accessories and other components that allow assembly in any pattern (*Figure 47*). The track lighting wiring is in the metal track rails. It can be either single-circuit or multiple-circuit wiring to allow two or more sets of lights to be independently controlled. The track is an extruded aluminum channel that holds two or more electrical connectors on opposite sides of the channel. The inside surfaces of the conductors are bare. When a track lighting fixture is snapped into the track, two terminals on the fixture contact the track conductors, supplying power to the fixture.

Track lighting systems are made so that the proper electrical polarity is always maintained. The polarity lines inside the track must be aligned when the sections of track are installed.

---

CONCRETE INSERT    T-BAR CLIP    ADJUSTABLE SWING HANGER    I-BEAM CLIP

BEAM CLAMP    FLANGE HANGERS    FLAT SURFACE HANGERS

203F44.EPS

*Figure 44*    Examples of hangers used to attach fixtures to building structural members.

The lighting fixtures will snap into the track in only one way.

> **NOTE**
> The *NEC®* requires that lighting fixtures, accessories, etc., used in a track system be specifically designed for the track on which they are to be installed.

Track sections are made in lengths of 2', 4', and 8'. These sections are fastened together using straight, flexible, X, L, or T connectors. End caps provide a finished look. Track lighting can be connected to a branch circuit outlet box in many ways, including the use of a floating power feed, end power feed, or X, L, or T power feed connector. Some are connected to a receptacle outlet using a cord and plug connector.

*NEC Sections 410.151 through 410.155* cover the installation requirements for track lighting. The track is mounted directly to the ceiling or wall using screws, toggle bolts, or other appropriate fasteners installed through holes in the track. When installed on a wall, the track must be at least 5' above the finished floor, except where protected from physical damage or when using low-voltage (less than 30V rms) track.

The *NEC®* requires that the track be fastened so that each fastener is suitable for supporting the maximum weight of the fixtures that can be installed. Unless identified for use with supports at greater intervals, a single section of track that is 4' or less in length must be supported in at least two places. If more track is being used, at least one additional support for each individual extension that is 4' long or less must be used.

A typical installation begins by first attaching the track electrical feed assembly to the designated outlet box. If using an end power feed for this purpose, the track adapter plate covers the outlet box and holds the track connector and electrical housing (*Figure 48*).

When using a wire-in connector, track clips that are spaced at the required distance apart are used to hold the track ¼" to ½" away from the ceiling or wall mounting surface. These clips are fastened in place using screws, toggle bolts, or other appropriate fasteners. It is a good idea to snap a chalkline from the center of the outlet box or center slot on the track connector to the proposed end of the track run for use as a track installation guide. Once the clips are installed, the first section

**Figure 45** Typical methods of support for suspended fluorescent fixtures.

203F45.EPS

of track is fitted solidly into the track connector, and then the track is snapped into the clips. Following this, setscrews in the sides of the clips are tightened to hold the track firmly in place. The remaining sections of track are installed in the same manner using the appropriate connectors and accessories. When it is necessary to cut a section of track to length, it can be cut using a hacksaw. Make sure to follow the manufacturer's instructions as to where and how to make the cut and how to close off the cut with a proper end cap.

### 7.5.0 Electrical Connection of Lighting Fixtures

Making the electrical connections between the lighting fixture and related outlet box wires is done at the appropriate point during the mechanical installation of the fixture. The specific way in which the supply wires in the outlet box are connected to the lighting fixture wires

varies depending on the design of the branch circuit and the way in which the branch circuit cable or raceway wiring is routed. Therefore, you must refer to the lighting plan to determine the exact connections.

#### 7.5.1 Conventional Hard-Wired Installations

The connection of the lighting fixture wiring to the outlet box supply wires should be done in accordance with the manufacturer's instructions. The general procedure is basically the same, no matter what type of fixture is being connected.

> **WARNING!**
>
> Before attempting to connect lighting fixtures to a branch circuit outlet box, make sure that the power is turned off to the branch circuit involved at the circuit breaker panel. Then, make sure to follow the prevailing lockout and tagout procedures.

CLIP FIXTURE TO GRID WHERE SHOWN AND PLUG INTO NEAREST RECEPTACLE

GRID

CORD WITH PLUG MOUNT; CORD WITH STRAIN RELIEF IN TOP OF J-BOX

TRACK CLIP (SECURE CLIP TO BOTTOM OF HEXAGONAL J-BOX WITH TWO #8 FLAT HEAD MACHINE SCREWS)

FIXTURE CANOPY WITH TRANSFORMER

ROOF DECK

4 × 4 BOX WITH COVER

¾" CONDUIT

SAFETY CABLE

PLUG

¼" FENDER WASHER

2½" BOLT

¼" LOCK WASHER

STRUT

NUT SECURES J-BOX TO STRUT

EMT CONDUIT

FIXTURE CANOPY

1½" DEEP HEXAGONAL J-BOX

14/3 CORD TIE WRAP TO CABLE 12" OC

CABLE COUPLER

AIRCRAFT CABLE

FIXTURE

SHEET METAL SCREW

TOP RAIL

¾" IMC

CONDUIT HANGER

THREAD SIZE REDUCER

CONNECTIONS MADE IN CONDULET

1" T-CONDULET

¾" RIGID STEEL NIPPLE

FIXTURE JUNCTION BOX

203F46.EPS

*Figure 46*   Examples of fixture supports for commercial/industrial applications.

relays that are activated or deactivated by the EMS control unit. *Figure 55* shows a typical lighting switching scheme involving the use of split-wired, multi-ballasted lighting fixtures. By split-wiring three-lamp and four-lamp lighting fixtures, multiple light intensities can be provided in a single zone. The relay-based control system provides full lighting for specific times of the day, while allowing a reduced lighting level and reduced power for those times when less lighting is needed.

*Figure 55(A)* shows a relay-controlled split-wiring system connected to two three-lamp lighting fixtures. This arrangement allows for four levels of lighting: 0%, 33⅓%, 66⅔%, and 100%. As shown, the relay to the inboard ballasts is closed, allowing two of the six lamps to be turned on, thus providing light at the 33⅓% level.

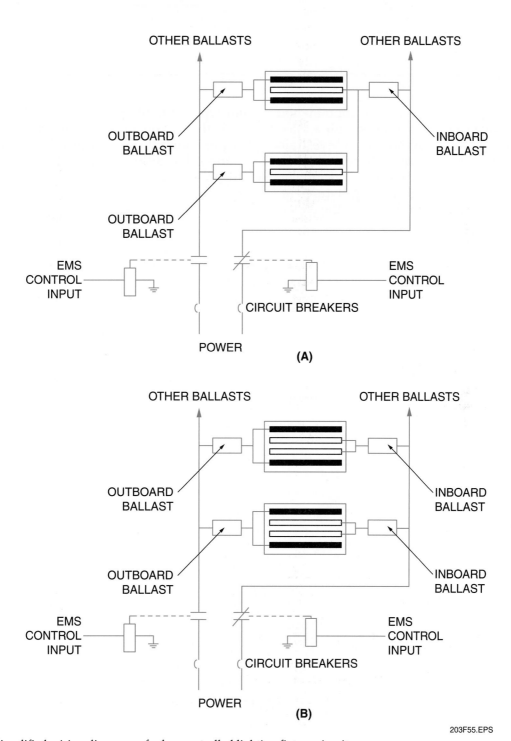

**Figure 55** Simplified wiring diagrams of relay-controlled lighting fixture circuits.

203F55.EPS

*Figure 55(B)* shows a similar relay-controlled split-wiring system connected to two four-lamp lighting fixtures. This arrangement allows for three levels of lighting: 0%, 50%, and 100%. As shown, the relay to the inboard ballasts is closed, allowing four of the eight lamps to be turned on, thus providing light at the 50% level.

## SUMMARY

Electric lighting serves to provide illumination for the performance of visual tasks with a maximum of comfort and a minimum of eyestrain and fatigue, allowing individuals to perform their daily living and work-related tasks.

The proper installation of lighting fixtures requires that you become familiar with the different types of lighting fixtures and the lamps used with them. You must acquire the skills needed to mechanically and electrically install lighting fixtures for different applications according to the prevailing building and electrical codes.

1. Light impulses are transferred from the eye to the brain by the part of the eye called the _____.

   a. rods
   b. cones
   c. pupil
   d. optic nerve

2. Light energy that is *not* absorbed by an object can be _____.

   a. reflected from the object
   b. transmitted through the object
   c. both reflected from and transmitted through the object
   d. dissipated as heat in the object

3. When all the wavelengths of the light spectrum are presented to the eye in nearly equal proportions, the color of the light is _____.

   a. red
   b. white
   c. blue
   d. green

4. In an incandescent lamp, the filament resistance is _____.

   a. much higher when the lamp is turned on
   b. lower when the lamp is turned on
   c. somewhat higher when the lamp is turned on
   d. the same whether the lamp is turned on or off

5. True or False? The filament of a tungsten halogen lamp is encased in a capsule containing iodine or bromine gas.

   a. True
   b. False

6. The diameter of a 4', T-12 fluorescent lamp is _____.

   a. 1"
   b. 1⅛"
   c. 1¼"
   d. 1½"

7. During operation, a high-output (HO) fluorescent lamp typically has a current draw of _____.

   a. 400mA to 600mA
   b. 800mA to 1,000mA
   c. 1,100mA to 1,300mA
   d. 1,500mA to 1,700mA

8. The type of high-intensity discharge (HID) lamp that uses the oldest HID technology is the _____.

   a. low-pressure sodium lamp
   b. high-pressure sodium lamp
   c. mercury vapor lamp
   d. metal halide lamp

9. A light that is considered warm in tone is produced by lamps with color temperatures of _____.

   a. 3,000K and less
   b. 3,500K and less
   c. 4,100K and less
   d. 5,000K and less

10. To improve both ballast efficiency and lamp efficacy, an electronic ballast typically operates the connected fluorescent lamps at a frequency of _____.

    a. 50Hz to 60Hz
    b. 10kHz to 20kHz
    c. 20kHz to 50kHz
    d. 60kHz to 100kHz

11. Most code requirements governing the installation of lighting fixtures, lampholders, and lamps are covered in _____.

    a. *NEC Article 411*
    b. *NEC Article 370*
    c. *NEC Article 410*
    d. *NEC Article 422*

12. Fixture whips used to connect from a junction box mounted above the ceiling to a fixture installed in a suspended ceiling cannot exceed a length of _____.

    a. 18"
    b. 2'
    c. 4'
    d. 6'

13. When cutting the branch circuit wires in an outlet box in preparation for connecting a light fixture, the amount of free conductor length that should be left after it emerges from the cable or raceway is _____.

    a. 3"
    b. 5"
    c. 6"
    d. 8"

14. All of the following sensors are normally used as occupancy sensors *except* a(n) _____.

    a. motion detector
    b. infrared sensor
    c. photosensor
    d. sound sensor

**Figure 1**

203RQ01.EPS

15. If the contacts of both relays are closed in the circuit shown in *Figure 1*, what percentage of lighting will be produced by the lighting fixtures?

    a. 0%
    b. 33⅓%
    c. 66⅔%
    d. 100%

1. The limiting factor that determines the minimum time required for the nerve terminals of the eye to respond to an image on the retina is the _____.
   a. intensity threshold
   b. contrast
   c. intensity angle
   d. time threshold

2. When light strikes an object, it can be _____, transmitted through the object and _____, and/or it can be _____ from the object.

3. When light passes through a prism, the spectrum ranges from _____ on the lower end to _____ on the upper end.

4. True or False? The amount of light a lamp produces is directly proportional to the amount of energy the lamp consumes to produce its rated light output.

5. Which of the following types of fluorescent lamp circuits requires the use of a separately installed starter?
   a. Instant start
   b. Rapid start
   c. Preheat
   d. Bi-pin

6. The type of HID lamp that produces light with the best color rendition is the _____ lamp.
   a. metal halide
   b. mercury vapor
   c. high-pressure sodium
   d. low-pressure sodium

7. True or False? When installing a bi-pin fluorescent lamp, it should be rotated 1/4 turn in its lampholder to seat and connect it.

8. Fluorescent troffers are a type of _____ lighting fixture.
   a. surface-mounted
   b. recessed
   c. suspended
   d. track-mounted

9. When opening a shipping carton containing a light fixture, make sure to open it _____ so as not to _____ the fixture.

10. True or False? When installing a surface-mounted fixture that weighs more than 50 pounds, it must always be supported independently of the outlet box unless listed and labeled for such use.

11. Recessed fixtures not labeled Type IC must be installed so that they are at a minimum distance of _____ from any adjacent insulation.

12. A fixture whip connected between a junction box mounted above the suspended ceiling and a fixture installed in a suspended ceiling must have a minimum length of _____ and be no longer than _____.

13. When fluorescent troffers and other fixtures are installed in a suspended ceiling, the ceiling framing members must be _____ _____ together and to the _____ structure.

14. True or False? Lighting fixtures and/or accessories for track lighting systems can be used interchangeably between systems made by different manufacturers.

15. When cutting the branch circuit wires in preparation for connecting a light fixture, you should leave at least _____ of free conductor length at the outlet box where the wire emerges from its cable sheath or raceway.
    a. 3"
    b. 5"
    c. 6"
    d. 8"

# Trade Terms Introduced in This Module

**Ballast:** A circuit component in fluorescent and HID lighting fixtures that provides the required voltage surge at startup and then controls the subsequent flow of current through the lamp during operation.

**Color rendering index (CRI):** A measurement of the way a light source reproduces color. The higher the index number (0–100), the closer colors are to how an object appears in full sunlight or incandescent light.

**Dip tolerance:** The ability of an HID lamp or lighting fixture circuit to ride through voltage variations without the lamp extinguishing and cooling down.

**Efficacy:** The light output of a light source divided by the total power input to that source. It is expressed in lumens per watt (LPW).

**Incandescence:** The self-emission of radiant energy in the visible light spectrum resulting from thermal excitation of atoms or molecules such as occurs when an electric current is passed through the filament in an incandescent lamp.

**Incident light:** The light emitted from a self-luminous object such as the sun or an electric source.

**Lumen (lm):** The basic measurement of light. One lumen is defined as the amount of light cast upon one square foot of the inner surface of a hollow sphere with a one-foot radius with a light source of one candela in its center.

**Lumen maintenance:** A measure of how a lamp maintains its light output over time. It may be expressed either numerically or as a graph of light output versus time.

**Lumens per watt (LPW):** A measure of the efficiency, or, more properly, the efficacy of a light source. The efficacy is calculated by taking the lumen output of a lamp and dividing by the lamp wattage. For example, a 100W lamp producing 1,750 lumens has an efficacy of 17.5 lumens per watt.

**Luminaire (fixture):** A complete lighting unit consisting of a lamp (or lamps) and ballasts (where applicable), together with the parts designed to distribute the light, position and protect lamps, and connect them to the power supply.

**Luminance:** The luminous intensity of any surface in a given direction per unit area of that surface as viewed from that direction. Measured in candela/m². The term *luminance* is commonly used to express brightness.

**Reflected:** The bouncing back of light waves or rays by a surface.

**Reflected light:** Light reflected from an object that does not have self-luminating properties but reflects light provided from another source.

**Refracted/refraction:** The bending of a light wave or ray as it passes obliquely from one medium to another of different density, or through layers of different density in the same medium.

**Starter:** A circuit component of certain fluorescent lighting fixtures that is used to heat the lamp electrodes before the lamps are lighted.

**Troffer:** A recessed lighting fixture installed with the opening flush with the ceiling.

# Appendix A

## Summary of Lighting Fixture Symbols

LUMINAIRE; POINT SOURCE, SURFACE-MOUNTED, (DRAWN TO SCALE OR LARGE ENOUGH FOR CLARITY)

LUMINAIRE; EXTENDED SOURCE, SURFACE-MOUNTED (DRAWN TO SCALE)

LUMINAIRE; STRIP-TYPE (LENGTH DRAWN TO SCALE)

LUMINAIRE; FLOOD-TYPE

LINEAR SOURCE; i.e., LOW-VOLTAGE STRIP, NEON, FIBER OPTIC, ETC. (LENGTH DRAWN TO SCALE)

EXIT SIGN; MOUNTING, NO. OF FACES AND ARROWS AS SHOWN

RECESSED

WALL-MOUNTED

PENDANT, CHAIN- OR STEM-MOUNTED

POLE-MOUNTED WITH ARM

POLE-MOUNTED ON TOP

ACCENT/DIRECTIONAL ARROW (DRAWN FROM PHOTOMETRIC CENTER IN DIRECTION OF OPTICS OR PHOTOMETRIC ORIENTATION)

DIRECTIONAL AIMING LINE (DRAWN FROM PHOTOMETRIC CENTER TO ACTUAL AIMING POINT)

TRACK-MOUNTED; LENGTH, LUMINAIRE TYPES AND QUANTITIES AS SHOWN (TRACK LENGTH DRAWN TO SCALE)

LUMINAIRE PROVIDING EMERGENCY ILLUMINATION (FILLED IN)

DOUBLE LINE FOR INDIRECT LUMINAIRES

LOUVERS

LUMINAIRE; BOLLARD-TYPE

203A01.EPS

RECESSED; 2' x 4', EMERGENCY

RECESSED; ACCENT OR WALL WASHER

CHAIN HUNG STRIPLIGHT

WALL-MOUNTED WALL WASHER @ 84" AFF TO
LIGHT CENTER

POLE-MOUNTED WITH PHOTOMETRIC ORIENTATION
TO SIDE

WALL-MOUNTED, INDIRECT

POLE-MOUNTED FLOODLIGHTS

WALL-MOUNTED, DOUBLE-FACE EXIT W/ARROWS

2' x 4' SURFACE LUMINAIRE WITH LOUVER
CONTROLLED BY SWITCHES a AND b

TRACK WITH FLOOD, ACCENT, AND DOWN LIGHTS

SURFACE-MOUNTED POINT SOURCE, TYPE B2

RECESSED; 2' x 2' CONTROLLED BY SWITCH a, TYPE C

WALL-MOUNTED FLOOD AT 96" WITH AIMING LINE

PENDANT HUNG, 1' x 4', INDIRECT, TYPE D2 ON
CIRCUIT 3 CONTROLLED BY SWITCH b

MOUNTING HEIGHT

LUMINAIRE IDENTIFIER; SEE LUMINAIRE SCHEDULE
(OR USE SUBSCRIPT AS BELOW)

SUBSCRIPTS ADJACENT TO LUMINAIRE ARE USED FOR
ADDITIONAL IDENTIFICATION, SUCH AS:
   1,2,3, ETC. – CIRCUIT OR AIMING SCHEDULE NUMBER
   a,b,c, ETC. – SWITCH IDENTIFICATION
   A,B1,C2, ETC. – LUMINAIRE IDENTIFIER

NIGHT LIGHT

ROADWAY LUMINAIRE – COBRA HEAD

LOWERING DEVICE

203A02.EPS

# Appendix B

## Testing Procedure and Checklist

### Electrical Testing Procedures

*Caution: High voltages, currents, and temperature are required to operate lamps. Therefore, shock and burn hazards exist, and testing or evaluating fixtures or components should be done only by qualified individuals.*

### A. Testing Lamps

The easiest method of troubleshooting a fixture is to try a **known** good lamp in the inoperative fixture. If the lamp being replaced exhibits any of the following conditions, replace with a new lamp.

1. **Sodium Leaker Lamp -** will have a brown/golden coating on the inside of the lamp envelope other than at the base of the lamp.

2. **Amalgam Leaker Lamp -** the lamp envelope will have a smoked bronze appearance on the inside of the envelope.

3. **Faulty Base to Lamp Envelope Seal -** a white powdery substance will appear at the base of the lamp where oxygen has leaked inside the lamp.

4. **End of Lamp Life -** the arc tube will be black on both ends or the entire length of the arc tube will be black.

5. **Broken Welds or Arc Tube Support Brackets -** mechanical breaks occasionally occur due to rough handling or internal thermal stresses. Broken welds in evacuated (HPS) lamps can also create a problem known as vacuum switching. Extremely high voltage surges occur in the lamp circuit if the weld opens while the lamp is operating. Secondary coil burnout, ignitor arcing and socket arcing can occur. Rewelding may occur and the lamp may **appear** to be satisfactory; however, if left in operation, failure of the ballast and/or ignitor is likely.

### B. Ignitors/Starters

The starter provides the necessary voltage and energy required to initiate the arc in the lamp. The easiest way to check the ignitor on 35W to 150W HPS units is to install a 120V incandescent lamp in the fixture. If the incandescent lamp operates but a known good HPS lamp will not ignite, replace starter. In 200W to 1000W HPS fixtures, install a mercury lamp of similar wattage. If the mercury lamp lights and the HPS lamp will not, replace the starter.

Do not operate incandescents or mercury lamps used to check the starter for extended periods of time (more than ½ hour).

### C. Capacitors

Testing capacitors may be accomplished by:

1. **Visual inspection for swollen capacitors.** If the capacitor is swollen or bulged on the sides or top where the terminals are located, remove and replace with a new one.

2. **Verify the correct microfarad rating** as specified on the ballast I.D. label.

3. **Using an ohmmeter to check capacitors:**
   • discharge capacitor by shorting between the terminals
   • disconnect capacitor from circuit
   • remove bleed resistor

4. **Set ohmmeter to highest resistance scale** and connect leads to capacitor terminals.
   • if resistance starts low and gradually increases, the capacitor is good.
   • if resistance starts low and doesn't increase, the capacitor is shorted and should be replaced.
   • if resistance is high and remains relatively the same, the capacitor is open and should be replaced.

### D. Ballast

Visual inspection of the coil for burned or charred windings is the easiest method for checking the ballast. If lamps, ignitors and capacitors test good, replace the ballast. Testing the voltage at the socket is another method of checking the ballast. However, to use this procedure you must know specific ballast/lamp voltage and amperage requirements. The starting aid (if present) should be disconnected prior to testing the voltage at the socket. Failure to remove the starting aid could damage the test equipment.

203A03.EPS

# Fluorescent Service Checklist

(Identify problem and test for cause in numerical sequence.)

| PROBLEM | End of Lp. Life | Incorr. Lamp | Def. Blst. | Incorr. Blst. | Incorrect Supply Voltage | No Supply Voltage | Incomp. Lamp Seating | (B) Dirty Lamps | Incorr. Fixt. Wiring | Loose Socket Connect. | (C) Low Amb Temp. | High Amb. Temp. |
|---|---|---|---|---|---|---|---|---|---|---|---|---|
| Failure to Start (A) | 1 | 5 | 4 | 10 | 6 | 7 | 8 | 2 | 9 | 11 | 3 | |
| Slow Starting | | 3 | | | 4 | | | 2 | | | 1 | |
| Blinking Off and On - Very Short Duration Cycles | 1 | | | | 2 | | | | | 4 | 3 | |
| Long Duration Cycles - Several Minutes to Hours (D) | | 3 | 2 | | 4 | | | | 5 | | | 1 |
| Short Lamp Life | | 2 | 1 | 3 | 4 | | 5 | | 6 | | 7 | |
| Lamp End Blackening - One End Only (A) | | | 4 | | | | 1 | | 2 | 3 | | |
| Lamp End Blackening - Both Ends (A) | 1 | 2 | 3 | 6 | 7 | | 4 | | 5 | | | |
| Lamp Ends Only Lighted | | 1 | 2 | 3 | | | | | 4 | | | |

A. Generally indicates lamp cathode heat is missing. Problem source must be identified and corrected.

B. In humid weather or when air conditioning systems are operated only during working hours, dust on the lamps may gather condensation and prevent reliable starting. Lamps and fixtures must be cleaned and the lamp waxed with a good silicone wax.

C. Energy saving lamp, ballast combinations will not reliably start near or below 60°F; all others 50°F unless low temperature systems are used.

D. Indicates ballast thermal protector is cycling. May be the result of high ambient temperature, fixture misapplication or restricted air circulation. With recessed fixtures, check to be sure insulation has not been placed directly on the recessed fixture body. With surface mounted fixtures mounted against insulated ceiling, make sure they are rated for such applications.

All service checklists detail the more probable problem causes. It is important to keep in mind that other elements may be involved or more than one cause present. All electrical service work must begin with a careful and detailed evaluation of the problem and thorough inspection of all components, paying special attention to all wiring connections.

The ability to field test ballasts is very limited. The normal verification is to substitute a known good ballast in the problem fixture.

**On rapid start systems only,** cathode heater voltage may be verified by a voltage reading between **the contacts on each socket.** (A voltage of 3.5 to 5 is considered normal.)

*Caution - Voltages in excess of 500V may be present between the lamp socket parts and ground.*

203A04.EPS

# Mercury Vapor and Metal Halide Service Checklist

**Identify problem and test for cause in numerical sequence.**

| PROBLEM | A End of Lamp Life | B Incorrect Lamp | C Open Cap. | D Def. Starter (if app.) | E Incorrect Supply Voltage | F Photo-Cont. | G New Lamp Repl. | H Line Voltage Dips | I Shorted. Cap. | J Burned Ballast Windings | K Incorrect Ballast |
|---|---|---|---|---|---|---|---|---|---|---|---|
| FAILURE TO OPERATE | 1 | 5 | 3 | 4 | 2 | | | | | | 6 |
| LAMP CYCLING | | 1 | | | 2 | 3 | | | | | |
| COLOR SHIFT | | | | | | | 1 | 2 | | | |
| LOW LIGHT OUTPUT | 2 | | | | 1 | | | | | | |
| TRIPPED BREAKER OR BLOWN FUSE | | | | | | | | | 1 | 2 | 3 |

## A. End of Lamp Life

At a lamp's end of life, the voltage requirements of the lamp exceeds the output ability of the ballast. The usual failure mode for mercury is low light output followed by failure to operate, and for metal halide low light levels, color shifts, and lamp operating instability (cycling). Replace end of life lamp as dictated by lamp testing procedure.

## B. Incorrect Lamp

Lamp wattage, voltage, burning position, and type must be checked against the fixture label to be sure the proper lamp has been installed.

## C. Open Capacitor

This is the usual result of electrical failure or mechanical damage. Very often the capacitor will be bulged or distorted. Where the capacitor is used in series with the lamp, the lamp will not operate. See tests for capacitors or replace with a known good capacitor.

## D. Defective Starter

The function of the starter is to provide a high voltage pulse to ignite the lamp. To test, replace with a known good lamp. (Also see Ignitor/Starter testing procedure)

## E. Incorrect Supply Voltage

When investigating problems of low light output or cycling, voltage readings first must be taken at the fixture to properly identify power distribution problems. In the case of multiple supply type ballasts, verification that the supply voltage is connected to the appropriate input lead is advised.

## F. Photocontrol

Problems may result from electrical failure, incorrect wiring, or from an incorrect amount of light reaching the cell. First cover the eye of the cell with electrical tape to verify fixture and cell operation. If the fixture fails to operate, the cell must be bypassed electrically to identify the problem source. Problems of incorrect amount of light usually can be resolved by repositioning the fixture or by using cell caps to regulate the light level.

## G. New Lamp Replacement

New lamps, when installed, go through a period of burn-in or seasoning which may extend for a period of 100 hours or more. The usual result is noticeable color variation between lamps. While metal halide lamps are noted for this, the system will stabilize as the burn-in period ends. It is important to understand that some variation in color may be noted between lamp manufacturers, or between old and new lamps.

## H. Line Voltage Dips

When investigating voltage dips, it is important to identify distribution system loading. The usual cause is the starting of large motors or the use of electric welding equipment. Line voltage recorders will usually identify the problem. It is important to understand that mild dips will cause a color shift, while severe dips will cause the lamp to go out. Dip tolerance depends on lamp type, lamp age and ballast type.

## I. Shorted Capacitor

This is the direct result of electrical failure or mechanical damage. The most common result of a shorted capacitor is ballast failure. In all cases of shorted capacitors, both capacitor and transformer should be replaced. See tests for capacitors or replace with a known good capacitor.

## J. Burned Ballast Coils

### 1. Burned Primary Coils

This is the usual result of fixture connection to incorrect supply voltage. Repeated failure often in conjunction with capacitor failure may indicate short duration high voltage spikes on the distribution system. The use of a scope along with power company assistance is generally required to identify these spikes. Equally important as a cause of failure is a shorted capacitor (See "I" above - Shorted Capacitor). See tests for ballasts on preceding page and replace if necessary.

### 2. Burned Secondary Coils

This may be caused by a short circuit in the lamp circuit wiring or by mechanical failure in the lamp. Carefully inspect all lamp circuit wiring. See tests for ballasts on preceding page and replace it if necessary. **In all cases, replace the lamp.**

## K. Incorrect Ballast

This requires checking only a new fixture or if a recurring problem is encountered. Carefully compare details on the transformer to the fixture label and lamp used in the circuit. Change components as required.

203A05.EPS

---

# High-Pressure Sodium Service Checklist

## Identify problem and test for cause in numerical sequence.

| PROBLEM | A End of Lamp Life | B Incorrect Supp. Voltage | C Incorrect Lamp | D Shorted Capacitor | E Photo-Control | F Line Volt. Dip | G Defective Lamp | H Defective Starter | I Open Capacitor | J Burned Ballast Windings |
|---|---|---|---|---|---|---|---|---|---|---|
| LAMP CYCLING SINGLE FIXTURE | 1 | 3 | 2 | | 4 | | | | | |
| LAMP CYCLING GROUP OF FIXTURES | | | | | 2 | 1 | | | | |
| FAILURE TO START | | 2 | 3 | | | | 4 | 1 | 5 | 6 |
| LOW LIGHT OUTPUT | | 2 | 3 | 1 | | | 4 | | | |

### A. End of Lamp Life

At a lamp's end of life, the operating voltage requirements of the lamp exceed the output ability of the ballast. This results in the lamp cycling off and on. It is important to understand that in the early stages of failure, the lamp may operate for several hours before cycling off. As the lamp nears total failure, the on time will decrease until the lamp fails to ignite at all. Cycling lamps should immediately be replaced to avoid starting aid damage. See lamp tests or replace with known good lamp.

### B. Incorrect Supply Voltage

When investigating problems of low light output, cycling, or failure to start, voltage readings must be taken *at the fixture* to properly identify distribution system problems. For multiple supply type ballasts, proper lead connection must be verified.

### C. Incorrect Lamp

Lamp wattage, voltage, burning position, and type must be checked against fixture label to be sure the proper lamp has been installed in the fixture.

### D. Shorted Capacitor

This is the direct result of electrical failure or mechanical damage. The most common result is low light output; cycling may also occur.

### E. Photocontrol

Problems may result from electrical failure or from an incorrect amount of light reaching the cell. First cover the eye of the cell with electrical tape, to verify fixture and cell operation. If the fixture fails to operate, the cell must be bypassed electrically to identify the problem source. Problems of incorrect amount of light usually can be resolved by repositioning the fixture or by using cell caps to regulate the light level. Occasionally, the cell will see light from the fixture reflected off a nearby object and turn itself off. Repositioning the cell or reflecting object may be required.

### F. Line Voltage Dip

When investigation voltage dips, it is important to identify distribution system loading. The usual cause is the starting of large motors or the use of electric welding equipment. Line voltage recorders will usually identify the problem. It is important to understand that lamps nearing end of life will be more susceptible to voltage dips than new lamps, and lamp operating on reactor ballasts are more sensitive to voltage dips than those on regulating ballasts.

### G. Defective Lamp

This is normally the result of some mechanical failure in the lamp. This can often be determined by brown or silver coating on the lamp outer jacket, or by deposits at the base of the lamp. See lamp testing procedures or replace with a known good lamp.

### H. Defective Starter

The function of the starter is to provide a high voltage pulse to ignite the lamp. Failure to operate generally results from electrical failure in the starter. See tests for starters or replace with a known good starter.

### I. Open Capacitor

This generally results from electrical failure or mechanical damage. Very often the capacitor will be bulged or distorted. When the capacitor is used in the secondary (lamp) circuit, the fixture will not operate. See tests for capacitors or replace with a known good capacitor.

### J. Burned Ballast Windings

This is often the result of fixture being connected to incorrect supply voltage. Repeated primary winding failures often in conjunction with capacitor failures may indicate short duration high voltage spikes on the distribution system. The use of a scope in conjunction with power company assistance is generally required to identify these spikes. See tests for ballasts and replace if necessary.

203A06.EPS

# Additional Resources

This module presents thorough resources for task training. The following resource material is suggested for further study.

*Lighting Handbook*, Latest Edition. New York: Illuminating Engineering Society of North America (IESNA).

*National Electrical Code® Handbook*, Latest Edition. Quincy, MA: National Fire Protection Association.

# Figure Credits

# Conduit Bending

**John Paul Jones Arena**

The Barton Malow Company won an Aon Build America Award for New Construction Management for its work with the University of Virginia's sports and entertainment complex in Charlottesville, Va.

**26204-11**

Trainees with successful module completions may be eligible for credentialing through NCCER's National Registry. To learn more, go to **www.nccer.org** or contact us at **1.888.622.3720.** Our website has information on the latest product releases and training, as well as online versions of our *Cornerstone* newsletter and Pearson's Contren® product catalog.

Your feedback is welcome. You may email your comments to **curriculum@nccer.org,** send general comments and inquiries to **info@nccer.org,** or use the User Update form at the back of this module.

 V.1 6/11

## Objectives

When you have completed this module, you will be able to do the following:

1. Describe the process of conduit bending using power tools.
2. Identify all parts of electric and hydraulic benders.
3. Bend offsets, kicks, saddles, segmented, and parallel bends.
4. Explain the requirements of the *National Electrical Code®* (*NEC®*) for bending conduit.
5. Compute the radius, degrees in bend, developed length, and gain for conduit up to six inches.

## Performance Tasks

Under the supervision of the instructor, you should be able to do the following:

1. Use an electric or hydraulic bender to bend a 1"conduit stub-up to an exact distance of 15¼" above the deck.
2. Make an offset in a length of conduit to miss a 10" high obstruction with a clearance between the obstruction and the conduit of not less than 1" nor more than 1½".
3. Make a saddle in a length of conduit to cross an 8" pipe with 1" clearance between the pipe and the conduit.

## Trade Terms

| | | |
|---|---|---|
| Approximate ram travel | Gain | Rise |
| Back-to-back bend | Inside diameter (ID) | Segment bend |
| Bending protractor | Kicks | Segmented bending shoe |
| Bending shot | Leg length | Springback |
| Concentric bending | Ninety-degree bend | Stub-up |
| Conduit | Offsets | Sweep bend |
| Degree indicator | One-shot shoe | Take-up (comeback) |
| Developed length | Outside diameter (OD) | |
| Elbow | Radius | |

## Required Trainee Materials

1. Pencil and paper
2. Appropriate personal protective equipment
3. Copy of the latest edition of the *National Electrical Code®*

Note: *NFPA 70®*, *National Electrical Code®*, and *NEC®* are registered trademarks of the National Fire Protection Association, Inc., Quincy, MA 02269. All *National Electrical Code®* and *NEC®* references in this module refer to the 2011 edition of the *National Electrical Code®*.

# Contents —————————————————————

*Topics to be presented in this module include:*

## Figures and Tables

## Figures and Tables (*continued*)

## 1.0.0 INTRODUCTION

The normal installation of intermediate metal conduit (IMC), rigid metal conduit (RMC), and electrical metallic tubing (EMT) requires many changes of direction in the conduit runs, ranging from simple offsets at the point of termination at outlet boxes and cabinets to complicated angular offsets at columns, beams, cornices, and so forth. This module explains how to use conduit bending machines for these tasks.

Unless the contract specifications dictate otherwise, such changes in direction, particularly in the case of smaller sizes, are made by bending the conduit or tubing as is required. In the case of 1¼" and larger sizes, right angle changes of direction are sometimes accomplished with the use of factory elbows or conduit bodies. In most cases, however, it is more economical to make these conduit bends in the field.

On-the-job conduit bends are also performed when multiple runs of the larger conduit sizes are installed. Truer parallel alignment of multiple runs is maintained by using on-the-job conduit bends rather than factory elbows. Such bends can all be made from the same center, using the bends of the largest conduit in the run as the pattern for all other bends. This is just one of the useful techniques covered in this module.

Exposed conduit work is one area of electricians' work that puts their skills on display. Exposed conduit directly reflects on the ability of the installer. With these thoughts in mind, it will benefit you to learn several methods of bending conduit that will ensure accurate and precisely bent conduit—conduit that you can step back and look at with pride because it was bent right the first time.

### 1.1.0 Safety Considerations

Always keep the following safety guidelines in mind when bending conduit:

- Read and understand the operating instructions before using any tool. Always use the right tool for the job; never use bending tools for any other purpose.
- Never operate tools with damaged or missing parts.
- Watch for pinch points on power bending tools.
- Use tools in a well-lit, uncluttered area, with enough space for the conduit to move while bending.
- Be prepared for the unexpected.

- Make sure all hydraulic connections are clean and tight. Replace damaged or worn hoses before using.
- Make sure tools are complete and properly assembled before operating.
- Bending shoes and follow bars should be treated as precision instruments. Never toss them into a tool chest or allow them to become damaged or bent out of shape. The quality of the final job will depend on using the proper shoes and follow bars in good condition.
- When working with PVC conduit, make sure adequate ventilation is provided to carry off fumes from joint cement or glue.

## 2.0.0 *NEC*® REQUIREMENTS

*NEC Chapter 3* contains the installation requirements for various types of conduit. *NEC Section 344.24* requires that all rigid metal conduit bends be made so that the conduit will not be damaged and the internal diameter of the conduit will not be effectively reduced. To accomplish this, the *NEC*® further specifies that the minimum radius (*Figure 1*) to the centerline of the conduit shall not be less than that listed in *Table 1*. There is a good reason for this rule. When bends are too tight, pulling becomes extremely difficult and the insulation on the conductors may be damaged.

The *NEC*® requirements for other bends are shown in *Table 2*.

### 2.1.0 Number of Bends per Run

Every change of direction in a conduit run adds to the difficulty of the pull. The *NEC*® specifically states that no more than four quarter bends (360° total) may be made in any one conduit run between outlet boxes, cabinets, panels, or junction boxes; that is, between pull points.

½"

4"

**(A)**

204F01A.EPS

**(B)**

204F01B.EPS

*Figure 1*   Inside radius requirements.

Some electricians believe that offsets, kicks, and saddles are not bends, especially in areas where the electrical inspectors are lax. These electricians count only those bends that are actually a quarter circle (90°). The misconception of this is quickly apparent when wires are pulled. Offsets and saddles add just as much resistance to pulling conductors as any 90° elbow. A 45° offset, for example, takes two 45° bends, which equal one 90° bend. A saddle may be as low as 60° or as high as 180°, depending on the types of bends used.

A 15° kick in a conduit run may seem insignificant, but after several of these are incorporated into the run, the difficulty of pulling wire becomes apparent. The number of degrees in each kick should be included in the total count, and in no case should the total number (number of bends × number of degrees in each bend) exceed 360°. This is the maximum number of degrees allowed. For example, you could have one 45° offset, a 90° saddle, and two elbows. Many electricians prefer to install pull boxes at closer intervals to reduce the number of bends, especially when the larger conductor sizes are being pulled. The additional cost of the pull boxes and the labor to install them is often offset by the labor saved in pulling the conductors.

**Table 1** *NEC® Minimum Requirements for Radius of Conduit Bends – One-Shot and Full-Shoe Benders (Data from NEC Chapter 9, Table 2)*

| Trade Size (Inches) | Radius to Center of Conduit (Inches) |
|---|---|
| ½ | 4 |
| ¾ | 4½ |
| 1 | 5¾ |
| 1¼ | 7¼ |
| 1½ | 8¼ |
| 2 | 9½ |
| 2½ | 10½ |
| 3 | 13 |
| 3½ | 15 |
| 4 | 16 |
| 5 | 24 |
| 6 | 30 |

Reprinted with permission from *NFPA 70®*, the *National Electrical Code®*. Copyright © 2010, National Fire Protection Association, Quincy, MA 02269. This reprinted material is not the complete and official position of the National Fire Protection Association on the referenced subject, which is represented only by the standard in its entirety.

**Table 2** *NEC® Minimum Requirements for Radius of Other Conduit Bends (Data from NEC Chapter 9, Table 2)*

| Trade Size (Inches) | Other Bends (Inches) |
|---|---|
| ½ | 4 |
| ¾ | 5 |
| 1 | 6 |
| 1¼ | 8 |
| 1½ | 10 |
| 2 | 12 |
| 2½ | 15 |
| 3 | 18 |
| 3½ | 21 |
| 4 | 24 |
| 5 | 30 |
| 6 | 36 |

Reprinted with permission from *NFPA 70®*, the *National Electrical Code®*. Copyright © 2010, National Fire Protection Association, Quincy, MA 02269. This reprinted material is not the complete and official position of the National Fire Protection Association on the referenced subject, which is represented only by the standard in its entirety.

## 3.0.0 TYPES OF BENDS

There are various types of conduit bends: elbows, offset bends, back-to-back bends, saddles, etc. A brief review of each follows:

- *Elbow* – An elbow, or ell, is a 90° bend that is used when a conduit must turn at a 90° angle. In single conduit runs when the larger sizes of conduit are being installed, factory elbows are frequently used to save labor on setting up a power bending machine, calculating and marking the conduit for bending, and finally, making the bend. However, in multiple conduit runs, a neater job will result if on-the-job sweep bends or concentric bends are properly calculated and installed. See *Figure 2*.

- *Offset* – An offset consists of two equal bends and is used when the conduit run must go over, under, or around an obstacle. An offset is also used at outlet boxes, cabinets, panelboards, and pull boxes. See *Figure 3*.

90° BENDS USING FACTORY ELBOWS

90° CONCENTRIC SEGMENTED BENDS MADE ON THE JOB

204F02.EPS

*Figure 2* Typical 90° bends.

OFFSET CONDUIT TO
MISS OBSTACLES

AIR DUCT

**(A)**

204F03A.EPS

**(B)**

204F03B.EPS

*Figure 3*   Applications of conduit offsets.

- *Saddle* – A saddle is used to cross a small obstruction or other runs of conduit. A saddle is made by marking the conduit at a point where the saddle is required and placing a bender a few inches ahead of this point. Bends are made as shown in *Figure 4*. Both three-bend and four-bend saddles can be used, depending on the type of obstruction.

- *Kick* – A kick is a single change in direction of a conduit run of less than 90°. It is used mostly where the conduit run will be concealed in deck work. The first bend in an offset, for example, is really a kick, as shown in *Figure 5*; another kick in the opposite direction transforms the bend into an offset.

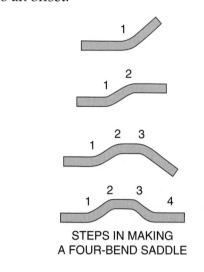

STEPS IN MAKING
A FOUR-BEND SADDLE

CONDUIT

SADDLE

AIR DUCT

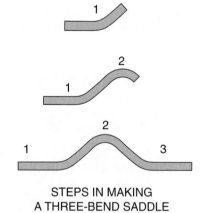

STEPS IN MAKING
A THREE-BEND SADDLE

CONDUIT

SADDLE

PIPE

204F04.EPS

*Figure 4*   Practical application of a saddle bend.

Figure 5 Kick.

## 4.0.0 THE GEOMETRY OF BENDING CONDUIT

Learning to bend conduit involves a thorough knowledge of some basic geometry (the science that deals with the properties of lines, angles, surfaces, and solids). The basic bends discussed previously will handle the majority of the electrician's needs, as these simple bends are merely combined to form more complex bends. When these basic skills are mastered, you will be able to calculate and bend conduit to fit most situations.

The 90° or right angle bend is probably the most basic of all and is used much of the time, regardless of the type of conduit being installed. All other bends are typically made with angles less than 90°.

### 4.1.0 Right Triangle

A right triangle, as shown in *Figure 6(A)*, is defined as any triangle with one 90° angle. The side directly opposite the 90° angle is called the hypotenuse and the side on which the triangle sits is the base. The vertical side is called the height or altitude. For offset bends, right triangle characteristics can also be applied because the offset forms the hypotenuse of a right triangle, as shown in *Figure 6(B)*. There are reference tables available for sizing offset bends based on these characteristics.

### 4.1.1 Trigonometry Fundamentals

Right triangles are used to develop trigonometric equations (*Figure 7*). These equations can be used to determine the rise or stub-up distance between bend points, the distance between bend joints, the distance a kick needs to be above the surface, or the amount of additional conduit required. There are six basic trigonometric functions that will be required:

- Sine
- Cosine
- Tangent
- Cotangent
- Secant
- Cosecant

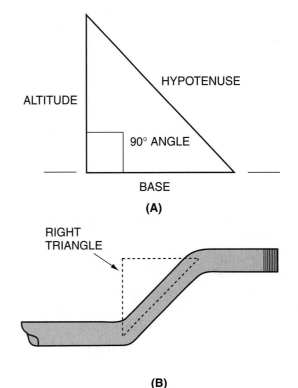

Figure 6 A right triangle and its relationship to a conduit offset.

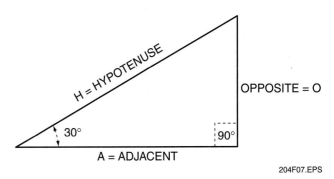

Figure 7 Trigonometry fundamentals of a right triangle.

The sine function can be computed by dividing the hypotenuse by the side opposite the angle being considered. This function can be represented by the following equation:

$$\text{Sine} = \frac{\text{opposite}}{\text{hypotenuse}} = \frac{O}{H}$$

The equations for the other functions are as follows:

$$\text{Cosine} = \frac{\text{adjacent}}{\text{hypotenuse}} = \frac{A}{H}$$

$$\text{Tangent} = \frac{\text{opposite}}{\text{adjacent}} = \frac{O}{A}$$

$$\text{Cotangent} = \frac{\text{adjacent}}{\text{opposite}} = \frac{A}{O}$$

$$\text{Secant} = \frac{\text{hypotenuse}}{\text{adjacent}} = \frac{H}{A}$$

$$\text{Cosecant} = \frac{\text{hypotenuse}}{\text{opposite}} = \frac{H}{O}$$

*Where:*

H = hypotenuse, side facing the right (90°) angle

O = side opposite the angle you are working with

A = side adjacent to the angle you are working with, but not the hypotenuse

Therefore, in *Figure 6(B)*, the hypotenuse or distance between bends could be computed by using the following cosecant trigonometric function:

$$\text{Cosecant} = \frac{\text{hypotenuse}}{\text{opposite}} = \frac{H}{O}$$

Let's say that the angle of the first bend is 30° and the rise or side opposite is 12". Therefore:

$$\text{Cosecant } 30° = \frac{H}{12''}$$

The cosecant of 30° is 2 (check using your calculator).

$$2 = \frac{H}{12''}$$

Cross multiply:

$$H = 2 \times 12''$$
$$H = 24''$$

Therefore, the hypotenuse or distance between bends would be 24".

Here is another example. In this case, a kick is to be made on the end of a piece of conduit (*Figure 8*). The angle of the kick is to be 30°. The hypotenuse of the kick is 20". Determine how far off the surface the end of the kick needs to be.

When calculating the hypotenuse of a bend, the cosecant of the angle is multiplied by the side opposite. If the angle and the hypotenuse are known, then the inverse can be used to determine the side opposite. Therefore, divide 20" by the cosecant of 30°, or:

$$20'' \div 2 = 10''$$

The end of the conduit needs to be brought 10" off the surface to acquire a 30° bend. This process eliminates the need for a protractor level.

### 4.2.0 Circle

A circle is defined as a closed curved line whose points are all the same distance from its center, as shown in *Figure 9(A)*. The distance from the center point to the edge of the circle is called the radius and thwe length of a straight line from one edge through the center to the other edge is called the diameter. The distance around the circle is called the circumference. A circle can be divided into four equal quadrants, as shown in *Figure 9(B)*. Each quadrant accounts for 90°, making a total of 360°. A 90° bend is based on ¼ of a circle, or one quadrant.

Concentric circles, shown in *Figure 9(C)*, are several circles that have a common center but different radii. The concept of concentric circles can be applied to concentric 90° bends in conduit.

204F08.EPS

*Figure 8*  Kick example.

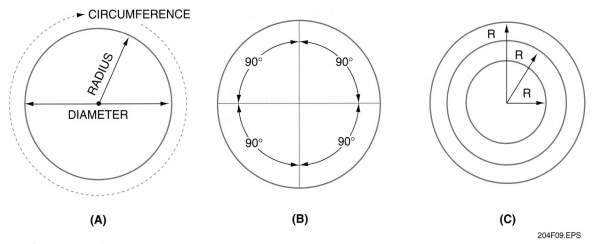

Figure 9 Characteristics of a circle.

Such bends have the same center point, but the radius of each is different. *Figure 10* shows how parts of a circle relate to a 90° conduit bend.

For bending conduit, it is necessary to understand the dynamics of the unit circle (*Figure 11*). The unit circle is a circle with a given radius of 1.

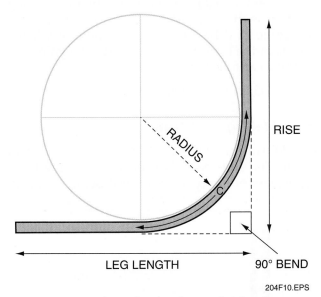

Figure 10 Parts of a circle related to conduit bending.

In calculating the circumference of a circle, the following formula is used:

$$C = 2\pi R$$

*Where:*

$$C = \text{circumference}$$
$$\pi = 3.14 \text{ (pi)}$$
$$R = \text{radius}$$

By taking the radius of the unit circle and substituting it into the formula, the result is:

$$C = 2\pi \times 1$$

Any number multiplied by the number 1 is equal to that number, or:

$$1 \times 2 = 2$$

Therefore, $2\pi \times 1 = 2\pi$. This means that in terms of pi, the circumference of the circle is $2\pi$; 360° is $2\pi$ and 180° = $\pi$. See *Figure 12*.

If you look at 90° in terms of $\pi$, 90° is half of 180° or half of $\pi$, so 90° is equal to ½ $\pi$ or $\pi \div 2$. Again, $\pi$ is a symbol for the numerical value 3.14 (rounded). $\pi \div 2$ is 3.14 ÷ 2 or 1.57. Therefore, 90° is represented by the numerical value 1.57. Looking

Figure 11 Unit circle.

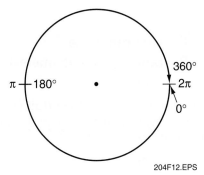

Figure 12 $\pi$ and $2\pi$.

back at the circumference formula shows that multiplying $2 \times \pi$ or 6.28 times the given radius of the circle gives the linear distance around the circle. With that in mind, if you multiply 1.57 times the given radius, it will provide the linear distance from 0° to 90° on the circle. This linear distance is known as the developed length in regard to the amount of conduit that is required to make a 90° sweep.

### 4.2.1 Types of Angular Representations

There are three basic ways that angles can be represented or specified. Angles can be stated in degrees, as already mentioned; radians; or gradients.

Angles are measured in degrees from 0° to 360°. They may also be measured in radians from 0 to 6.28 or 0 to $2\pi$. The third way is to specify angles in gradients, from 0 to 400 gradients (grads). These measurements can be equated as follows:

$$90° = \pi \div 2 = 1.57 \text{ radians} = 100 \text{ grads}$$
$$180° = \pi = 3.14 \text{ radians} = 200 \text{ grads}$$
$$270° = 3\pi \div 2 = 4.71 \text{ radians} = 300 \text{ grads}$$
$$360° = 2\pi = 6.28 \text{ radians} = 400 \text{ grads}$$

### 4.3.0 Equations

To calculate conduit bends accurately, you need to use some basic equations. The equations given here are the most relevant ones for the electrician. Examples are provided for clarification. To calculate the circumference of a circle, use the following equation:

$$C = \pi D$$

*Where:*

C = circumference
D = diameter
$\pi$ = 3.14

As discussed earlier, another equation for finding the circumference of a circle is as follows:

$$C = 2\pi R$$

*Where:*

C = circumference
R = radius (½ the diameter)
$\pi$ = 3.14

If the radius in a circle (and conduit bend) measures two feet, the circumference of the circle may be found as follows:

$$C = 2\pi R$$
$$C = 2 \times 3.14 \times 2' = 12.56'$$

The arc of a quadrant equals ¼ the circumference of the circle, and its length is found as follows:

$$\text{Length of arc} = (0.25)(2\pi R)$$
$$\text{Length of arc} = (0.25)(2 \times 3.14 \times R)$$
$$\text{Length of arc} = 1.57R$$

Therefore, the length of the arc in the quadrant, if the radius is 2', may be found as follows:

$$\text{Length of arc} = 1.57R$$
$$\text{Length of arc} = 1.57 \times 2' = 3.14'$$

If the radius of the bend and the outside diameter of the conduit are known, then the distance C (the length of the arc) can be calculated. This length, as related to conduit bending, is known as the developed length; it is the actual length of the bend.

Conduit bends with the circumference of the circle (*Figure 13*) and not at right angles. Therefore, the length of the conduit needed for a bend will not equal the right angle distances A and B. **Gain** is the difference between the right angle distances A and B and the shorter distance C, the length of conduit actually needed for the bend.

The gain for a 90° bend is found by multiplying the radius of the bend × 0.43 (rounded up from the value of 0.4292 in *Table 3*). Therefore, if the radius of the bend in *Figure 13* is 2', the gain may be found as follows:

$$\text{Gain} = 2' \times 0.43 = 0.86' = 10.32''$$

### 4.3.1 Making a 90° Bend

The 90° stub bend is probably the most basic bend of all. The stub bend is used much of the time, regardless of the type of conduit being installed.

GAIN = DISTANCE SAVED

204F13.EPS

*Figure 13* Gain.

**Table 3** Gain Factors

| | — | 1° | 2° | 3° | 4° | 5° | 6° | 7° | 8° | 9° |
|---|---|---|---|---|---|---|---|---|---|---|
| 0° | 0 | 0 | 0 | 0 | 0 | 0 | 0.0001 | 0.0001 | 0.0003 | 0.0003 |
| 10° | 0.0005 | 0.0006 | 0.0008 | 0.001 | 0.0013 | 0.0015 | 0.0018 | 0.0022 | 0.0026 | 0.0031 |
| 20° | 0.0036 | 0.0042 | 0.0048 | 0.0055 | 0.0062 | 0.0071 | 0.0079 | 0.009 | 0.01 | 0.0111 |
| 30° | 0.0126 | 0.0136 | 0.015 | 0.0165 | 0.0181 | 0.0197 | 0.0215 | 0.0234 | 0.0254 | 0.0276 |
| 40° | 0.0298 | 0.0322 | 0.0347 | 0.0373 | 0.04 | 0.043 | 0.0461 | 0.0493 | 0.0527 | 0.0562 |
| 50° | 0.06 | 0.0637 | 0.0679 | 0.0721 | 0.0766 | 0.0812 | 0.086 | 0.0911 | 0.0963 | 0.1018 |
| 60° | 0.1075 | 0.1134 | 0.1196 | 0.126 | 0.1327 | 0.1397 | 0.1469 | 0.1544 | 0.1622 | 0.1703 |
| 70° | 0.1787 | 0.1874 | 0.1964 | 0.2058 | 0.2156 | 0.2257 | 0.2361 | 0.247 | 0.2582 | 0.2699 |
| 80° | 0.2819 | 0.2944 | 0.3074 | 0.3208 | 0.3347 | 0.3491 | 0.364 | 0.3795 | 0.3955 | 0.4121 |
| 90° | 0.4292 | — | — | — | — | — | — | — | — | — |

Before beginning to make the bend, you need to know two measurements:

- The desired rise or stub-up
- The **take-up (comeback)** distance of the bender

The desired rise is the height of the stub-up. The take-up is the amount of conduit the bender will use to form the bend. Take-up distances are usually listed in the manufacturer's instruction manual. Once the take-up has been determined, subtract it from the stub-up height. Mark that distance on the conduit (all the way around) at that distance from the end. The mark will indicate the point at which you will begin to bend the conduit. Line up the starting point on the conduit with the starting point on the bender. Most benders have a mark, such as an arrow, to indicate the starting point.

> **NOTE**
> When bending conduit, the conduit is placed in the bender and the bend is made facing the end of the conduit from which the measurements were taken.

As discussed above, if the radius of the bend in *Figure 13* is 2', the gain is 10.32".

*Table 4* shows the equivalent fractions for many decimals. For 10.32", search the decimal column for 0.32. The closest decimal in the table is 0.3125, equivalent to ⁵⁄₁₆. Therefore, the number in question becomes 10⁵⁄₁₆".

A decimal may also be mathematically converted to a fraction. A decimal whose denominator is contained in the numerator without a remainder can easily be converted to a fraction by removing the decimal point from the numeral, which then becomes the numerator (the top numeral of the fraction). The denominator is always one plus as many zeros as there are decimal places in the decimal. For example:

$$0.75 = {}^{75}\!/_{100} = {}^{3}\!/_{4}$$

*or:*

$$0.375 = {}^{375}\!/_{1000} = {}^{3}\!/_{8}$$

> **NOTE**
> Decimals may also be converted to fractions by using a calculator, provided the calculator has a fraction key. The exact procedure will vary with the different models of calculators.

Gain factors for 0° to 90° bends are shown in *Table 3*. To demonstrate the use of this table, assume that it is desired to find the gain on a 45° conduit bend with a 15" centerline radius. Referring to *Table 3*, look in the left-hand column; glance down the column until the number 40° is found. Since the bend is 45°, read to the right in this row until the column titled 5° is found; note the figure, 0.043. Therefore, the gain factor for a 45° bend is 0.043.

Multiply the gain factor by the centerline radius (15") to obtain the full gain of a 45° bend.

$$0.043 \times 15" = 0.645"$$

To convert this figure to a readable figure on the foot rule, convert the decimal to a common fraction: ⁶⁴⁵⁄₁₀₀₀ = approximately ⁵⁄₈. Thus, the full gain of the 45° bend is ⁵⁄₈".

For example, the gain for a given bend is 2½". In *Figure 14*, there are two back-to-back 90° bends. The rise for the first 90° bend is 2' and the rise for the second 90° bend is 3'. The back-to-back measurement for these bends is 4'. You need to

**Table 4** Decimal Equivalents of Some Common Fractions

| Fraction | Decimal | MM | Fraction | Decimal | MM |
|---|---|---|---|---|---|
| 1/64 | 0.015625 | 0.397 | 33/64 | 0.515625 | 13.097 |
| 1/32 | 0.03125 | 0.794 | 17/32 | 0.53125 | 13.494 |
| 3/64 | 0.046875 | 1.191 | 35/64 | 0.546875 | 13.891 |
| 1/16 | 0.0625 | 1.588 | 9/16 | 0.5625 | 14.288 |
| 5/64 | 0.078125 | 1.984 | 37/64 | 0.578125 | 14.684 |
| 3/32 | 0.09375 | 2.381 | 19/32 | 0.59375 | 15.081 |
| 7/64 | 0.109375 | 2.778 | 39/64 | 0.609375 | 15.478 |
| 1/8 | 0.125 | 3.175 | 5/8 | 0.625 | 15.875 |
| 9/64 | 0.140625 | 3.572 | 41/64 | 0.640625 | 16.272 |
| 5/32 | 0.15625 | 3.969 | 21/32 | 0.65625 | 16.669 |
| 11/64 | 0.171875 | 4.366 | 43/64 | 0.671875 | 17.066 |
| 3/16 | 0.1875 | 4.763 | 11/16 | 0.6875 | 17.463 |
| 13/64 | 0.203125 | 5.159 | 45/64 | 0.703125 | 17.859 |
| 7/32 | 0.21875 | 5.556 | 23/32 | 0.71875 | 18.256 |
| 15/64 | 0.234375 | 5.953 | 47/64 | 0.734375 | 18.653 |
| 1/4 | 0.25 | 6.35 | 3/4 | 0.75 | 19.05 |
| 17/64 | 0.265625 | 6.747 | 49/64 | 0.765625 | 19.447 |
| 9/32 | 0.28125 | 7.144 | 25/32 | 0.78125 | 19.844 |
| 19/64 | 0.296875 | 7.54 | 51/64 | 0.796875 | 20.241 |
| 5/16 | 0.3125 | 7.938 | 13/16 | 0.8125 | 20.638 |
| 21/64 | 0.32812 | 8.334 | 53/64 | 0.828125 | 21.034 |
| 11/32 | 0.34375 | 8.731 | 27/32 | 0.84375 | 21.431 |
| 23/64 | 0.359375 | 9.128 | 55/64 | 0.859375 | 21.828 |
| 3/8 | 0.375 | 9.525 | 7/8 | 0.875 | 22.225 |
| 25/64 | 0.390625 | 9.922 | 57/64 | 0.890625 | 22.622 |
| 13/32 | 0.40625 | 10.319 | 29/32 | 0.90625 | 23.019 |
| 27/64 | 0.421875 | 10.716 | 59/64 | 0.921875 | 23.416 |
| 7/16 | 0.4375 | 11.113 | 15/16 | 0.9375 | 23.813 |
| 29/64 | 0.453125 | 11.509 | 61/64 | 0.953125 | 24.209 |
| 15/32 | 0.46875 | 11.906 | 31/32 | 0.96875 | 24.606 |
| 31/64 | 0.484375 | 12.303 | 63/64 | 0.984375 | 25.003 |
| 1/2 | 0.5 | 12.7 | 1 | 1 | 25.400 |

determine the length of straight conduit required to accomplish this task. The conduit is to be cut to length and threaded before bending.

The gain is the amount of conduit that is saved by the radius of the bend. The amount of conduit saved is 2½" per 90° bend. In this situation, there are two 90° bends, or a savings of 5". The total straight lengths add up to 9' of conduit. Therefore:

$$9' - 5" = 8'\text{-}7"$$

## 5.0.0 MECHANICAL BENDERS

Conduit bends are normally made in the smaller sizes of conduit and tubing by hand with the use of hickeys or EMT bending tools. However, on many projects, an advantage can be gained by the use of mechanical bending equipment with suitable adjustable stops and guides. With the use of such equipment, the exact bend can be duplicated in quantity with a minimum of effort. The angle of the bend and the location of the bend in relation to the end of the length of conduit are preset.

A popular mechanical bender is shown in *Figure 15*. This type of bender was originally called the Chicago bender, as it was made by the Chicago Equipment and Manufacturing

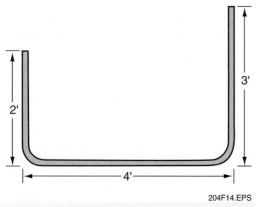

204F14.EPS

*Figure 14* Back-to-back 90° bends.

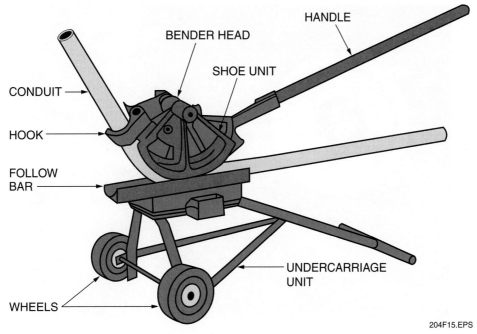

*Figure 15*  Typical mechanical bender.

CONDUIT

BENDER HEAD

HANDLE

SHOE UNIT

HOOK

FOLLOW BAR

UNDERCARRIAGE UNIT

WHEELS

204F15.EPS

Company. Today, however, this type of bender is manufactured by several different companies and the correct name is *portable mechanical conduit bender*. In any event, you may still hear the term *Chicago bender* on many jobs.

This type of mechanical bender is very popular with many electrical contractors. To use it, a length of conduit is placed in position and secured in place; a long bending handle is then pulled around and the bend completed. This type of bender may be used as a one-shot bender for the smaller sizes of conduit (bypassing the ratchet mechanism). The ratchet mechanism, however, is usually activated when bending the larger sizes of conduit to make the work easier. It is suitable for making bends in conduit sizes up to 2" EMT, 1¼" IMC, and 1½" rigid steel conduit, provided the proper bending accessories are used (e.g., bending shoes, follow bars, etc.).

Bending shoes and follow bars are designed to form a particular radius bend for a certain type and size of conduit, that is, EMT, IMC, or rigid conduit. These accessories should be treated as precision instruments; any damage will result in inaccurate bends, kinks, and so on. The first consideration is to use only the proper shoe and follow bar for the type and size of conduit being bent. For example, never use an EMT shoe for bending rigid conduit or a 3" shoe for bending 2½" conduit. In general, make certain that the bending shoes and follow bar are compatible with the type and size of conduit to be bent; to do otherwise may damage the tool and result in inaccurate bends.

The ratchet feature is normally engaged for the larger sizes of conduit, whereas a spring-loaded pawl engages the ratchet for easier bending in segments. For the smaller sizes of conduit, however, the ratchet may be bypassed so that the bend can be made in one shot.

A bending gauge with an adjustable pointer on the bender is helpful when making multiple bends at the same angle. This pointer is set at the desired angle and then the setscrew is tightened. As the bend is being made, the handle is operated until the pointer reaches the index mark. To ensure the correct angle of bend, the first bend should be checked with a bending protractor (*Figure 16*) and any necessary adjustments made to the bending gauge pointer before continuing. All successive bends will be exactly the same as the first.

For example, follow this process for charting a bender. This chart will contain minimum size, gain, and centerline distances from the arrow on the bender to the center of the bend for 15°, 30°, 45°, and 60° bends. This is to be accomplished with one scrap piece of conduit.

For this discussion, the scrap piece will be ½" conduit and will be 3' in length (*Figure 17*).

Place a mark on the conduit at a given distance from the end of the conduit. For this discussion, the mark will be 10" in from the end of the conduit (*Figure 18*).

Take a ½" conduit bender and place the arrow, which represents the take-up or minimum rise of the bender, on the mark 10" in from the end of the bender.

Figure 16 Bending protractor.

Figure 17 Conduit.

Figure 18 Conduit with 10" mark.

Figure 19 Kick of 15°.

Figure 20 Conduit and straightedge.

Figure 21 Conduit and horizontal straightedge.

Place a protractor level on the conduit in front of the bender. This will be on the piece of conduit marked for 10" back from the end. Bend the 10" portion until the protractor level reads 15°. Remove the bender from the conduit (*Figure 19*).

Now take a straightedge, such as a ruler or torpedo level, and lay it on the inside of the bend so the straightedge lies across the bend and rests against the straight portion of the conduit (*Figure 20*).

With a sharp pencil, scribe a line across the bend of the conduit.

Now take the straightedge and lay it across the conduit so the straightedge is against the side adjacent to the previous position. It should extend across the bend once again (*Figure 21*).

Now scribe a line across the bend of the conduit. The two pencil lines should cross to form an X on the conduit. This represents the center of the bend (*Figure 22*).

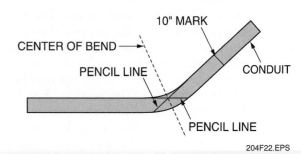

Figure 22 Center of bend.

NCCER — *Contren® Learning Series*  26204-11

Now take a tape measure and measure from the 10" mark to the centerline of the bend. Record this distance as follows:

15° – 1" (assuming 1" is the measured distance)

Now place the bender back on the conduit so that the take-up arrow is on the 10" mark on the conduit. Place the protractor level back on the conduit. Bend the conduit until the protractor level reads 30°. Take the bender off the conduit and repeat the line crossing process that was discussed for 15°, using the same straightedge. Once again, measure from the 10" mark to the point where the pencil lines cross in the center of the bend. Record this measurement.

15° – 1"

30° – 1½" (assuming this is the measured distance)

Repeat the previous process for 45°. Record this measurement.

15° – 1"

30° – 1½"

45° – 2" (assuming this is the measured distance)

Repeat the previous process for 60°. Record this measurement.

15° – 1"

30° – 1½"

45° – 2"

60° – 2¼" (assuming this is the measured distance)

The centerline distances for the different bends from the take-up mark of the bend have now been recorded. This portion of the chart is now completed.

The next step is to determine the minimum rise of the bender. Place the bender back on the conduit so the take-up mark of the bender is on the 10" mark on the conduit. Place the protractor level back on the conduit and bend the conduit until the protractor level reads 90°. Take the bender off the conduit. Measure from the back of the conduit to the 10" mark. The reading on the tape is 5". This is the minimum 90° stub length (*Figure 23*). Record this measurement.

15° – 1"

30° – 1½"

45° – 2"

60° – 2½"

Minimum rise – 5" (assuming this is the measured distance)

The next step is to determine the gain of the bender for 90°. Measure the length of both sides of the scrap piece of conduit (*Figure 24*).

*Figure 23*   90° stub-up.

*Figure 24*   90° elbow.

Add the two measured stub lengths together.

15" + 1'-11⅝" = 3'-2⅝"

Subtract the original length of the conduit, which was 3', from 3'-2⅝".

3'-2⅝" – 3' = 2⅝"

This is the gain for this particular ½" conduit bender. Record this information.

15° – 1"

30° – 1½"

45° – 2"

60° – 2½"

Minimum rise – 5"

Gain – 2⅝" (assuming this is the measured distance)

The ½" bender has now been charted. This same process would need to be repeated for a ¾" bender, 1" bender, 1¼" bender, 1½" bender, etc.

This charting process should help you to understand how the manufacturer of a bender comes up with the marks that determine the ability to bend exact 90° bends of any length, which is the minimum rise mark. The star mark seen on many benders is used for back-to-back bends. The star mark is 2⅝" back from the minimum rise mark (take-up). In other words, it is the measured gain distance back from the minimum rise mark. The centerline marks are used in lining up the centers of bends for various sizes of conduit. The centerline marks can also be used in lining up saddles on the centerlines of I-beams and process piping. This information is good for charting any type of wrap-around bender.

## 6.0.0 MECHANICAL STUB-UPS

Stub-ups are quickly and easily made with mechanical benders. A deduct decal is provided on many benders, but sometimes these decals become damaged, making them difficult to read. Therefore, backup charts should be provided on all jobs. With this deduct chart on hand, the following is an example of making a 90° stub-up to a given height.

Assume that you are working on a deck job and need a number of 1" rigid stub-ups with a rise of 15" each. When you check the deduct chart on the bender for a 15" stub-up using 1" rigid conduit, you note that 11" should be deducted from the total rise of 15". Since 15" − 11" = 4", measure back from the end of the conduit by 4" and make a mark. Encircle the entire conduit at this point so you will not lose the mark once the conduit is placed in the bender. Many electricians like to use a black felt-tip marker for marking conduit.

Load the conduit into the bender with the mark lined up with the front of the bender hook. Engage the ratchet and start pumping the bender handle until the bender pointer reaches the preset index mark for 90°. Move the bender handle forward, then remove the conduit from the bender and check its height. It should be exactly 15". If the height of the bend is slightly off, make the necessary adjustments before continuing. Once the correct height is reached, the remaining bends will also be correct.

## 7.0.0 MECHANICAL OFFSETS

The decal chart on the bender that provides deduct information for the stub-ups also contains data for making offsets that require 20°, 30°, and 45° bends. This chart is necessary to make perfect offsets every time using the mechanical bender.

Assume that you are running a raceway system with ½" rigid conduit and an air duct must be bypassed, requiring the conduit run to be offset. After taking measurements on the job, you find that an offset of 12" is needed to clear the air duct.

Measure the distance from the end of the conduit to the start of the first bend; mark the conduit as before. Referring to the chart on the bender

## Making Accurate Offsets

Be sure to use the same angle for both bends of an offset. To avoid dog legs, make sure your first bend lines up evenly with the rest of the conduit, the handle, and the bender. Take your time, and be sure of proper alignment before making the second bend. A No-Dog® is a simple, pocket-sized device that may be used to prevent crooked bends. It is screwed onto the end of the conduit and has a built-in level to ensure straight bends.

204SA02.EPS

with offset information, you decide to make the offset with two 45° bends. The chart indicates that the distance between bends is 16⁵⁄₁₆". Therefore, measure and mark this distance back from the first mark.

Insert the conduit into the bender and line up the first mark with the front of the bender hook. The ratchet may be used, but for ½" conduit, the ratchet override on the front of the bender shoe is normally employed. Make the first 45° bend. Move the bender handle forward to release the conduit. Now slide the conduit forward through the bender hook until the second mark lines up with the front of the bender hook, and then turn the bend over so the end of the conduit is pointing downward toward the deck. Also, make sure that the first bend lines up with the next bend to be made to prevent a crooked bend (commonly referred to as a dog leg) in the conduit. Once everything is aligned, engage the bender handle and make another 45° bend. The height of the offset should be exactly 12" (see *Figure 25*).

DISTANCE BETWEEN MARKS
WILL VARY ACCORDING TO
TYPE AND SIZE OF CONDUIT –
CHECK BENDING CHART

A

SECOND
MARK

FIRST
MARK

A

OFFSET

204F25.EPS

*Figure 25*    Bending offsets in conduit.

The distances between marks for offset bends will vary depending upon the size and type of conduit being bent, so always check the offset information on the bender.

### 7.1.0 General Tips for Bending with the Chicago Bender

The following are some general tips to keep in mind when bending with the portable mechanical conduit bender:

- An engineer's rule marked in hundredths will simplify formula bending by eliminating the need to convert to fractions.
- When minimum length stubs are being bent, the shoe tends to creep and deforms the end of the conduit and threads. Screwing a coupling onto the pipe stops the shoe from creeping forward and protects the threads.
- When bending offsets, the front of the bender can be temporarily elevated for clearance requirements.
- Most bender shoes are made of cast aluminum and are easily pitted and gouged if foreign material gets between the shoe and the pipe. For longer shoe life, keep the pipe clean and the shoes wiped down.
- When the remaining pipe length is not long enough to reach the roller or pipe support, a larger diameter conduit can be slid over the pipe being bent to complete the bend; or, if the pipe has threads, screw on a coupling and a short piece of scrap pipe.

- Segment and concentric bending of smaller sizes of pipe can be performed with this bender. Bend a scrap piece of pipe and measure from the center of the bend to the front of the bending shoe. Use this measurement to adjust the start mark using the segment and concentric bending procedures.
- To make matching bends in two different sizes of conduit when using a mechanical bender, make both bends using the larger shoe.

> NOTE
>
> For matching bends in sizes ½", ¾", and 1" rigid conduit and IMC, bend all pipes with the 1" shoe.

### 8.0.0 ELECTRIC AND HYDRAULIC CONDUIT BENDERS

Bends in larger sizes of conduit (over 2" EMT, 1¼" IMC, or 1½" rigid conduit) are normally made using hydraulic benders. In some instances, bending tables have been developed for use with bending tools or hydraulic benders to simplify making bends to certain dimensions. Various adaptations of benders have been developed to serve certain specific purposes. For example, one type of hydraulic bender is designed for making bends in a section of conduit that has been installed in a raceway system.

### 8.1.0 Electric Conduit Benders

Electric conduit benders operate on the same basic principle as the mechanical benders described previously except that the bending is accomplished by a gear motor rather than manual power. There are several types of electric benders on the market. A typical electric unit is shown in *Figure 26*. This bender will make stub-ups, offsets, and saddles quickly and easily at any location where a 120V receptacle is available. Furthermore, this bender will make 90° one-shot bends in ½" through 2" EMT, IMC, or rigid conduit.

Bending charts are often included in the form of decals attached to the bender for quick reference on the job. These include deduct and springback figures along with information for making offsets. Springback is when the conduit relaxes to a slightly looser bend angle when the bender is released.

Many electric benders are sold as a power unit and the bending shoes for various types of conduit are sold as accessories. These shoes quickly

**Figure 26** Typical electric bender.

204F26.EPS

snap into place on the power unit, ready for use in seconds.

To describe the operation of the example bender, assume that you need a 16" stub-up in a piece of 1¼" rigid conduit. Measure and mark a piece of conduit 16" from one end. When you check the deduct chart on the bender for this size conduit, the figure is 12¾". Therefore, 12¾" must be deducted from 16", which leaves 3¼". So a mark is made at this distance from the end of the conduit. Encircle the conduit with this mark so it will be plainly visible during the bend.

Insert the conduit in the bender with the mark lined up with the front of the bender hook. For this particular bend, the mark is 3¼" from the end of the conduit.

The three-position operating switch on the example bender is attached to a flexible cord. The center position de-energizes the machine, the Up position (forward) is for bending, and the Down position (reverse) is for unloading the conduit after the bend has been made. When released, the switch automatically springs to the center or Off position. A pointer on the shoe indicates the degree of bend as the bend is being made.

The machine is jogged or inched up until the shoe pointer lines up with the zero mark on the degree scale. The switch is then pressed upward and held in this position to start the bend. As the pointer approaches the 90° mark, refer to the springback chart for the size and type of conduit being bent. In this case, 1¼" rigid conduit, the chart indicates 95°. Therefore, the pointer should

pass the 90° mark and stop at 95° to allow for springback when the conduit is removed from the bender; the stub-up is ready for installation.

An offset is made in the electric bender similar to the method described for the mechanical bender; that is, the first mark is located on the conduit and then offset information is obtained from the bender chart. In the case of a 16" offset in a length of 1¼" rigid conduit, the chart indicates a distance of 22⅝" between marks for a 16" offset using 45° bends.

Mark the conduit and insert the conduit into the bending shoe, positioning the mark so that it is even with the front of the bender hook. Start the bend as discussed previously until the pointer reaches the 45° mark on the scale; allow for any springback as indicated in the chart. Reverse the motor until the conduit is loose, then turn the conduit upside down. Position the second mark at the front of the bender hook, making certain the conduit is aligned to prevent a dog leg in the bend. Start the second bend until the pointer reaches the 45° mark, reverse the bender, and unload the conduit. A magnetic torpedo level placed on the side of the conduit will help align bends for offsets. An anti-dog device can also be used for this.

Finally, check the bends for accuracy. Although rebending is possible, it is not a good practice. Rebending puts considerable strain on the conduit, and while it may not break, the coating may crack and cause corrosion.

### 8.1.1 Speed Benders

A speed bender operates basically the same as the standard bender except the speed bender utilizes remote digital control with easy-to-use bending charts and instructions to ensure fast, accurate, and consistent bends in conduit sizes from ½" to 2".

This bender can be operated upright or laid on its back for large offsets and saddles. In operation, rather than holding the switch up in the bend position as with the standard bender, the operator sets the desired bend via digital controls. The bender automatically bends to that degree once the conduit is placed in the bending shoe and the start control is activated. Otherwise, the conduit is marked and bent in the same way as described for the standard bender.

### 8.2.0 Hydraulic Conduit Benders

The hydraulic bender is indispensable for bending large conduit. Only by the addition of the hydraulic pump and cylinder to a bender frame is the necessary power available for bending rigid conduit up to 6" trade size.

The typical hydraulic bender (*Figure 27*) provides extended bending capability, but also requires additional responsibility. Hydraulic bending equipment requires a large initial investment, and the bender must be properly cared for if it is to last and give the service necessary to justify the cost.

The bender frame, shoes, follow bars, and saddles will require little more than occasional wiping down to remove the dirt and oil film. The pump, cylinder, and hydraulic hoses, however, will demand more attention.

The pump should have its fluid level checked periodically. Low fluid pressure will keep the pump from developing its rated hydraulic pressure. When the level is low and fluid must be added, use only approved hydraulic oil. Ordinary oil will cause damage to the pump assembly.

Hydraulic benders have two types of bending shoes, one-shot and segment.

The one-shot shoe has a full 90° radius. The conduit can be formed around the shoe to a full 90° bend without collapsing the walls of the pipe. One-shot benders require no new bending techniques. The take-up is figured to the center of the bend and the bender shoe will have an indicating mark at its center. If take-up values are not listed on the bender frame or bender storage case, scrap

pipe can be bent and take-up figures found. All the methods and layout techniques discussed for EMT and rigid metal conduit can be used with a hydraulic bender and one-shot shoes. Offsets will have to be adjusted (less angle of bend) so the spacing between the bend marks is far enough apart to allow the first bend to be rolled 180° and advanced enough to clear the shoe.

Segment shoes are shorter and have a radius that is far less than 90°. A ninety-degree bend cannot be made in one operation, as the conduit walls would collapse. Bends, then, must be made in several steps (as few as four and as many as 30) to form a smooth radius. The segmented bending shoe allows pipe to be bent to larger size radii.

Segmented shoes are used for concentric bending (bending several conduits with increasing or decreasing radii). One-shot shoes can be used for a segment bend, but are not as convenient. Concentric bending is covered in detail later in this module.

Accurate bending of large conduit is possible but requires practice, patience, and ability. With few exceptions, all formulas and bending techniques discussed to this point will apply.

### 8.2.1 Bending Tips for Rigid Aluminum

Rigid aluminum conduit is available in all trade sizes from ½" through 6". It is lightweight, corrosion resistant, and has low ground impedance. It is, however, difficult to bend consistently and accurately. Two pipes out of the same bundle will act differently when bent. Even if two pipes are bent using the same layout, they do not always come out the same. Do not be discouraged by this; it is the nature of the metal and it cannot be helped.

Another disadvantage of aluminum is that a one-shot bending shoe will dig in and score the pipe. Also, where the pipe rides on the shoe, it is prone to wrinkling and scoring. Applying petroleum jelly or a lubricant such as WD-40® to the

FRAME UNIT

HYDRAULIC RAM

CONDUIT SUPPORT

MOTOR

204F27.EPS

*Figure 27*  Hydraulic conduit bender.

shoe will allow the pipe to slide without the shoe digging in. Petroleum jelly will also make it easier to remove the conduit when the bend is complete.

### 8.2.2 One-Shot Bending

Accurate one-shot stub-ups are easily made on hydraulic benders by applying a little basic geometry in making calculations, and then knowing the operating principles of the bender. *Figure 28* shows the reference points of a common 90° bend. To make one-shot 90° bends, first determine the leg length and rise, the gain, the radius of the bend, and the half-gain.

Use the following procedure for laying out accurate stub-ups:

*Step 1* Determine lengths A and B.

*Step 2* Add lengths A and B. Subtract X for the length of pipe required.

*Step 3* Subtract Y from length A or B to get the center of the bend.

*Step 4* Calculate the developed length and the length of conduit required.

*Step 5* Determine the center of the bend. This can be done by taking the half-gain value from the distance A or B. See *Table 5*.

### 8.2.3 90° Segment Bends

When bending conduit in segments with a hydraulic bender, the following factors must be determined:

- The size of conduit to be bent
- The radius of the bend
- The total number of degrees in the bend
- The developed length
- The gain of the bend

To determine the developed length for a 90° bend, multiply the radius by 1.57. The next step is to locate the center of the bend. Most benders have the center mark indicated on the bending shoes. Once the center mark on the conduit is found, it is easy to locate the other bend marks (see *Figure 29*).

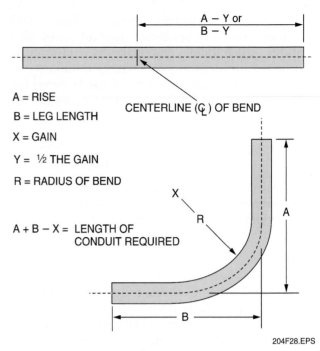

A = RISE
B = LEG LENGTH
X = GAIN
Y = ½ THE GAIN
R = RADIUS OF BEND

A + B − X = LENGTH OF CONDUIT REQUIRED

CENTERLINE (℄) OF BEND

204F28.EPS

*Figure 28*  Laying out stub-ups.

**Table 5**  Dimensions of Stub-Ups for Various Sizes of Conduit

| Pipe and Conduit Size | Radius of Bend R | Minimum Developed Length 90° | Gain X | ½ Gain Y |
|---|---|---|---|---|
| ½" | 4" | 6⁵⁄₁₆" | 1¹¹⁄₁₆" | ²⁷⁄₃₂" |
| ¾" | 4½" | 7¹⁄₁₆" | 1¹⁵⁄₁₆" | ³¹⁄₃₂" |
| 1" | 5¾" | 9" | 2½" | 1¼" |
| 1¼" | 7¼" | 11³⁄₈" | 3⅛" | 1⁹⁄₁₆" |
| 1½" | 8¼" | 13" | 3½" | 1¾" |
| 2" | 9½" | 14¹⁵⁄₁₆" | 4¹⁄₁₆" | 2¹⁄₃₂" |

You must now determine the number of bending shots that will make the bend to suit the requirements, preferably an odd number so that there are an equal number of bends on each side of the center mark. Next, calculate the width of the spaces for each segment bend and make the layout on the conduit. Make an equal number of spaces on each side of the center mark. The gain need only be determined if the bend is being fitted between two existing conduit runs or junction boxes.

To determine the bending data for a 90° bend using 3" conduit with a rise of 48", a leg length of 46", and a centerline radius of 30", proceed as follows (see *Figure 30*).

*Step 1* Multiply the radius by 1.57 to determine the developed length.

30" × 1.57 = 47.10" (47⅛") developed length

Bend the second sweep and your offset is built. The final measurements should be as follows:

Developed length of each sweep = 15.7"

Distance between centerline of each sweep = 60"

Distance between shot marks in each sweep = 2.62"

Degrees per shot in each sweep = 4.29°

## 9.1.0 Concentric Bending

If two or more parallel runs of conduit must be bent in the same direction, as shown in *Figure 39*, the best results will be obtained by using concentric bends. When laying out concentric bends, the bend for the innermost conduit is calculated first. In the example in *Figure 39*, the first bend has a radius of 20". If this dimension is multiplied by 1.57, it yields a value of 31¹³⁄₃₂". This is the developed length of the shortest radius bend.

The second radius is found by adding the **outside diameter (OD)** size of the first pipe to the radius of the first pipe and the desired spacing between pipes. In this example, the radius of the second bend would be equal to 24⅜" (see *Figure 39*).

Radius of first pipe = 20"

OD size of first pipe = 2⅜"

Spacing between pipes = 2"

Radius of second pipe:

20" + 2⅜" + 2" = 24⅜" or 24.375"

Once the developed length of the second radius is found, the spacing between marks can be determined. The start mark must be the same distance from the end of the second pipe as it was for the first pipe (*Figure 39*). This will be true for each additional pipe as well. The spacing will increase

*Figure 39* Principles of concentric bending.

204F39.EPS

between marks as the radius of the pipes increases. This is due to increasing developed length. The pipes in *Figure 39* are all laid out for 15 shots.

The equation for the developed length (DL) is $1.57 \times R$ (radius). The developed length for the second bend is found by multiplying $1.57 \times 24.375" = 38.27"$.

The radius and developed length of each successive bend are found in a similar manner. After determining the developed length, you must establish the number of segment bends needed to form each 90° bend. In concentric bending, every bend must receive the same number of segment bends to maintain concentricity. As illustrated, 15 segment bends of 6° each will total 90°.

> **NOTE**
>
> The radius change from one bend to another affects the spacing of segment bends. To find the segment bend spacing, divide the developed length of each bend by 15. For the first bend illustrated, the spacing for each segment bend is 2³⁄₃₂"; for the second bend, it is 2⁹⁄₁₆", and so on, as shown in *Figure 39*.

If the legs of the bends have to be a certain length, the gain must be considered just as in any segment bending procedure. When the runs of conduit are not the same size, the radius of each successive bend can be found as follows:

**Step 1** Determine the radius of the innermost bend.

**Step 2** Calculate one-half the outside diameter of the innermost conduit and of the next adjacent conduit.

**Step 3** Note the distance between the two runs of conduit.

**Step 4** Add these quantities.

### 9.2.0 Offset Bends

Many situations require a conduit to be bent so that it can pass by or over objects such as beams and other conduit or to enter panelboards and junction boxes. The bends used for this purpose are called offsets. To produce an offset, two equal bends of less than 90° are required at a specified distance apart. This distance is determined by the angle of the two bends and can be calculated by using the following procedure and the table in *Figure 40*.

First, determine the offset needed, then find the degree of bend. Next, multiply the offset measurements by the figure directly under the degree of bend (see *Figure 40*). This applies to all sizes of

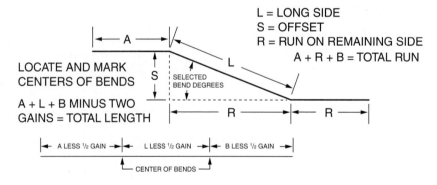

| TO FIND UNKNOWN | KNOWN | TIMES CORRESPONDING MULTIPLIER EQUALS | | | | | | | | UNKNOWN |
| --- | --- | --- | --- | --- | --- | --- | --- | --- | --- | --- |
| | | TABLE OF MULTIPLIER FOR SELECTED DEGREES OF BEND | | | | | | | | |
| | | 5⅝° | 11¼° | 15° | 22½° | 30° | 37½° | 45° | 60° | |
| L | S | 10.207 | 5.126 | 3.864 | 2.613 | 2.00 | 1.643 | 1.414 | 1.155 | L |
| S | L | 0.098 | 0.195 | 0.259 | 0.383 | 0.50 | 0.609 | 0.707 | 0.866 | S |
| R | S | 10.158 | 5.027 | 3.732 | 2.414 | 1.732 | 1.303 | 1.00 | 0.577 | R |
| S | R | 0.098 | 0.199 | 0.268 | 0.414 | 0.577 | 0.767 | 1.00 | 1.732 | S |
| L | R | 1.005 | 1.02 | 1.035 | 1.082 | 1.155 | 1.260 | 1.414 | 2.00 | L |
| R | L | 0.995 | 0.981 | 0.966 | 0.933 | 0.866 | 0.793 | 0.707 | 0.50 | R |
| GAIN PER BEND | RADIUS OR SHOE | 0.0002 | 0.0006 | 0.0015 | 0.0051 | 0.0124 | 0.0212 | 0.0430 | 0.1076 | GAIN PER BEND |

204F40.EPS

*Figure 40*   Conduit offset bending table.

conduit. For example, to form an 18" offset with two 45° bends, first make the following calculation to determine the distance between bends:

$$18" \times 1.414 = 25\frac{1}{2}"$$

This is the distance between bends and is labeled side $L$ (see *Figure 40*). To connect the two ends of an offset to two pieces of conduit that are already in place, it is necessary to know the total or overall length (OL) of the offset from end to end before bending.

Note the following equation:

$$OL = (A + L + B) - (2 \times gain)$$

*Where:*

$$A = 36"$$
$$L = 25\frac{1}{2}"$$
$$B = 48"$$
$$2 \times gain = 1\frac{9}{32}"$$

The gain is calculated by multiplying the shoe radius by the decimal figures shown on the last line under Degrees of Bend in *Figure 40*. For example, to calculate how an offset might be made in a length of 3" conduit using 45° bends:

$$OL = (A + L + B) - (2 \times gain)$$

**NOTE**

This offset uses a 15" radius.

$$OL = 109\frac{1}{2}" - 1\frac{9}{32}" = 108\frac{7}{32}"$$

$$15" \times 0.0430 \text{ (from the table in } \textit{Figure 40}\text{)} = 0.645°$$

For two gains, you have:

$$1.290 = 1\frac{9}{32}$$

Since you already know the long dimension of the offset (25"), to find the amount of offset, refer to the table in *Figure 40*, second line down

## Think About It

### Concentric Bends

What does concentric mean? Why can't you just bend all your conduit sections to the same radius and lay them side by side?

204SA04.EPS

(to find S), then move across the row until you come to the 45° column. Note that the multiplier is 0.707.

Therefore:

$$25'' \times 0.707 = 17.675''$$

This is the height of the offset. To make the offset, the conduit is positioned in the hydraulic bender, as shown in *Figure 41*. The bending shoe makes one 45° bend on the first center mark, the conduit is reversed in the bender, and the next 45° bend is made at the second mark. When making bends, always refer to the bending charts that accompany the bender being used for the operation.

For example, an offset is to be built to go over an I-beam that is 24" in height. The distance from the end of the conduit run to the J box on the opposite side of the I-beam is 8' (see *Figure 42*).

The task is to determine the distance between bend points where a 45° bend is being used. Determine the exact length of conduit needed so threading can occur before bending. The conduit will have no clearance over the top of the I-beam.

To find the distance between bend points, multiply the amount of offset required by the cosecant of the bending angle (in this case, 45°):

$$24'' \times 1.41 = 33.84''$$

The straight distance from the end of the conduit run to the J box is 8'. To find the amount of conduit to add to this because of the 45° bend, use the cosine formula (C = A ÷ H). The hypotenuse

Figure 42 24" offset.

or the distance between bend points is 33.84". The side adjacent is found by converting C = A ÷ H to A = C × H, or:

$$A = 0.707 \times 33.84'' = 23.92''$$

Subtract 23.92" from 33.84" and this gives the amount of extra conduit needed when making this 45° bend:

$$33.84'' - 23.92'' = 9.92'' \text{ or } 9^{92}/_{100}''$$

Add the 9.92" to 8' and the amount of straight conduit required has been determined:

$$8' + 9.92'' = 8'\text{-}9.92'' \text{ of straight conduit}$$

The final measurements should be as follows:

Distance between bend points = $33^{21}/_{25}''$ or 33.84"

Length of straight conduit = 8'-9.92" or 8'-9$^{92}/_{100}$"

Now let us complete another example. In this case, an offset is to be made without the use of a

Figure 41 Position of conduit in bender for making offsets.

NCCER — *Contren® Learning Series* 26204-11

protractor level. The offset is to be 20" in height (*Figure 43*). The bend angle is 45°. Explain the process required to accomplish this.

The cosecant for 45° is 1.41. The distance between bend points is 1.41 × 20". Therefore:

$$1.41 \times 20" = 28.2"$$

The first mark on the conduit is placed 14.1" back from the end of the conduit. The end of the conduit needs to be brought up 10" to accomplish a 45° bend. The second mark on the conduit is placed 28.2" down the conduit from the first mark. The bender is placed on the second mark. The conduit is bent until the two surfaces are parallel. At this point, a 20" offset has been created.

## 9.3.0 Concentric Offsets

To make concentric offsets, use the same procedure as for 90° stubs. Increase the radius of the second pipe by an amount equal to the radius of the first pipe plus the OD size of the first pipe plus the spacing desired between pipes. Although the examples used to explain offset bending had several shots per bend, fewer shots can be used. Successful offsets can be made with only two or three shots per bend. However, with offsets, 90° bends, or any other segment bends, the

extra time spent bending smoother radius bends (more shots) is returned when the conductors are pulled into the completed raceway.

### 9.3.1 Saddle Bends

As mentioned previously, saddle bends are used to cross small obstructions or other runs of conduit. Saddle bends in conduit up to four inches **inside diameter (ID)** may be made in hydraulic conduit benders. Always refer to the charts that accompany the bender. For example, to make a saddle bend on a length of 2" rigid conduit so that it can pass over a 3" water pipe with ¼" clearance, three bending operations are required, as shown in *Figure 44*.

The bend is calculated by referring to appropriate bending charts. The one shown in *Table 6* gives measurements for 2" conduit with ¼" clearance. Therefore, look in the left column under straight-run conduit and go down the column until the 3" row is reached. Moving to the right in this row, it can be seen that the spacing between bends is 15⅝" on center, and bends No. 1 and No. 2 will be 15°, while the third bend is 30°.

Using the information found in the bending table, mark the conduit accordingly. Insert the conduit in the bender and make bends No. 1 and No. 2 first, both at 15°. Back off the pump pressure, reverse the conduit in the bender, and then make the third bend at 30°. Release the pump pressure, remove the conduit from the bender, and check the saddle for accuracy.

For another example, suppose a saddle is to be built to go over an I-beam that is 36" in height (*Figure 45*). The distance from the end of the conduit to the center of the I-beam is 5'.

*Figure 43*   20" offset.

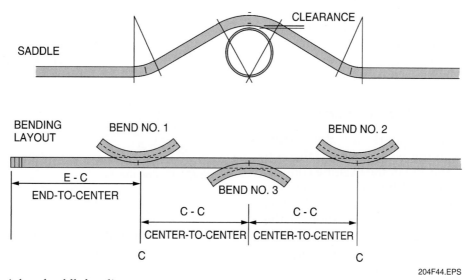

*Figure 44*   Principles of saddle bending.

**Table 6** Saddle Table

| Straight-Run Conduit | Minimum Length | Bend Spacing | Bend Degrees | Bend Degrees | Bend Degrees |
|:---:|:---:|:---:|:---:|:---:|:---:|
| — | E-C | C-C | No. 1 | No. 2 | No. 3 |
| 1" | 20" | 16" | 6 | 6 | 12 |
| 1¼" | 20" | 16" | 7 | 7 | 14 |
| 1½" | 20" | 16" | 8 | 8 | 16 |
| 2" | 20" | 15⅞" | 10 | 10 | 20 |
| 2½" | 20" | 15¾" | 12½ | 12½ | 25 |
| 3" | 20" | 15⅝" | 15 | 15 | 30 |
| 3½" | 20" | 15½" | 18 | 18 | 36 |
| 4" | 20" | 15½" | 20 | 20 | 40 |

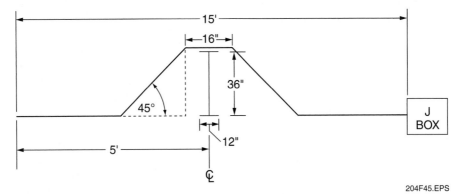

204F45.EPS

*Figure 45* 36" saddle.

The task is to determine the distance between bend points when a 45° bend is being used. Determine the exact length of conduit needed so threading can be completed before bending. The conduit is to clear the top of the I-beam by 1". Determine the centerline layout of the saddle (*Figure 46*).

The 16" between bend points in this problem is an arbitrary measurement. It is determined by the person doing the bending. Enough conduit should be used to extend past the outside dimensions of the 12" I-beam so the radius of each bend does not run into the I-beam.

The distance between bend points on the hypotenuse of the imaginary triangle is calculated by multiplying the total height of the I-beam (36" + 1" clearance = 37") by the cosecant of the 45° angle of the bend or 1.41 × 37":

$$1.41 \times 37" = 52^{17}/_{100}" \text{ or } 52.17"$$

The centerline layout mark for the saddle would be 5' if the conduit were to be installed on a straight line from the end of the existing conduit run to the center of the I-beam. However, the conduit is not being installed on a straight line. Therefore, the measurement in question is the amount of additional conduit needed because of the 45° angle of deviation (*Figure 47*). The amount of additional conduit is equal to the side adjacent subtracted from the hypotenuse.

204F46.EPS

*Figure 46* Saddle.

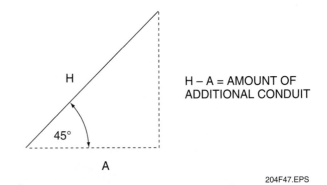

H – A = AMOUNT OF ADDITIONAL CONDUIT

204F47.EPS

*Figure 47* Cosine function.

The side adjacent (A) first needs to be calculated, then that measurement can be subtracted from the hypotenuse (52.17"). To calculate the side adjacent, use the cosine formula (C = A ÷ H).

NCCER — *Contren®  Learning Series*  26204-11

## Saddle Bends

Whenever possible, a three-bend saddle should not be sharper than 30°. If the object you are trying to saddle is more than 6" in diameter, use a four-bend saddle rather than a three-bend saddle. In this installation, two three-bend saddles were required.

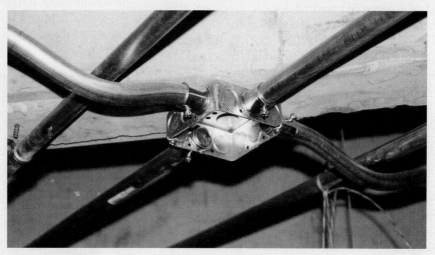

204SA05.EPS

Because you are looking for A, convert the formula to $A = C \times H$, or:

$$A = 0.707 \times 52.17"$$
$$A = 36.88"$$

Subtract 36.88 from 57.17, or:

$$52.17" - 36.88" = 15.29"$$

That is how much extra conduit is needed to get the center of the saddle to the center of the I-beam. Therefore, the layout mark for the saddle is:

$$5' + 15.29" = 6'-3.29"$$

The layout of the conduit is shown in *Figure 48*.

To find the total amount of conduit to be used from the end of the existing conduit to the J box, simply take the straight line distance of 15' and add the difference between the hypotenuse and side adjacent to 15' two times, or:

$$15' + 15.29" + 15.29" = 15' + 30.58" = 17'-6.58"$$

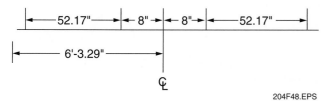

*Figure 48*   Conduit layout.

The final measurements should be as follows:

Distance between bend points on the top portion of the saddle = 16"

Distance between bend points that form each hypotenuse = $52^{17}/_{100}$" or 52.17"

Distance from the end of the conduit to the centerline layout mark of the saddle = 6'-3.29" or $6'-3^{29}/_{100}$"

Overall length of the conduit required to reach the J box = $17'-6^{58}/_{100}$" or 17'-6.58"

### 10.0.0 Tricks of the Trade

There are many tricks of the trade that you will learn during your career as an electrician. Some of these will be handed down to you by experienced workers; others will be learned through experience. Most professionals are constantly seeking new methods to improve their efficiency and to make the work go smoother and faster without sacrificing workmanship.

One of the handiest personal tools for bending conduit is the small, magnetic torpedo level with a 35° and 45° bubble. This tool should be in the tool pouch of every electrician. In many cases, it can make the difference between a good job and a poor job; that is, obtaining level and plumb conduit runs or not.

Workers have also designed their own custom tools to help in conduit installations. One tip is to take a pair of vice grips and weld a small, flat piece of iron on the top of the jaws for use with the magnetic level. During multiple bends, the vice grips are positioned at the desired point on the end of a piece of conduit, locked in place, and then the magnetic level is placed on the flat plate so the bubble may be watched during the bend. There are also some new commercial tools that are designed for leveling conduit bends; keep up with these developments by reading trade journals.

### 10.1.0 Eliminating Dog Legs

One of the problems that exists when bending an offset or saddle is the creation of a dog leg. A dog leg occurs when the centerline that runs the length of the conduit is not maintained for both bends of an offset and for the four bends of a saddle. Dog legs can be eliminated using an anti-dog device, as discussed earlier, or by following the procedure described here.

See *Figure 49*. When the conduit is snugged up by the bender, take a piece of Unistrut® and clamp it to the end of the conduit farthest from the bending shoe. Before tightening the conduit strap, level the Unistrut®. Once this is done, make your first bend. Release the conduit from the bending shoe and move the conduit forward to the next bending mark. Roll the conduit 180° and re-level the Unistrut®. Tighten the conduit in the bender and again make sure that the Unistrut® is level. Bend the second bend and you will have an offset with no dog leg.

### 10.2.0 Using a Table Corner or Plywood Sheet for Calculating Added Length

The task is to use a table corner or the corner of a sheet of plywood and demonstrate how to come up with the amount of additional conduit needed to go from one point to another once a bend has been made. The distance between points is 8'. The amount of offset to be made is 25" (*Figure 50*). The degree of bend is 30°.

## Trapeze Hangers

Trapeze hangers are often used to support multiple overhead conduit runs. This reduces the number of individual supports and also reduces the number of bends, as the conduit can be run below obstacles as shown here.

204SA06.EPS

See *Figure 51*. Measure 25" up one side of the sheet of plywood or the corner of a table. The hypotenuse of the triangle will be 2 × 25" or 50" (2 is the cosecant for 30°). Measure from the 25"

*Figure 50*   25" offset.

*Figure 49*   Dog leg elimination.

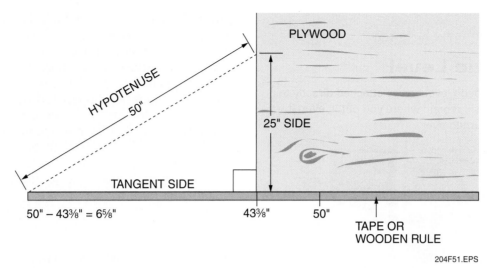

**Figure 51** Extra conduit.

mark to the surface that is tangent to the side that is 25" long, until the 50" mark on the tape measure crosses the tangent side. Mark this 50" point. Then, place the end of the ruler on the 50" mark and let the 50" point on the ruler extend beyond the end of the plywood or table. The amount of tape that extends beyond the plywood or table edge is the amount of extra conduit that is needed to accomplish this bend. This is 6⅝" more conduit than if going on a straight line for 8' or 8'-6⅝" of conduit required to accomplish the task.

## 11.0.0 PVC CONDUIT INSTALLATIONS

Rigid nonmetallic conduit and fittings (PVC electric conduit) may be used in direct earth burial; in walls, floors, and ceilings of buildings; in cinder fill; and in damp and dry locations. Its use is prohibited by the *NEC*® in certain hazardous locations, for support of fixtures or other equipment, where subject to physical damage, and under certain other conditions. See *NEC Article 352*.

PVC conduit can be cut easily at the job site without special tools, although PVC cutters help in cutting square ends. Sizes from ½" through 1½" can be cut with a fine-tooth saw. For sizes 2" through 6", a miter box or similar saw guide should be used to keep the conduit steady and ensure a square cut. To ensure satisfactory joining, care should be taken not to distort the end of the conduit when cutting.

After cutting, deburr the pipe ends and wipe clean of dust, dirt, and plastic shavings. Deburring is accomplished easily using a pocket knife or file.

One of the important advantages of PVC conduit, in comparison with other rigid conduit materials, is the ease and speed with which solvent-cemented joints can be made. The following steps are required for a proper joint:

*Step 1*  The conduit should be wiped clean and dry.

> **WARNING!**
>
> The cement used with PVC can be hazardous. Provide adequate ventilation, avoid skin contact, and always refer to the MSDS and follow the manufacturer's usage and safety instructions.

*Step 2*  Apply a full, even coat of PVC cement to the end of the conduit and the fitting. The cement should cover the area that will be inserted in the fitting.

*Step 3*  Push the conduit and fitting firmly together with a slight twisting action until it bottoms and then rotate the conduit in the fitting (about a half turn) to distribute the cement evenly. Avoid cement buildup inside the conduit. The cementing and joining operation should not exceed more than 20 seconds.

*Step 4*  When the proper amount of cement has been applied, a bead of cement will form at the joint. Wipe the joint with a brush to remove any excess cement. The joint should not be disturbed for 10 minutes.

# Magnetic Level

A magnetic torpedo level can be attached directly to the conduit. It will remain where you put it, even if the conduit is turned over.

204SA07.EPS

## 12.0.0 BENDING PVC CONDUIT

Most manufacturers of PVC conduit offer various radius bends in a number of segments. Where special bends are required, PVC conduit is easy to bend on the job. Stub-ups, saddles, concentric bends, offsets, and kicks are all possible with PVC conduit, just as with metallic conduit.

PVC conduit is bent with the aid of a heating unit. The PVC must be heated evenly over the entire length of the curve. Heating units are available from various sources that are designed specifically for the purpose in sizes to accommodate all conduit diameters (*Figure 52*). While some heaters use gas for the heat source, most employ

204F52.EPS

*Figure 52*  PVC heating units.

infrared heat energy, which is more quickly absorbed in the conduit. Small sizes are ready to bend after a few seconds in the hotbox. Larger diameters require two or three minutes, depending on the conditions. Other methods of heating PVC conduit for bending include heating blankets and hot air blowers. The use of torches or other flame-type devices is not recommended. PVC conduit exposed to excessively high temperatures may take on a brownish color. Sections showing evidence of such scorching should be discarded.

> **WARNING!**
>
> Always wear gloves when working with heat.

If a number of identical bends are required, a template can be helpful (*Figure 53*). A simple template can be made by sawing a sheet of plywood to match the desired bend. Nail to a second sheet of plywood. The heated conduit section is placed in the template, sponged with water to cool, and is then ready to install. Care should be taken to fully maintain the ID of the conduit when bending.

If only a few bends are needed, scribe a chalk line on the floor or workbench. Then, match the heated conduit to the chalk line and cool. The conduit must be held in the desired position until it is relatively cool since the PVC material will tend to revert to its original shape. Templates are also available for many bends.

Another method is to take the heated conduit section to the point of installation and form it to fit the actual installation (*Figure 54*). This method is especially effective for making blind bends or compound bends using smaller sizes of PVC. After bending, wipe a wet rag over the bend (*Figure 55*) to cool it.

Bends in small-diameter PVC conduit (½" to 1½") require no filling for code-approved radii. When bending PVC of 2" or larger diameter, there is a risk of wrinkling or flattening the bend. To help eliminate this problem, a plug set is used. A plug is inserted in each end of the piece of PVC being bent, and then a hand pump is used to pressurize the conduit before bending it. The pressure is about three to five pounds per square inch.

Place airtight plugs (*Figure 56*) in each end of the conduit section before heating. The retained air will expand during the heating process and hold the conduit open during bending. Do not remove the plugs until the conduit has cooled.

*Figure 53* Plywood template.

204F53.EPS

204F54.EPS

*Figure 54* Some PVC bends may be formed by hand.

204F56.EPS

*Figure 56* Typical plug set.

are factors, an O-ring expansion coupling should be installed near the fixed end of the run, or fixture, to take up any expansion or contraction that may occur. Confirm the expansion and contraction lengths available in these fittings as they may vary by manufacturer. Charts are available that indicate what expansion can be expected at various temperature levels. The coefficient of linear expansion of PVC conduit is 0.0034" per 10' per °F. *Table 7* lists the expansion rates for various temperatures.

Expansion couplings are seldom required in underground or slab applications. Expansion

204F55.EPS

*Figure 55* After the bend is formed, wipe a wet rag over the bend to cool it.

In applications where the conduit installation is subject to constantly changing temperatures and the runs are long, precautions should be taken to allow for expansion and contraction of PVC conduit. When expansion and contraction

Conduit Bending

**Table 7** PVC Expansion Rates

| Temperature Change in °F | Length Change in Inches per 100 Ft of PVC Conduit | Temperature Change in °F | Length Change in Inches per 100 Ft of PVC Conduit |
|---|---|---|---|
| 5 | 0.2 | 105 | 4.2 |
| 10 | 0.4 | 110 | 4.5 |
| 15 | 0.6 | 115 | 4.7 |
| 20 | 0.8 | 120 | 4.9 |
| 25 | 1.0 | 125 | 5.1 |
| 30 | 1.2 | 130 | 5.3 |
| 35 | 1.4 | 135 | 5.5 |
| 40 | 1.6 | 140 | 5.7 |
| 45 | 1.8 | 145 | 5.9 |
| 50 | 2.0 | 150 | 6.1 |
| 55 | 2.2 | 155 | 6.3 |
| 60 | 2.4 | 160 | 6.5 |
| 65 | 2.6 | 165 | 6.7 |
| 70 | 2.8 | 170 | 6.9 |
| 75 | 3.0 | 175 | 7.1 |
| 80 | 3.2 | 180 | 7.3 |
| 85 | 3.4 | 185 | 7.5 |
| 90 | 3.6 | 190 | 7.7 |
| 95 | 3.8 | 195 | 7.9 |
| 100 | 4.1 | 200 | 8.1 |

and contraction may generally be controlled by bowing the conduit slightly or burying the conduit immediately. After the conduit is buried, expansion and contraction cease to be factors. Care should be taken, however, in constructing a buried installation. If the conduit should be left exposed for an extended period of time during widely-variable temperature conditions, an allowance should be made for expansion and contraction.

In above-ground installations, care should be taken to provide proper support of PVC conduit because of its semi-rigidity. This is particularly important at high temperatures. The distance between supports should be based on temperatures encountered at the specific installation. Charts are available that clearly outline at what intervals support is required for PVC conduit at various temperature levels.

### Think About It

## Putting It All Together

Examine the visible conduit bends in your home or workplace. Are they neat and professionally made? If not, what might you have done differently to improve the installation?

## SUMMARY

Many conduit installations are visible; that is, they run exposed. Consequently, electricians must take special care to ensure that all exposed conduit runs are parallel, level, and plumb. Nothing else will do. This is one phase of the electrical construction industry where electricians have an opportunity to show off. In fact, an expert installation of a conduit system is similar to a work of art.

Learn the basics of conduit bending and installations, put your knowledge to practical use, and take pride in your abilities to perform a professional conduit installation. Make your work second to none. Of course, contractors and clients want speed, but if you have a good basic knowledge of conduit bending and then put this knowledge to use, your craftsmanship will let you bend conduit smarter and faster, giving a good-looking installation along with speed to satisfy your employer, the building owners, and all concerned.

1. The minimum bending radius for 3" rigid conduit is _____.

   a. 8"
   b. 10"
   c. 11"
   d. 13"

2. The minimum bending radius for ½" rigid conduit is _____.

   a. 4"
   b. 6"
   c. 8"
   d. 10"

3. The maximum number of 90° bends allowed between pull points in a conduit system is _____.

   a. one
   b. two
   c. three
   d. four

4. A saddle bend is counted as _____.

   a. 90°
   b. 120°
   c. 180°
   d. It depends on the type of bends used.

5. When referring to the two bends required to make an offset in a length of conduit, which of the following is always true?

   a. The degree of bend for each must be exactly 45°.
   b. The degree of bend for each must be exactly 30°.
   c. The degree of bend for each must be equal.
   d. The degree of bend for each must be unequal.

6. The reason for making a saddle bend in a run of conduit is to _____.

   a. change direction in the height of a conduit run
   b. make a 90° bend
   c. cross a small obstacle
   d. make a conduit termination in an outlet box

7. The equation used to calculate the circumference of a circle is _____.

   a. $C = \pi D$
   b. $C = 1\pi R$
   c. $C = \pi R^2$
   d. $C = \pi^2$

8. The fractional equivalent of the decimal 0.015625 is _____.

   a. ¹⁄₆₄
   b. ¹⁄₃₂
   c. ³⁄₆₄
   d. ¹⁄₁₆

9. The rise in a conduit bend is best described as the _____.

   a. horizontal distance that runs from the point of bend and parallel with the deck
   b. leg length
   c. radius
   d. height of the stub-up

10. The best type of bending shoe for making concentric bends is the _____.

   a. concentric shoe
   b. one-shot shoe
   c. segmented shoe
   d. long shoe

11. Rigid aluminum conduit is available in standard sizes from _____.

   a. ½" to 4"
   b. ½" to 6"
   c. ¾" to 6"
   d. ¾" to 4"

12. Two or more parallel bends that are bent in the same direction are known as _____.

   a. stubs
   b. perpendicular bends
   c. concentric bends
   d. axial bends

13. Which of the following formulas should be used to find the side adjacent?

   a. $S = O \times H$
   b. $S = O \div H$
   c. $C = A \div H$
   d. $C = A \times H$

14. An important safety precaution to use when working with PVC conduit is to _____.
    a. provide proper ventilation to carry off fumes from the joint cement
    b. use lengths of PVC of not more than 8'
    c. use inside diameters of 4" or less
    d. never wear gloves that may adhere to the joint cement

15. To prevent flattening when bending larger diameter PVC pipe, use _____.
    a. plugs and air pressure
    b. sand fill
    c. water fill
    d. a higher temperature

1. In what chapter of the *NEC*® will you find the installation requirements for conduit?

   a. Two
   b. Three
   c. Four
   d. Five

2. The maximum number of degrees allowed between pull points in a conduit run is _____.

3. Offsets and saddles _____.

   a. do not need to be counted if less than 15°
   b. must be included when calculating total degrees between pull points
   c. add no resistance
   d. all count as 30°

4. A conduit run with _____ is acceptable to pull wire.

   a. two 90° bends, four 45° bends, and one box offset
   b. two 90° bends, three 45° bends, and one saddle
   c. two 90° bends, two 45° bends, and two 30° bends
   d. four 90° bends and two box offsets

5. The maximum size rigid steel conduit that can be bent using a portable mechanical bender is _____.

6. The maximum size IMC that can be bent using a portable mechanical conduit bender is _____.

7. All of the following may be bent using a mechanical bender *except* _____.

   a. EMT
   b. IMC
   c. PVC
   d. rigid conduit

8. The maximum size rigid conduit that can be bent with a hydraulic bender is _____.

9. When making segment bends, _____ must be determined.

   a. the radius of the bend
   b. the total number of degrees in the bend
   c. the developed length
   d. All of the above.

10. When making a saddle bend using 4" conduit with a minimum length (E-C) of 20" and bend spacing (C-C) of 15½", the angle of the third bend will be _____.

    a. 36°
    b. 20°
    c. 40°
    d. 30°

11. The code requires a minimum radius for rigid conduit bends because _____

    _____

12. A bend used for a change in direction of less than 90° is called a(n) _____.
    a. kick
    b. saddle
    c. stub-up
    d. offset

13. A mechanical bender may be used to make bends in _____.
    a. 2" EMT
    b. 4" rigid conduit
    c. 1½" PVC
    d. IMC conduit only

14. To make matching bends in both ½" and 1" IMC when using a mechanical bender, _____.
    a. adjust the deduct length for the ½" conduit
    b. use the ½" shoe for the ½" conduit and the 1" shoe for the 1" conduit
    c. use the 1" shoe for both conduits
    d. get matched pairs of conduit shoes from the manufacturer

15. Speed benders can bend conduit up to _____.

16. True or False? The one-shot shoes used with hydraulic benders can make 90° bends in a single operation.

17. One method of bending segment is to use _____.
    a. an elapsed time method to keep track of each bend
    b. a stress meter to monitor each bend
    c. a level to monitor each bend
    d. the amount of travel method to monitor each bend

18. To make a 20" offset using 45° angles, a multiplier of 1.414 is used. If the first bend is marked at 12" from the end of the conduit, the second bend would be marked at _____ from the end.
    a. 16.96"
    b. 28.28"
    c. 40.28"
    d. 32"

19. Airtight plugs are installed in the ends of large sizes of PVC conduit during bending to _____.
    a. prevent the conduit from wrinkling or flattening the bend
    b. hold the heat inside the conduit to keep it flexible
    c. prevent condensation within the conduit
    d. prevent deadly PVC gases from escaping

20. One of the disadvantages of aluminum conduit is that _____.
    a. it cannot be bent with one-shot benders
    b. it is difficult to bend consistently and accurately
    c. special bending shoes must be used
    d. aluminum kinks when bent

# Trade Terms Introduced in This Module

**Approximate ram travel:** The distance the ram of a hydraulic bender travels to make a bend. To simplify and speed bending operations, many benders are equipped with a scale that shows ram travel. Using a simple table (supplied with many benders), the degree of bend can easily be converted to inches of ram travel. This measurement, however, can only be approximated because of the variation in springback of the conduit being bent.

**Back-to-back bend:** Any bend formed by two 90° bends with a straight section of conduit between the bends.

**Bending protractor:** Made for use with benders mounted on a bending table and used to measure degrees; also has a scale for 18, 20, 21, and 22 shots when using it to make a large sweep bend.

**Bending shot:** The number of shots needed to produce a specific bend.

**Concentric bending:** The process of making 90° bends in parallel runs of conduit. This requires increasing the radius of each conduit from the inside of the bend toward the outside.

**Conduit:** Piping designed especially for pulling electrical conductors. Types include rigid, IMC, EMT, PVC, aluminum, and other materials.

**Degree indicator:** An instrument designed to indicate the exact degree of bend while it is being made.

**Developed length:** The amount of straight pipe needed to bend a given radius. Also, the actual length of the conduit that will be bent.

**Elbow:** A 90° bend.

**Gain:** The amount of pipe saved by bending on a radius as opposed to right angles. Because conduit bends in a radius and not at right angles, the length of conduit needed for a bend will not equal the total determined length. Gain is the difference between the right angle distances A and B and the shorter distance C—the length of conduit actually needed for the bend.

**Inside diameter (ID):** The inside diameter of conduit. All electrical conduit is measured in this manner. The outside dimensions, however, will vary with the type of conduit used.

**Kicks:** Bends in a piece of conduit, usually less than 90°, made to change the direction of the conduit.

**Leg length:** The distance from the end of the straight section of conduit to the bend, measured to the centerline or to the inside or outside of the bend or rise.

**Ninety-degree bend:** A bend in a piece of conduit that changes its direction by 90°.

**Offsets:** Two equal bends made to avoid an obstruction blocking the run of the conduit.

**One-shot shoe:** A large bending shoe that is designed to make 90° bends in conduit.

**Outside diameter (OD):** The size of any piece of conduit measured on the outside diameter.

**Radius:** The relative size of the bent portion of a pipe.

**Rise:** The length of the bent section of conduit measured from the bottom, centerline, or top of the straight section to the end of the bent section.

**Segment bend:** Any bend formed by a series of bends of a few degrees each, rather than a single one-shot bend.

**Segmented bending shoe:** A smaller type of shoe designed for bending segmented bends only (always less than 15°).

**Springback:** The amount, measured in degrees, that a bent conduit tends to straighten after pressure is released on the bender ram. For example, a 90° bend, after pressure is released, will pull back about 2° to 88°.

**Stub-up:** Another name for the rise in a section of conduit.

**Sweep bend:** A 90° bend with a radius larger than that produced by a standard one-shot shoe.

**Take-up (comeback):** The amount that must be subtracted from the desired stub length to make the bend come out correctly using a point of reference on the bender or bending shoe.

# Additional Resources

This module presents thorough resources for task training. The following resource material is suggested for further study.

*Benfield Conduit Bending Manual*, 2nd Edition. KS: EC&M Books.

*National Electrical Code® Handbook*, Latest Edition. Quincy, MA: National Fire Protection Association.

*Tom Henry's Conduit Bending Package* (includes video, book, and bending chart). Winter Park, FL: Code Electrical Classes, Inc.

# Figure Credits

# CONTREN® LEARNING SERIES — USER UPDATE

NCCER makes every effort to keep its textbooks up-to-date and free of technical errors. We appreciate your help in this process. If you find an error, a typographical mistake, or an inaccuracy in NCCER's Contren® materials, please fill out this form (or a photocopy), or complete the online form at www.nccer.org/olf. Be sure to include the exact module number, page number, a detailed description, and your recommended correction. Your input will be brought to the attention of the Authoring Team. Thank you for your assistance.

*Instructors* – If you have an idea for improving this textbook, or have found that additional materials were necessary to teach this module effectively, please let us know so that we may present your suggestions to the Authoring Team.

### NCCER Product Development and Revision
3600 NW 43rd Street, Building G, Gainesville, FL 32606

**Fax:** 352-334-0932
**Email:** curriculum@nccer.org
**Online:** www.nccer.org/olf

☐ Trainee Guide   ☐ AIG   ☐ Exam   ☐ PowerPoints   Other _____

Craft / Level: _____ Copyright Date: _____

Module Number / Title: _____

Section Number(s): _____

Description: _____

_____

_____

_____

Recommended Correction: _____

_____

_____

_____

_____

Your Name: _____

Address: _____

_____

Email: _____ Phone: _____

# Pull and Junction Boxes

## Zachry Corporate Headquarters

Zachry Construction Corporation won the Aon Build America Award for its work on its own Corporate Headquarters in San Antonio, Texas. Zachary constructed a new conference and employment center using the U.S. Green Building Council (USGBC) Leadership in Energy and Environmental Design (LEED) Programs.

26205-11

Trainees with successful module completions may be eligible for credentialing through NCCER's National Registry. To learn more, go to **www.nccer.org** or contact us at **1.888.622.3720.** Our website has information on the latest product releases and training, as well as online versions of our *Cornerstone* newsletter and Pearson's Contren® product catalog.

Your feedback is welcome. You may email your comments to **curriculum@nccer.org,** send general comments and inquiries to **info@nccer.org,** or use the User Update form at the back of this module.

# PULL AND JUNCTION BOXES

## Objectives

When you have completed this module, you will be able to do the following:

1. Describe the different types of nonmetallic and metallic pull and junction boxes.
2. Properly select, install, and support pull and junction boxes and their associated fittings.
3. Describe the *National Electrical Code®* (*NEC®*) regulations governing pull and junction boxes.
4. Size pull and junction boxes for various applications.
5. Understand the NEMA and IP classifications for pull and junction boxes.
6. Describe the purpose of conduit bodies and Type FS boxes.

## Performance Tasks

Under the supervision of the instructor, you should be able to do the following:

1. Identify various NEMA boxes.
2. Properly select, install, and support pull and junction boxes over 100 cubic inches in size.
3. Identify various conduit bodies and fittings.

## Trade Terms

Conduit body
Explosion-proof
Handhole
Junction box

Mogul
Pull box
Raintight
Waterproof

Watertight
Weatherproof

## Required Trainee Materials

1. Pencil and paper
2. Copy of the latest edition of the *National Electrical Code®*
3. Appropriate personal protective equipment

Note: *NFPA 70®*, *National Electrical Code®*, and *NEC®* are registered trademarks of the National Fire Protection Association, Inc., Quincy, MA 02269. All *National Electrical Code®* and *NEC®* references in this module refer to the 2011 edition of the *National Electrical Code®*.

# Contents

*Topics to be presented in this module include:*

## Figures and Tables

# 1.0.0 INTRODUCTION

Pull boxes and junction boxes (*Figure 1*) are provided in an electrical installation to facilitate the installation of conductors, or to provide a junction point for the connection of conductors, or both. In some instances, the location and size of pull boxes are designated on the drawings. In most cases, however, the electricians on the job will have to determine the proper number, location, and sizes of pull or junction boxes to facilitate conductor installation.

> **NOTE**
> Smaller conductors may use a regular octagon or square device box with a blank cover as a junction box.

Pull and junction boxes must be sized, installed, and supported to meet the requirements of *NEC Article 314*. Since the *NEC*® limits the number of conductors allowed in each box according to its volume, you must install boxes that are large

205F01.EPS

*Figure 1*    Pull and junction boxes.

enough to accommodate the number of conductors that must be spliced in the box or fed through it. Therefore, a knowledge of the various types of boxes and their volumes is essential.

Besides being able to calculate the required box sizes, you must also know how to select the proper type of box for any given application. For example, metallic boxes used in concrete deck pours are different from those used as pull and junction boxes in residential or commercial buildings. Boxes for use in hazardous locations must be rated as explosion-proof.

You must also know what fittings are available for terminating the various wiring methods in these boxes.

## 1.1.0 Boxes for Damp and Wet Locations

In damp or wet locations, boxes and fittings must be placed or equipped to prevent moisture or water from entering and accumulating within the box or fitting. It is recommended that approved boxes of nonconductive material be used with nonmetallic sheathed cable or approved nonmetallic conduit when the cable or conduit is used in moisture-prone locations. Boxes installed in wet locations must be approved for the purpose per *NEC Section 314.15*.

A wet location is any location subject to saturation with water or other liquids, such as locations exposed to weather or water, washrooms, garages, and interiors that might be hosed down. Underground installations or those in concrete slabs or masonry in direct contact with the earth must be considered to be wet locations. Raintight, waterproof, or watertight equipment (including fittings) may satisfy the requirements for weatherproof equipment. Boxes with threaded conduit hubs and gasketed covers will normally prevent water from entering the box except for condensation within the box.

A damp location is a location subject to some degree of moisture. Such locations include partially protected outdoor locations—such as under canopies, marquees, and roofed open porches.

# Outdoor Boxes

Outdoor wiring must be able to resist the entry of water. Outdoor boxes are either driptight, which means sealed against falling water from above, or watertight, which means sealed against water from any direction. Driptight boxes simply have lids that deflect rain; they are not waterproof. Watertight boxes are sealed with gaskets to prevent the entry of water from any angle.

It also includes interior locations subject to moderate degrees of moisture—such as some basements, some barns, and cold storage warehouses.

## 1.2.0 NEMA and IP Enclosure Classifications

Like switch enclosures, pull and junction boxes are also classified according to the degree of protection they provide from the elements. The National Electrical Manufacturers Association (NEMA) classifications for enclosures are as follows:

- *NEMA Type 1: General Purpose* – This enclosure is primarily intended to prevent accidental contact with the enclosed apparatus. It is suitable for general-purpose applications indoors where it is not exposed to unusual service conditions. A NEMA Type 1 enclosure serves as protection against dust and light and indirect splashing, but is not dust-tight.
- *NEMA Type 3: Dust-Tight, Raintight* – This enclosure is intended to provide suitable protection against specified weather hazards. A NEMA Type 3 enclosure is suitable for outdoor applications, such as construction work. It is also sleet-resistant.
- *NEMA Type 3R: Rainproof, Sleet Resistant* – This enclosure protects against interference in operation of the contained equipment due to rain and resists damage from exposure to sleet. It is designed with conduit hubs and external mounting as well as drainage provisions.
- *NEMA Type 4: Watertight* – A watertight enclosure is designed to meet a hose test, which consists of a stream of water from a hose with a 1" nozzle, delivering at least 65 gallons per minute. The water is directed on the enclosure from a distance of not less than 10' for a period of five minutes. During this period, it may be pointed in one or more directions, as desired. There should be no leakage of water into the enclosure under these conditions.
- *NEMA Type 4X: Watertight, Corrosion-Resistant* – These enclosures are generally constructed along the lines of NEMA Type 4 enclosures except that they are made of a material that is highly resistant to corrosion. For this reason, they are ideal in applications such as meat packing and chemical plants, where contaminants would ordinarily destroy a steel enclosure over a period of time.
- *NEMA Type 7: Hazardous Locations, Class I* – These enclosures are designed to meet the application requirements of the *NEC®* for Class I hazardous locations. Class I locations are those in which flammable gases or vapors are or may be present in the air in sufficient quantities to produce explosive or ignitable mixtures. In this type of equipment, the circuit interruption occurs in the air.
- *NEMA Type 9: Hazardous Locations, Class II* – These enclosures are designed to meet the application requirements of the *NEC®* for Class II hazardous locations. Class II locations are those that are hazardous because of the presence of combustible dust. The letter or letters following the type number indicate the particular group or groups of hazardous locations (as defined in the *NEC®*) for which the enclosure is designed. The designation is incomplete without a suffix letter or letters.
- *NEMA Type 12: Industrial Use* – This type of enclosure is designed for use in those industries where it is desired to exclude such materials as dust, lint, fibers and flyings, oil seepage, or coolant seepage. There are no conduit openings or knockouts in the enclosure, and mounting is by means of flanges or mounting feet.
- *NEMA Type 13: Oil-Tight, Dust-Tight* – NEMA Type 13 enclosures are generally made of cast iron, gasketed, or permit use in the same environments as NEMA Type 12 devices. The essential difference is that due to its cast housing, a conduit entry is provided as an integral part of the NEMA Type 13 enclosure, and mounting is by means of blind holes rather than mounting brackets.

*Table 1* provides an explanation of the ingress protection (IP) classification system. Each enclosure is designated IP followed by a two-digit number. For example, a dust-tight enclosure that is suitable for continuous immersion would be an IP68. *Table 2* is a cross reference of NEMA and IP enclosure classifications.

**Table 1**  IP Classification System

| First Number<br>Degree of Protection against Solid Objects | Second Number<br>Degree of Protection against Water |
|---|---|
| 0  Not protected.<br>1  Protected against a solid object greater than 50 mm, such as a hand.<br>2  Protected against a solid object greater than 12 mm, such as a finger.<br>3  Protected against a solid object greater than 2.5 mm, such as a wire or tool.<br>4  Protected against a solid object greater than 1.0 mm, such as wire or thin strips of metal.<br>5  Dust-protected. Prevents ingress of dust sufficient to cause harm.<br>6  Dust-tight. No dust ingress. | 0  Not protected.<br>1  Protected against water dripping vertically, such as condensation.<br>2  Protected against dripping water when tilted up to 15°.<br>3  Protected against water when spraying at an angle of up to 60°.<br>4  Protected against water splashing from any direction.<br>5  Protected against jets of water from any direction.<br>6  Protected against heavy seas or powerful jets of water. Prevents ingress sufficient to cause harm.<br>7  Protected against harmful ingress of water when immersed between a depth of 150 mm to 1 m.<br>8  Protected against submersion. Suitable for continuous immersion in water. |

**Table 2**  NEMA and IP Enclosures

| NEMA Enclosure Type | IP Classification |
|---|---|
| 1 | IP20 |
| 2 | IP21 |
| 3 | IP54 |
| 3R | IP24 |
| 3S | IP54 |
| 4,4X | IP56 |
| 5 | IP52 |
| 6,6P | IP67 |
| 12,12K | IP52 |
| 13 | IP54 |

## 2.0.0 SIZING PULL AND JUNCTION BOXES

Pull boxes should be as large as possible. Workers need space within the box for both hands and in the case of the larger wire sizes, workers will need room for their arms to feed the wire. *NEC Section 314.16* specifies that pull and junction boxes must provide adequate space and dimensions for the installation of conductors.

Long runs of conductors should not be made in one pull. Pull boxes, installed at convenient intervals, will relieve much of the strain on the conductors. The length of the pull, in many cases, is left to the judgment of the workers or their supervisor, and the conditions under which the work is installed.

The installation of pull boxes may seem to cause a great deal of extra work and trouble, but they save a considerable amount of time and hard work when pulling conductors. Properly placed, they eliminate many bends and elbows and do away with the necessity of fishing from both ends of a conduit run.

If possible, pull boxes should be installed in a location that allows electricians to work easily and conveniently. For example, in an installation where the conduit comes up a corner of a wall and changes direction at the ceiling, a pull box that is installed too high will force the electrician to stand on a ladder when feeding conductors, and will allow no room for supporting the weight of the wire loop or for the cable-pulling tools.

Unless the contract drawings or project engineer state otherwise, it is just as easy for the pull boxes to be placed at a convenient height that allows workers to stand on the floor with sufficient room for both wire loop and tools.

In some electrical installations, a number of junction boxes must be installed to route the conduit in the shortest, most economical way. The *NEC*® requires all junction boxes to be accessible. This means that a person must be able to get to the conductors inside the box without removing plaster, wall covering, or any other part of the building.

Junction boxes detract from the decorative scheme of a building. Therefore, where such boxes will be used in areas open to the public or in other areas where the boxes will be unattractive, they should be installed above suspended ceilings, in closets, or at least in corners of the room or area.

Junction boxes or pull boxes must be securely fastened in place on walls or ceilings or adequately suspended.

While certain sizes of factory-constructed boxes are available with knockouts, it may be necessary to have them custom built to meet the job requirements. When it is not possible to accurately anticipate the raceway entrance requirements, it will be necessary to cut the required knockouts on the job. In the case of large pull boxes and troughs, shop drawings should be prepared prior to the construction of these items with all required knockouts accurately indicated in relation to the conduit run requirements.

## 2.1.0 Sizing Pull and Junction Boxes for Systems Under 600V

For raceways containing conductors of No. 4 or larger, and for cables containing conductors of No. 4 or larger but carrying less than 600V, the minimum dimensions of pull boxes, junction boxes, and conduit bodies must meet the following requirements:

- In straight pulls, the length of the box or conduit body shall not be less than eight times the trade diameter of the largest raceway.
- Where angle or U pulls are made, the distance between each raceway entry inside the box or conduit body and the opposite wall of the box or conduit body shall not be less than six times the trade diameter of the largest raceway in a row. This distance shall be increased for additional entries by the amount of the sum of the diameter of all other raceway entries in the same row on the same wall of the box. Each row shall be calculated individually, and the single row that provides the maximum distance shall be used.
- The distance between raceway entries enclosing the same conductor shall not be less than six times the trade diameter of the larger raceway.
- When transposing cable size into raceway size, the minimum trade size raceway required for the number and size of conductors in the cable shall be used.

*Figure 2* shows a junction box with several runs of conduit. Since this is a straight pull, and 4"

conduit is the largest size in the group, the minimum length required for the box can be determined by the following calculation:

Trade size of largest conduit × 8 [per *NEC Section 314.28(A)(1)*] = minimum length of box

*or:*

$$4" \times 8 = 32"$$

Therefore, this particular pull box must be at least 32" in length. The width of the box, however, need only be of sufficient size to enable locknuts and bushings to be installed on all the conduits or connectors entering the enclosure.

Junction or pull boxes in which the conductors are pulled at an angle (*Figure 3*) must have a distance of not less than six times the trade diameter of the largest conduit [*NEC Section 314.28(A)(2)*]. The distance must be increased for additional conduit entries by the amount of the sum of the diameters of all other conduits entering the box on the same

*Figure 3*　Pull box with conduit runs entering at right angles.

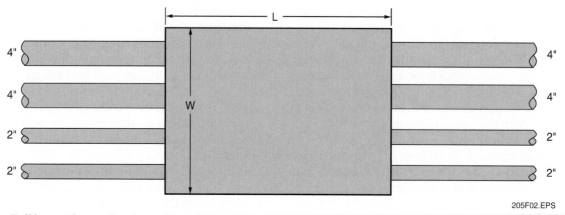

*Figure 2*　Pull box with two 4" and two 2" conduit runs.

side. The distance between raceway entries enclosing the same conductors must not be less than six times the trade diameter of the largest conduit.

Since the 4" conduit is the largest size in this case:

$$L_1 = 6 \times 4" + (3 + 2) = 29"$$

Since the same conduit runs are located on the adjacent wall of the box, $L_2$ is calculated in the same way; therefore, $L_2 = 29"$.

The distance $(D) = 6 \times 4"$ or 24". This is the minimum distance permitted between conduit entries enclosing the same conductor.

The depth of the box need only be of sufficient size to permit locknuts and bushings to be properly installed. In this case, a 6"-deep box would suffice.

Power distribution blocks are permitted to be installed into pull and junction boxes rated over 100 in³. These blocks must be listed, and the installation must comply with *NEC Section 314.28(E)*. This requires proper bending space and protection from live parts. In addition, the terminals must be unobstructed.

### 2.2.0 Sizing Pull and Junction Boxes for Systems over 600V

*NEC Article 314, Part IV* covers requirements for pull and junction boxes, conduit bodies, and handholes used on systems over 600V. Because the conductors used with these systems are larger and heavier than those used on lower-voltage systems, the *NEC®* requires additional pulling and

bending space. The *NEC®* requirements for these systems include the following:

- For straight pulls, the length of the box must be at least 48 times the outside diameter of the largest shielded or lead-covered conductor entering the box. See *NEC Section 314.71(A)*. When the cable is unshielded, the length may be reduced to 32 times the outside diameter of the largest nonshielded conductor entering the box.

- For angle or U pulls, the distance between each cable or conductor entry inside the box and the opposite wall of the box must be at least 36 times the diameter, over sheath, of the largest cable or conductor. See *NEC Section 314.71(B)(1)*. The distance must be increased by the sum of the outside diameters, over sheath, of all other cables or conductors entering through the same wall of the box. When the cable is unshielded and not lead covered, the length may be reduced to 24 times the outside diameter of the largest nonshielded conductor entering the box.

- The distance between a cable or conductor entry and its exit must be at least 36 times the outside diameter, over sheath, of that cable or conductor. See *NEC Section 314.71(B)(2)*. When the cable is both unshielded and not lead covered, the distance may be reduced to 24 times the outside diameter.

- In addition to a suitable cover, pull boxes containing conductors over 600V must be provided with at least one removable side.

- Per *NEC Section 314.72(E)*, boxes housing conductors over 600V must be completely enclosed and permanently marked DANGER – HIGH VOLTAGE – KEEP OUT.

### 3.0.0 CONDUIT BODIES

Conduit bodies, also called condulets, are defined in *NEC Article 100* as a separate portion of a conduit or tubing system that provides access through a removable cover to the interior of the system at a junction of two or more sections of the system or at a terminal point of the system. Conduit bodies are usually used with RMC and IMC. The cost of conduit bodies, because they are cast, is significantly higher than the stamped steel boxes. Splicing in conduit bodies is typically not recommended; however, it is permitted under certain conditions as specified in *NEC Section 314.16(C)(2)*.

As an electrical trainee, you will hear such terms as LL, LR, as well as other letters to distinguish between the various types of conduit bodies. To identify certain conduit bodies, an old trick of the trade is to hold the **conduit body** like a pistol (*Figure 4*). When doing so, if the oval-shaped

TYPE LR – OPENING ON RIGHT

TYPE LL – OPENING ON LEFT

205F04.EPS

*Figure 4*  Identifying conduit bodies.

opening of the conduit body is to your left, it is called an LL—the first L stands for elbow and the second L stands for left. If the opening is on your right, it is called an LR—the R stands for right. If the opening is facing upward, this type of conduit body is called an LB—the B stands for back. If there is an opening on both sides, it is called an LRL—for both left and right. The other popular shapes are named for their letter look-alikes, that is, T and X. The only exception is the C conduit body. Let us take a closer look at each of these.

### 3.1.0 Type C Conduit Body

A Type C conduit body (*Figure 5*) may be used to provide a pull point in a long conduit run or a conduit run that has bends totaling more than 360° (see *NEC Sections 342.26, 344.26, and 358.26*). In this application, the Type C conduit body is used as a pull point.

### 3.2.0 Type L Conduit Bodies

A Type L conduit body (*Figure 6*) is used as a pulling point for conduit that requires a 90° change in direction. (Again, the letter L is short for elbow.) To use a Type L conduit body, the cover is removed, the wire is pulled out and coiled on the ground (or floor), and then it is reinserted into the other opening and pulled. Type L conduit bodies are available with the cover on the back (Type LB), on the sides (Type LL and Type LR), or

205F06.EPS

*Figure 6*  Type L conduit bodies.

on both sides (Type LRL). The cover and gasket for conduit body fittings must be ordered separately; do not assume that these parts come with the conduit body when it is ordered.

### 3.3.0 Type T Conduit Body

A Type T conduit body, also known as a tee, is used to provide a junction point for three intersecting conduits and low point drains (*Figure 7*). Tees are used extensively in rigid conduit systems. The cost of a tee conduit body is more than twice that of a standard 4" square box with a cover. Therefore, the use of Type T conduit bodies with EMT is limited.

205F05.EPS

*Figure 5*  Type C conduit body.

205F07.EPS

*Figure 7*  Type T conduit body.

NCCER — *Contren® Learning Series*  26205-11

### 3.4.0 Type X Conduit Body

A Type X conduit body is used to provide a junction point for four intersecting conduits. The removable cover provides access to the interior of the X so that wire pulling and splicing may be performed (see *Figure 8*).

205F08.EPS

*Figure 8*    Type X conduit body.

### 3.5.0 FS and FD Boxes

FS boxes are cast boxes available in single-gang, two-gang, and three-gang configurations. They are sized to permit the installation of switches and receptacles. Covers for switches and receptacles are available for FS boxes that have formed openings much like switch and receptacle plates. FD boxes are similar to FS boxes. The letter D in the FD box indicates it is a deeper box (2½" deep versus 1⅝" deep for an FS box). Neither FS nor FD boxes are considered by the *NEC*® to be conduit bodies. FS and FD boxes may be used in environments defined by NEMA 1 (dry, clean environments); NEMA 3R (outdoor, wet environments); and NEMA 12 (dusty, oily environments).

Engineers specify and electricians install FS and FD boxes for a reason. Never alter these boxes by drilling mounting holes in them. Most are provided with cast-in mounting eyes for this purpose. Mount these boxes only as recommended by the manufacturer.

### 3.6.0 Pulling Elbows

Pulling elbows are used exclusively for pulling wire at a corner point of a conduit run. The volume of a pulling elbow is too low to permit splicing wire. See *Figure 9(A)*.

### 3.7.0 Entrance Ells (SLBs)

An entrance ell, or SLB, is built with an offset so that it may be attached directly to the surface that is to have a conduit penetration. A cover on the back of the SLB permits wire to be pulled out and reinserted into the conduit that penetrates the support surface. See *Figure 9(B)*.

(B) ENTRANCE ELL

(A) PULLING ELBOW

205F09.EPS

*Figure 9*  Elbows.

### 3.8.0 Moguls

**Mogul** conduit bodies (*Figure 10*) are available in the same types as standard conduit bodies (Type L, Type T, and so on). They have raised covers to provide better access to large conductors during conductor installation or maintenance. Moguls also allow right angle bends where splices, pulls, and taps are needed in a weatherproof chamber. Like regular boxes, they must comply with the bending space requirements of *NEC Section 314.28*. Larger moguls may contain built-in cable-pulling rollers to facilitate installation.

### 4.0.0 HANDHOLES

According to *NEC Article 100*, a handhole (*Figure 11*) is "an enclosure identified for use in underground systems, provided with an open or closed bottom, and sized to allow personnel to reach into, but not enter, for the purpose of installing, operating, or maintaining equipment or wiring or both." Handholes are often used with underground PVC conduit for the installation of landscape lighting, light poles, traffic lights, and other applications. Because handholes are exposed to various levels of foot traffic and/or vehicular loading, it is critical to select the correct enclosure for the application. *NEC Section 314.30,*

205F10.EPS

*Figure 10*  Mogul.

*Informational Note* refers to *ANSI/SCTE 77-2002, Specification for Underground Enclosure Integrity,* for additional information on loading for underground enclosures. Some of the *NEC®* requirements for handholes are as follows:

- Like boxes and conduit bodies, handholes must be installed in such a way that the enclosed wiring remains accessible without removing any portion of the building, such as wall coverings, or without requiring excavation, such as for enclosures serving underground wiring installations. See *NEC Section 314.29.*
- Per *NEC Section 314.30(A),* handhole enclosures must be sized in accordance with *NEC Section 314.28(A)* for conductors operating at 600 volts or below, and in accordance with *NEC Section 314.71* for conductors operating at over 600 volts.
- Underground raceways and cable assemblies entering a handhole enclosure must extend into the enclosure, but they are not required to be mechanically connected to the enclosure per *NEC Section 314.30(B).*
- Where handhole enclosures without bottoms are installed, all enclosed conductors and any splices or terminations, if present, must be listed as suitable for wet locations per *NEC Section 314.30(C).*
- Handhole enclosure covers shall have an identifying mark or logo that prominently identifies the function of the enclosure, such as ELECTRIC (*Figure 11*) or TRAFFIC SIGNALS (*Figure 12*). To discourage unauthorized entry, handhole enclosure covers must either weigh over 100 pounds or require the use of tools to open.
- Per *NEC Section 314.30(D),* metal covers and other exposed conductive surfaces must be bonded in accordance with *NEC Sections 250.92 and 250.96(A).*

### 4.1.0 Handhole Construction

Handhole enclosures may be constructed of a variety of materials, including precast portland cement concrete, precast polymer concrete, thermoplastic, fiberglass-reinforced resin, steel, and aluminum. The type of material selected depends on the environment and the expected load.

> **CAUTION**
>
> Some enclosures use dissimilar materials for the body and cover of the enclosure. Dissimilar materials are likely to have different responses to temperature and pressure variations, which may result in cracking or other performance failures.

205F11.EPS

*Figure 11*   Handhole.

### 4.2.0 ANSI/SCTE Requirements

According to *ANSI/SCTE 77-2002,* handhole enclosures must be constructed in such a way that they withstand all loads likely to be imposed and remain safe and reliable for the intended

205F12.EPS

*Figure 12*   Handhole containing traffic signal wiring.

an adequate number of concentric knockouts. However, some may not have precut knockouts or the ones that are available may not be in the required location. In these cases, a knockout punch must be used to make a hole for the conduit connection. A hand-operated knockout punch is shown in *Figure 22*.

To use the knockout punch, the center of the hole is located in the box or enclosure and marked with a center punch. A pilot hole is then drilled to accept the threaded drive bolt of the knockout punch. The punch is separated from the drive screw, the screw is then placed through the pilot hole with the die on one side of the box wall, and the punch is screwed onto the drive screw on the opposite side of the wall. The punch is then aligned and screwed onto the drive screw hand-tight, in which case the punch should lightly bite into the wall of the enclosure. Finally, a wrench is then used to tighten the drive nut until the punch is drawn through the enclosure wall, making a neat circular opening.

Where many such openings must be made, or when holes for the larger sizes of conduit must be cut, power knockout tools are often used to speed up the process. The battery-powered knockout kit shown in *Figure 23* can be used to punch up to 2" conduit, while the hydraulic kit in *Figure 24* can be used to punch up to 4" conduit.

## Step Bits

Step bits can be used to quickly create holes for smaller size conduit.

205SA05.EPS

205F23.EPS

*Figure 23*   Battery-powered knockout kit.

205F22.EPS

*Figure 22*   Knockout punch kit.

205F24.EPS

*Figure 24*   Hydraulic knockout kit.

# Summary

Pull and junction boxes play an important role in conductor installation and maintenance. They are installed at various locations in a conduit run to provide additional pull points to facilitate conductor installation or to provide access for system service and maintenance. *NEC Article 314* provides details on the proper sizing and selection of pull and junction boxes. The rules for systems carrying over 600V are much more stringent than those for systems operating at lower voltages. Always refer to the latest edition of the *NEC®* as well as the manufacturer's instructions for the proper application and installation of pull and junction boxes.

Conduit bodies provide access to the electrical system at junction and termination points.

Moguls are a special type of conduit body with a raised cover to provide extra space when using larger conductors.

Handholes are underground enclosures that typically serve outdoor lighting systems and traffic signal wiring. Because they are subject to varying load conditions and may become energized if they fail, it is essential to select and install them for the specific environment to which they will be exposed.

Fittings are required to connect a raceway to a pull or junction box. The type of fitting selected depends on the type of conduit used and the application.

1. Where in the *NEC*® will you find the most information on pull and junction boxes?

   a. *NEC Article 240*
   b. *NEC Article 250*
   c. *NEC Article 314*
   d. *NEC Article 320*

2. When calculating the pull box size for straight pulls on systems operating at less than 600V, the length of the box must not be less than _____ times the trade diameter of the largest raceway.

   a. two
   b. four
   c. six
   d. eight

3. If the largest trade diameter of a raceway entering a pull box is 3", and it is a straight pull for a system operating at less than 600V, the minimum size box allowed is _____.

   a. 20"
   b. 24"
   c. 30"
   d. 36"

4. For a straight pull, the length of a pull box housing unshielded conductors carrying over 600V must be _____ times the outside diameter of the largest unshielded conductor entering the box.

   a. 24
   b. 32
   c. 48
   d. 60

5. For a straight pull, the length of a pull box housing shielded conductors carrying over 600V must be _____ times the outside diameter of the largest shielded conductor entering the box.

   a. 24
   b. 32
   c. 48
   d. 60

6. True or False? In addition to a suitable cover, pull boxes containing conductors over 600V must be provided with at least one removable side.

   a. True
   b. False

7. A conduit body that has openings on four different sides plus an access opening is called a _____.

   a. Type X
   b. Type C
   c. Type T
   d. Type LL

8. Which of the following best describes how Type FS boxes should be installed?

   a. Holes should be drilled in back of the box for mounting screws.
   b. Holes should be drilled on the sides of the box only for mounting screws.
   c. No holes should be drilled in the box for mounting.
   d. Holes may be drilled only in existing installations.

9. Which of the following enclosures is best suited for splicing conductors?

   a. Pulling ell
   b. SLB
   c. Conduit body
   d. Mogul

10. The combination coupling in *Figure 1* is used to connect _____.

   a. two sections of flexible conduit
   b. flexible conduit to EMT
   c. EMT to rigid conduit
   d. flexible conduit to rigid conduit

205RQ01.EPS

**Figure 1**

1. A Type T conduit body is used to provide a junction point for _____ intersecting conduits.

2. All boxes must be sized, installed, and supported to meet the current _____ requirements.

3. Class 1 locations contain flammable _____ or _____.

4. Conduit bodies, also called _____, provide access through a removable cover or covers to the interior of the system.

5. To distinguish between Type LL and Type LR conduit bodies, hold the fitting as if it were a(n) _____ and see where the cover is situated.

6. For a straight pull, the minimum length of the box must be at least _____ times the trade diameter of the largest raceway.

7. The minimum length for a junction box in which the conductors are pulled at an angle and that contains two 3" conduits and one 4" conduit entering each side is _____.

8. A(n) _____ locknut may be required if bonding jumpers are used inside of the box or enclosure.

9. The minimum length for a junction box with one 4" conduit on each side in which the conductors are pulled straight through the box is _____.

10. When sizing pull and/or junction boxes for systems over 600V, you need to know the outside diameter of the _____.

# Trade Terms Introduced in This Module

**Conduit body:** A separate portion of a conduit or tubing system that provides access through a removable cover (or covers) to the interior of the system at a junction of two or more sections of the system or at a terminal point of the system. Boxes such as FS and FD or larger cast or sheet metal boxes are not classified as conduit bodies.

**Explosion-proof:** Designed and constructed to withstand an internal explosion without creating an external explosion or fire.

**Handhole:** An enclosure used with underground systems to provide access for installation and maintenance.

**Junction box:** An enclosure where one or more raceways or cables enter, and in which electrical conductors can be, or are, spliced.

**Mogul:** A type of conduit body with a raised cover to provide additional space for large conductors.

**Pull box:** A sheet metal box-like enclosure used in conduit runs to facilitate the pulling of cables from point to point in long runs, or to provide for the installation of conduit support bushings needed to support the weight of long riser cables, or to provide for turns in multiple-conduit runs.

**Raintight:** Constructed or protected so that exposure to a beating rain will not result in the entrance of water under specified test conditions.

**Waterproof:** Constructed so that moisture will not interfere with successful operation.

**Watertight:** Constructed so that water will not enter the enclosure under specified test conditions.

**Weatherproof:** Constructed or protected so that exposure to the weather will not interfere with successful operation.

# Additional Resources

This module presents thorough resources for task training. The following resource material is suggested for further study.

*National Electrical Code® Handbook*, Latest Edition. Quincy, MA: National Fire Protection Association.

# Figure Credits

# CONTREN® LEARNING SERIES — USER UPDATE

NCCER makes every effort to keep its textbooks up-to-date and free of technical errors. We appreciate your help in this process. If you find an error, a typographical mistake, or an inaccuracy in NCCER's Contren® materials, please fill out this form (or a photocopy), or complete the online form at www.nccer.org/olf. Be sure to include the exact module number, page number, a detailed description, and your recommended correction. Your input will be brought to the attention of the Authoring Team. Thank you for your assistance.

*Instructors* – If you have an idea for improving this textbook, or have found that additional materials were necessary to teach this module effectively, please let us know so that we may present your suggestions to the Authoring Team.

### NCCER Product Development and Revision
3600 NW 43rd Street, Building G, Gainesville, FL 32606

**Fax:** 352-334-0932
**Email:** curriculum@nccer.org
**Online:** www.nccer.org/olf

☐ Trainee Guide    ☐ AIG    ☐ Exam    ☐ PowerPoints    Other _____

Craft / Level: _____ Copyright Date: _____

Module Number / Title: _____

Section Number(s): _____

Description: _____

_____

_____

_____

_____

Recommended Correction: _____

_____

_____

_____

_____

Your Name: _____

Address: _____

_____

Email: _____ Phone: _____

# Conductor Installations

## U.S. Department of Transportation Headquarters

Clark Construction Group built the new headquarters for the U.S. Department of Transportation (DOT) in Washington, D.C. on an 11-acre parcel. The 2,000,000-square-foot office complex spans two city blocks.

## 26206-11

Trainees with successful module completions may be eligible for credentialing through NCCER's National Registry. To learn more, go to **www.nccer.org** or contact us at **1.888.622.3720.** Our website has information on the latest product releases and training, as well as online versions of our *Cornerstone* newsletter and Pearson's Contren® product catalog.

Your feedback is welcome. You may email your comments to **curriculum@nccer.org,** send general comments and inquiries to **info@nccer.org,** or use the User Update form at the back of this module.

# CONDUCTOR INSTALLATIONS

## Objectives

When you have completed this module, you will be able to do the following:

1. Explain the importance of communication during a cable-pulling operation.
2. Plan and set up for a cable pull.
3. Set up reel stands and spindles for a wire-pulling installation.
4. Explain how mandrels, swabs, and brushes are used to prepare conduit for conductors.
5. Properly install a pull line for a cable-pulling operation.
6. Explain how and when to support conductors in vertical conduit runs.
7. Describe the installation of cables in cable trays.
8. Calculate the probable stress or tension in cable pulls.

## Performance Tasks

Under the supervision of the instructor, you should be able to do the following:

1. Prepare multiple conductors for pulling in a raceway system.
2. Prepare multiple conductors for pulling using a wire-pulling basket.

## Trade Terms

| | | |
|---|---|---|
| Basket grip | Conductor support | Setscrew grip |
| Cable grip | Conduit piston | Sheave |
| Capstan | Fish line | Soap |
| Clevis | Fish tape | |

## Required Trainee Materials

1. Paper and pencil
2. Appropriate personal protective equipment
3. Copy of the latest edition of the *National Electrical Code*®

Note: *NFPA 70*®, *National Electrical Code*®, and *NEC*® are registered trademarks of the National Fire Protection Association, Inc., Quincy, MA 02269. All *National Electrical Code*® and *NEC*® references in this module refer to the 2011 edition of the *National Electrical Code*®.

# Contents

*Topics to be presented in this module include:*

## Figures and Tables

# 1.0.0 INTRODUCTION

In most cases, the installation of conductors in raceway systems is merely routine. However, there are certain practices that can reduce labor and materials and help prevent damage to the conductors. The use of modern equipment, such as vacuum fish tape systems, is one way to reduce labor during this phase of the wiring installation.

There are three types of fish tape: steel, nylon, and fiberglass. They also come in different weights for various applications. The proper size and length of the fish tape, as well as the type, should be one of the first considerations. For example, if most of the runs between branch circuit outlets are 20' or less, a short fish tape of 25' will easily handle the job and will not have the weight and bulk of a larger tape. When longer runs are encountered, the required length of the fish tape should be enclosed in one of the metal or plastic fish tape reels. This way, the fish tape can be rewound on the reel as the pull is being made to avoid having an excessive length of tape lying around on the floor or deck.

> **WARNING!**
>
> Never fish a steel tape through or into enclosures or raceways containing an energized conductor.

When several bends are present in the raceway system, the insertion of the fish tape may be made easier by using flexible fish tape leaders on the end of the fish tape.

The combination blower and vacuum fish tape systems are ideal for use on long runs and can save time. Basically, the system consists of a tank and air pump with accessories. An electrician can vacuum or blow a line or tape in any size conduit from ½" through 4", or even up to 6" conduit with optional accessories.

After the fish tape is inserted in the raceway system, the conductors must be firmly attached by some approved means. On short runs, where only a few conductors are involved, all that is necessary is to strip the insulation from the ends of the wires, bend these ends around the hook in the fish tape, and securely tape them in place. Where several wires are to be pulled together, the wires should be staggered and the fish tape securely taped at the point of attachment so that the overall diameter is not increased any more than is absolutely necessary.

Basket grips (*Figure 1*) are available in many sizes for almost any size and combination of conductors. They are designed to hold the conductors firmly to the fish tape and can save time and trouble that would be required when taping wires.

In all but very short runs, the wires should be lubricated with wire lubricant prior to attempting the pull, as well as during the pull. Some of this

---

**On Site**

## Nonconductive Fish Tape

When fishing conductors in a conduit or raceway that already contains existing energized (live) conductors, the safest method is to turn off and lock out/tag the power sources for all the live conductors. However, in some rare instances, fishing a conductor through a conduit or raceway containing other live conductors may be unavoidable. In this case, always request your supervisor's approval before proceeding and always use a nonconductive fish tape made of nylon or fiberglass.

206SA01.EPS

---

lubricant should also be applied to the inside of the conduit.

Wire dispensers are great aids in keeping the conductors straight and facilitating the pull. Many different types of wire dispensers are available to handle virtually any size spool of wire or cable. Some dispensers are stationary, while others have casters that make it easy to move heavy spools to the pulling location (*Figure 2*). Wheeled dispensers are sometimes called wire caddies.

206F01.EPS

*Figure 1*   Basket grip.

**CABLE DISPENSER**

**WIRE CADDY**

206F02.EPS

*Figure 2*   Wire dispensers.

## 2.0.0 PLANNING THE INSTALLATION

The importance of planning any wire-pulling installation cannot be over-stressed. Proper planning will make the work go easier and labor will be saved.

Large sizes of conductors are usually shipped on reels, involving considerable weight and bulk. Consequently, setting up these reels for the pull, measuring cable run lengths, and similar preliminary steps will often involve a relatively large amount of the total cable installation time. Therefore, consideration must be given to reel setup space, proper equipment, and moving the cable reels into place.

Whenever possible, the conductors should be pulled directly from the shipping reels without pre-handling them. This can usually be done through proper coordination of the ordering of the conductors with the job requirements. While doing so requires extremely close checking of the drawings and on-the-job measurements (allowing for adequate lengths of conductors in pull boxes, elbows, troughs, connections, and splices), the extra effort is well worth the time to all involved.

When the lengths of cable have been established, the length of cable per reel can be ordered so that the total length per reel will be equal to the total of a given number of raceway lengths, and the reel so identified.

In most cases, the individual cables of the proper length for a given number of runs are reeled separately onto two or more reels at the factory, depending on the number of conductors in the runs.

When individual conductors are shipped on separate reels, it is necessary to set up for the same number of reels as the number of conductors to be pulled into a given run, as shown in *Figure 3*.

As an extra precaution against error in calculating the lengths of conductors involved, it is a good idea to actually measure all runs with a fish tape before starting the cable pull, adding for makeup to reach the terminations, and accounting for discarding the cable underneath the pulling sleeve or pulling wrap, as it will have been excessively stressed during the pull. Check these totals against the totals indicated on the reels. Under normal cable delivery schedules, when the feeder raceways have been installed at a relatively early stage of the overall building construction, it may not delay the final completion of the electrical installation to delay ordering the cables until the raceways can actually be measured.

When pulling conductors directly from the reels, care must be taken that each given run be

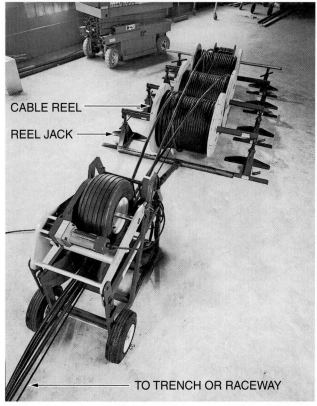

CABLE REEL

REEL JACK

TO TRENCH OR RACEWAY

206F03.EPS

*Figure 3*   Multi-cable pull.

cut off from the reel so that there is a minimum amount of waste. In other words, preclude the possibility of the final run of cable taken from the reel being too short for that run.

## 2.1.0 Pulling Location

Each job will have to be judged separately as to the best location of pulling setups; the number of setups should be reduced to a minimum in line with the best direction of the pull. The pulling location for a particular job is determined by the weight of the cable, height of the pull, practicality of moving the equipment to the pulling location, number of setups required, as well as the location of any bends. Also, a separate setup might have to be made at the top of each rise, whereas a single setup might be made at a ground floor pull box location from which several feeders are served with the same size and number of conductors.

### Think About It

# Multiple Reels

Why are multiple reel pulling setups used?

The location of the pulling equipment determines the number of workers required for the job. A piece of equipment that can be moved in and set up on the first floor by four workers in an hour's time may require six workers working two hours when set up in basements or parking levels of buildings. Therefore, when planning a cable pull, make certain enough workers are on hand to adequately handle the installation.

It is a simple operation for a few workers to roll cable reels from a loading platform to a first floor setup, whereas moving them to upper floors involves more handling and usually requires a crane or other hoists. In addition, the reel jacks have to be moved to the setup point when a downward pull is made, and after the pulling operation is completed, they must be taken back to the first floor.

## 2.2.0 Cable Pull Operations

These operations are performed to a lesser or greater degree in almost all cable pulls with larger sizes of conductors:

*Step 1*   Measure or re-check runs and establish communications between both ends of the pull.

*Step 2*   Provide pulling equipment.

*Step 3*   Receive and unload pulling equipment.

*Step 4*   Move pulling equipment to the pulling location.

*Step 5*   Set up and anchor pulling equipment.

*Step 6*   Remove to another location or move to a loading platform and load on a truck or forklift.

*Step 7*   Receive and store cable; may be moved directly to setup location if job conditions permit.

*Step 8*   Move to setup point.

*Step 9*   Move reel jacks and mandrel to setup point.

*Step 10*   Set up reels, then identify and tag the cables.

*Step 11*   Prepare cable ends.

*Step 12*   Install fish tape.

*Step 13*   Install pulling line or cable.

*Step 14*   Connect pulling line.

*Step 15*   Lubricate with proper lubricants.

*Step 16*   Pull cable.

***Step 17*** Disconnect pulling line.

***Step 18*** Remove reels.

***Step 19*** Permanently identify both ends of the cables/conductors.

***Step 20*** Rack cables in pull boxes and troughs.

***Step 21*** Splice or connect cables.

***Step 22*** Check and test.

In some instances, the following additional operations are involved, depending upon the exact details of the project. Other items of importance will be discussed later in this module.

***Step 23*** Remove lagging or other protective covering from reels.

***Step 24*** Unreel cable and cut it to length.

***Step 25*** Re-reel cable for pulling.

***Step 26*** Replace lagging on reels.

***Step 27*** Operate such auxiliary equipment as guide-through cabinets or pull boxes and signal systems between reel and pulling setups.

Each of these pulling operations is discussed in detail in this module (also see *Figure 4*).

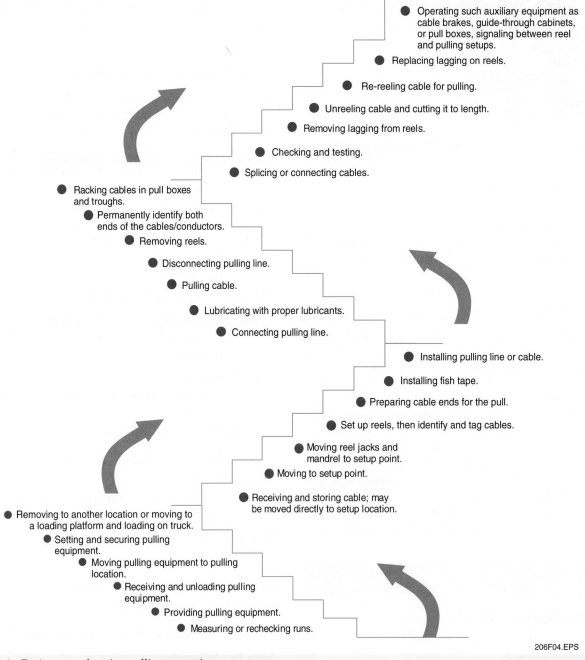

- Operating such auxiliary equipment as cable brakes, guide-through cabinets, or pull boxes, signaling between reel and pulling setups.
- Replacing lagging on reels.
- Re-reeling cable for pulling.
- Unreeling cable and cutting it to length.
- Removing lagging from reels.
- Checking and testing.
- Splicing or connecting cables.
- Racking cables in pull boxes and troughs.
- Permanently identify both ends of the cables/conductors.
- Removing reels.
- Disconnecting pulling line.
- Pulling cable.
- Lubricating with proper lubricants.
- Connecting pulling line.
- Installing pulling line or cable.
- Installing fish tape.
- Preparing cable ends for the pull.
- Set up reels, then identify and tag cables.
- Moving reel jacks and mandrel to setup point.
- Moving to setup point.
- Receiving and storing cable; may be moved directly to setup location.
- Removing to another location or moving to a loading platform and loading on truck.
- Setting and securing pulling equipment.
- Moving pulling equipment to pulling location.
- Receiving and unloading pulling equipment.
- Providing pulling equipment.
- Measuring or rechecking runs.

206F04.EPS

*Figure 4* Basic steps of a wire-pulling operation.

## Measuring Tape

A waterproof polyester tape with permanent markings every foot from 0' to 3,000' is available on reels for use in measuring conduit/raceway runs. It can be fished through the run manually or using a power fishing system.

206SA02.EPS

## 3.0.0 SETTING UP FOR WIRE PULLING

As mentioned previously, much planning is required for pulling the larger sizes of conductors in raceways. There are several preliminary steps required before the actual pull begins.

The proper use of appropriate equipment is crucial to a successful cable installation. The equipment needed for most installations is shown in the checklist in *Figure 5*. Some projects may require all of these items, while others may require only some of them. Each cable-pulling project must be taken on an individual basis and analyzed accordingly. Seldom will two pulls require identical procedures.

> **NOTE**
> Think of everything that can go wrong and take every precaution.

### 3.1.0 Setting Up the Cable Reels

When reels of cable arrive at the job site, it is best to move them directly to the setup location if at all possible. This prevents having to handle the reels more than necessary. However, if this is not practical, arrangements must be made for storage until the cable is needed.

The exact method of handling reels of cable depends on their size and the available tools and equipment. In many cases, the reels may be rolled to the pulling location by one or more workers. For reels up to 24" wide × 40" in diameter, a cable reel transporter can be used to transport the cable reel; it also acts as a dispenser during the pulling operation. When available, a forklift is ideal for lifting and transporting cable reels. See *Figure 6*.

### EQUIPMENT CHECKLIST

| | |
|---|---|
| ❑ PORTABLE ELECTRIC GENERATOR | ❑ SEVERAL WIRE ROPE SLINGS OF VARIOUS LENGTHS |
| ❑ EXTENSION CORDS AND GFCI | ❑ SHACKLES/ROPE CLEVIS |
| ❑ PUMP, DIAPHRAGM | ❑ GANG ROLLERS WITH AT LEAST 4' EFFECTIVE RADIUS |
| ❑ MAKEUP BLOWER AND HOSE | ❑ HAND WINCHES |
| ❑ MANHOLE COVER HOOKS | ❑ MANHOLE EDGE SHEAVE |
| ❑ WARNING FLAGS, SIGNS | ❑ PULLING ROPE |
| ❑ ELECTROSTATIC kV TESTER | ❑ SWIVELS |
| ❑ ELECTRIC SAFETY BLANKETS AND CLAMPS | ❑ BASKET GRIP PULLERS |
| ❑ RADIOS OR TELEPHONES | ❑ 0-1/5/10 KIP DYNAMOMETER |
| ❑ GLOVES | ❑ REEL ANCHOR |
| ❑ FLOOD LAMPS | ❑ REEL JACKS |
| ❑ FISH TAPE OR STRING BLOWER/VACUUM | ❑ CABLE CUTTERS |
| ❑ HAND LINE | ❑ LINT-FREE RAGS |
| ❑ DUCT-CLEANING MANDRELS | ❑ CABLE-PULLING LUBRICANTS |
| ❑ DUCT-TESTING MANDRELS | ❑ PRELUBING DEVICES |
| ❑ CAPSTAN-TYPE PULLER | ❑ PLYWOOD SHEETS |
| ❑ SNATCH BLOCKS | ❑ DIAMETER TAPE |
| ❑ SHORT ROPES FOR TEMPORARY TIE-OFFS | ❑ 50' MEASURING TAPE |
| ❑ GUIDE-IN FLEXIBLE TUBING (ELEPHANT TRUNKS) | ❑ SILICONE CAULKING (TO SEAL CABLE ENDS) |

206F05.EPS

*Figure 5* Wire-pulling equipment checklist.

CABLE REEL TRANSPORTER                    FORKLIFT

206F06.EPS

*Figure 6*  Two methods of transporting cable reels.

However, for very large reels (48" or more in diameter), a crane or similar hoisting apparatus is usually necessary for lifting the reels onto reel jacks supported by jack stands to acquire the necessary height. *Figure 7* shows a summary of proper and improper ways to transport reels of wire or cable on the job site.

*Figure 8* shows several types of reel jacks, including the spindle. For a complete setup, two stands and a spindle are required for each reel. The reel jacks or stands are available in various sizes from 13" to 54" high to accommodate reel diameters up to 96". Extension stands used in conjunction with reel stands can accommodate larger reels.

Reel-stand spindles are commonly available in diameters from 2⅜" to 3½" and from 59" to 100" in length for carrying reel loads up to 7,500 pounds. However, some heavy-duty spindles are rated for loads up to 15,000 pounds.

### 3.2.0 Preparing Raceways for Conductors

Another preliminary step prior to pulling conductors in raceway systems is to inspect the raceway itself. Few things are more frustrating than to pull four 1,000 kcmil conductors through a conduit and find out when the pull is almost done that the conduit is blocked or damaged. Such a situation usually requires pulling the conductors back out, repairing the fault, and starting all over again.

These problems are not too serious with exposed banks of conduit, as the conduit can usually be separated at the fault, the fault corrected, and another piece of conduit installed, using unions if necessary. However, in underground conduit runs—especially those encased in concrete—the situation can be both time-consuming and costly.

A test pull will detect any hidden obstructions in the conduit prior to the pull (*Figure 9*). Go and no-go steel and aluminum mandrels are available for pulling through runs of conduit before the cable installation. Mandrels should be approximately 80% of the conduit size (twice the 40% fill factor). If any obstructions are found, they can be corrected before wasting time on an installation that might result in conductor damage and the possibility of having to re-pull the conductors.

*Figure 10* shows several devices used to inspect raceway systems, as well as to prepare raceways for easier and safer conductor pulls. The conduit swab in *Figure 10* may be used ahead of the conductors during a pull. Its main purpose is to swab out debris from the raceway and spread a uniform film of pulling compound inside the conduit for easier pulling. The flexible mandrel is also used to clean the conduit and check the run for obstructions. The flexible discs make it easier to pull around tight bends.

The conduit brush in *Figure 10* helps clean and polish the interior of conduit before pulling the cable. Such brushes remove sand and other light obstructions. Note that this brush has a pulling eye on one end and a twisted eye on the opposite end, enabling it to be pushed or pulled through the conduit.

ON THE RIMS OF THE SPOOL (MOVING
EQUIPMENT DOES NOT COME INTO
CONTACT WITH CABLE)

ON THE FLAT SIDE OF THE SPOOL OR
ON THE CABLE (MOVING EQUIPMENT
COMPRESSES INSULATION AND MAY
DAMAGE CABLE)

206F07.EPS

*Figure 7*  Proper and improper ways of transporting cable reels.

SPINDLE

SCREW-TYPE
REEL STANDS

RATCHET-TYPE
REEL STAND

206F08.EPS

*Figure 8*　Typical reel stands.

GO OR NO-GO
MANDREL

BACKHOE
TOOTH MARK

RACEWAY

FLATTENED
BEND

CONCRETE
SEDIMENT

206F09.EPS

*Figure 9*　Faults that may be detected with a conduit mandrel.

LUBE SPREADER
SWAB

CONDUIT BRUSH

FLEXIBLE MANDREL

206F10.EPS

*Figure 10*　Devices used to inspect, clean, and lubricate raceway systems.

One final step before starting the pull is to measure the length of the raceway, including all turns in junction boxes and the like. A fish tape may be pushed through the raceway system and a piece of tape used to mark the end. When it is pulled back out, a tape measure may be used to measure the exact length. An easier way, however, is to use a power fishing system to push or pull a measuring tape through the conduit run. Details of this operation are explained in the next section.

When measuring the conductor length, be sure to allow sufficient room where measurements are made through a pull box. Conductors should enter and leave pull boxes in such a manner as to allow the greatest possible sweep for the conductors. Large conductors are especially difficult to bend but with proper planning, you can simplify the feeding of these conductors from one conduit to another.

For example, if a conduit run makes a right-angle bend through a pull box, the conduit for a given feeder should come into the box at the lower left-hand corner and leave diagonally opposite at the upper right-hand corner, as shown in *Figure 11*. This gives the conductors the greatest possible sweep with the box, eliminating sharp bends and consequent damage to the conductor insulation.

Runs should also be calculated to allow for splices and terminations in junction boxes, panelboards, motor-control centers, the cable discarded under the sleeve or pulling wrap, etc.

### 3.3.0 Installing the Pull Line

At one time, pull-in wires were frequently placed in conduit runs as the raceways were installed. However, in recent times, with modern cable-fishing equipment, this practice is seldom used.

Pull lines are sometimes manually fished in with a steel fish tape, but much time can be saved by using a blower/vacuum fish tape system. In general, a **conduit piston**—sometimes referred to as a mouse or missile—is blown with air pressure or vacuumed through the run. The foam piston is sized to the conduit and has a loop on both ends. In most cases, **fish line** or measuring tape is attached to the piston as it is blown or vacuumed through the conduit run. The measuring tape serves two purposes: it provides an accurate measurement of the conduit run, and the tape is used to pull the cable-pulling rope into the conduit run. In some cases, if the run is suitable, the pulling rope is attached directly to the piston and vacuumed into the run. *Figure 12* shows a blower/vacuum fish tape system being used to vacuum a pull line in a conduit while *Figure 13* shows the same apparatus blowing the piston through. Most of these units

provide enough pressure to clean dirt or water from conduit during the fishing operation. *Figure 14* shows two types of pistons used with this system. The one on the right utilizes air-guide vanes to prevent the piston from tumbling inside the larger sizes of conduit.

### 3.4.0 Preparing Cable Ends for Pulling

The pulling-in line or cable must be attached to the cable or conductors in such a manner that it cannot part from the cable during the pull. Two common methods include direct connection with the cable conductors themselves and connection by means of pulling grips or baskets placed over the cable or group of conductors. The use of the proper type of grip or basket will facilitate the pull, but in many cases—especially on long pulls—workers prefer to use three- or four-hole **cable grips** with setscrews that secure each conductor to the pulling block. *Figure 15* shows several types of pulling grips.

Most pulling blocks have a rope **clevis** as an integral part of the block. However, when using pulling grips or baskets in 2" or larger conduit, a rope clevis is normally used to facilitate connecting the pulling rope to the wire grip. Two types are currently used: the straight clevis and the swivel clevis.

#### 3.4.1 Stripping the Cable Ends

When the type of grip being used requires that the ends of the cable insulation be stripped from the conductors, conventional methods are used—the same as for terminating conductors for splices or connections to terminal lugs in panelboards, switchgear, etc.

In general, the ends of conductors should first be trimmed. Cable cutters capable of cutting conductors through 1,000 kcmil save workers time over using a hacksaw. There are also cable strippers, adjustable from 1/0 AWG through 1,000 kcmil, that handle midspan and termination stripping of THHN, THWN, XHHW, and similar insulation. These tools are excellent for stripping conductors for use in setscrew clamp-type pulling grips.

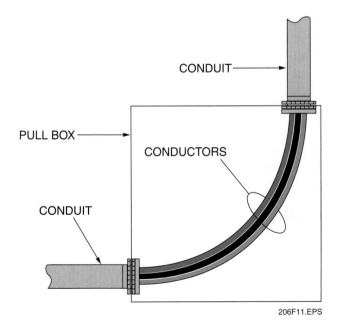

Figure 11 Obtaining the greatest possible conductor sweep in a pull box.

206F11.EPS

---

206F12A.EPS

206F12B.EPS

206F12C.EPS

*Figure 12* Power fishing system.

206F13.EPS

*Figure 13* Blower/vacuum fish tape system used to blow a pull line in conduit.

FLEXIBLE FOAM PISTON
FOR AIRTIGHT SEAL

PISTONS ARE AVAILABLE
IN SIZES FROM 1/2" TO 6"

FINS ARE SOMETIMES UTILIZED
ON PISTONS FOR LARGER SIZES
OF CONDUIT TO KEEP THE
PISTONS FROM TUMBLING

206F14A.EPS

206F14B.EPS

*Figure 14*  Types of pistons in common use.

To use a stripping tool, first mark the required distance from the ends of the conductors, using the pulling grip as a gauge. Close the jaws of the stripping tool on the cable and twist. These self-feeding devices ensure positive progression down the cable to any position desired. To stop stripping, apply back pressure to the stripper until a full circle has been completed.

Once the conductor ends have been stripped, insert one conductor at a time into the setscrew grip. Make sure the end of the bare conductor is firmly in place, and then tighten the setscrews with a hex wrench. Continue on to the next conductor, and so on, until all conductors are secured in the pulling grip.

### 3.4.2 Safety with Cutters

When using a stripping tool for the first time, make sure that you read and understand all instructions and warnings before using the tool.

The following should also be observed when working with cable terminators:

- Wear eye protection.
- Inspect tools before using, and replace damaged, missing, or worn parts.
- Be prepared for the unexpected. Make sure your footing and body position are such that you will not lose your balance.
- Use only the type and size material in the stated capacity.
- Do not use the tool on or near live circuits.

### 3.5.0 Types of Pulling Lines

Wire-pulling ropes have come a long way from the hemp ropes used by electricians a couple of decades ago. The most common wire-pulling ropes on the market include the following:

- Nylon
- Polypropylene

## Fish Poles

Rigid fish poles can be used in areas such as over drop-in ceilings where traditional fish tape might get caught up on ductwork or the ceiling grid. Glow-in-the-dark fish poles are also available.

FISH POLE IN USE

GLOW-IN-THE-DARK
FISH POLES

206SA03.EPS

- Multiplex polyester
- Double-braided polyester composite

The type of rope selected will depend mainly on the pulling load; that is, the weight of the cable, the length of the pull, and the total resistance to the pull. For example, Greenlee's Multiplex cable-pulling rope is designed for low-force cable pullers. It has a low stretch characteristic that makes it suitable for pulls up to 2,000 pounds. Lengths are available from 100 to 1,200 feet. Greenlee's double-braided composite rope for high-force cable pullers is designed for pulls up to 6,500 pounds.

**CAUTION**

Any equipment associated with the pull must have a working load rating in excess of the force applied during the pull. All equipment must be used and mounted in strict accordance with the manufacturer's instructions.

Care must be used in selecting the proper rope for the pull, and then every precaution must be taken to make sure that the cable-pulling force does not exceed the rope capacity. There are several reasons for this, but the main one is safety. For example, think of a 100' length of nylon rope with a 10,000-pound breaking strength. Such ropes can stretch 40' before breaking, releasing

*Figure 15* Various types of pulling grips used during conductor installation.

206F15.EPS

200,000 foot-pounds of energy in the process. Think of the damage this amount of energy could do to a raceway system and to nearby workers if the rope broke under this amount of force.

In some cases, a power blower/vacuum fish tape system is used to vacuum the piston through the raceway with the pulling rope attached. In other cases, a line is first blown or vacuumed

## Grips and Swivels

Conductors can be attached to a pulling line using various methods, including setscrew grips or basket grips. No matter what method is used, always insert a swivel of some sort between the pulling line or fish tape and the conductors to alleviate twisting of the conductors. Breakaway swivels are available that release at a specified tension to avoid damage to the conductors if excess force is applied or if the conductors get hung up during the pull.

SETSCREW GRIP

BASKET GRIP

BREAKAWAY SWIVELS

206SA04.EPS

through the conduit and then the pulling rope is attached to the line for pulling it through the raceway system. In either case, there are certain precautions that should be taken when using a power fish tape system:

- Read and understand all instructions and warnings before using the tool.
- Never fish in runs that might contain live power.
- Be prepared for the unexpected. Make sure your footing and body position are such that you will not lose your balance in any unexpected event.
- Use blower/vacuum systems only for specified light fishing and exploring the raceway system.
- Never use pliers or other devices that are not designed to pull a fish tape. They can kink or nick the tape, creating a weak spot.

### 4.0.0 CABLE-PULLING EQUIPMENT

Except for short cable pulls, hand-operated or power-operated cable pullers or winches are used to furnish the pulling power. In general, cable reels are set in place at one end of the raceway system, and the cable puller is set up at the opposite end. One end of the previously-installed pulling rope is attached to a clevis, basket, or other cable grip to which the conductors are attached. The other end of the rope (at the cable puller) is wrapped around the rotating drum capstan on the cable puller (*Figure 16*).

Wire-pulling lubricant—sometimes referred to as soap—is inserted into the empty conduit as well as wiped thoroughly onto the front of the cable. One or more operators are on hand to feed the cable, while one worker is usually all that is required on the pulling end.

**WARNING!**

Always use a wire-pulling lubricant that is compatible with the type of cable being pulled. Failure to do so can result in unsafe pulling forces and cable damage. Check with the cable manufacturer for their recommendations, and always contact the lubricant manufacturer about the compatibility of their products with specific cables. Also check the product's MSDS for any applicable safety requirements.

The number of wraps on the puller drum decides the amount of force applied to the pull. For example, the operator needs to apply only 10 pounds of force to the pulling rope in all cases. With this amount of force applied by the

# Communications

What type of communications equipment would most likely be used during cable-pulling operations?

# Monitored Cable Pullers

Many power cable pullers include force gauges with an automatic over-tension shutoff device. These units prevent you from accidentally exceeding the maximum recommended pulling tension for the conductor(s) being pulled.

206SA06.EPS

4,800 pounds. Also check the rope carefully for wear or damage prior to the pull. Remember that a rope is only as strong as its weakest point.

- A decision must be made as to the best end of the raceway for pulling. The reel setup must also be carefully placed. In general, conductors that are to be installed downward should be fed off the top of the reel and where conductors are to be fed upward, the best method is to feed from the bottom of the reel. This eliminates sharp kinks or bends in the conductors.
- The appropriate pulling equipment must be used. The equipment must be of the proper capacity for the job. Space consideration for the equipment is also important, as is the particular type of mounting.
- Having enough workers is critical. Never be caught short. This is where experienced workers earn their pay. In high-force wire pulling, experience is the best teacher.
- Safety must be foremost in everyone's mind.

When planning the cable pull, make sure that enough cable is on hand for the run. Cable may be verified while still on the reel by using a cable-length meter. Of course, the conduit run length should have already been checked, as discussed previously. Do not proceed further until everyone is assured that enough cable is on hand for all bends, sweeps in junction boxes, etc. If the cable falls short, take steps to correct the situation before continuing.

Equipment with feeding sheaves will help provide a smooth guide for the cable. Also, have sufficient lubricant on hand and use it both before and during the pull.

Select and install proper cable grips on the cable ends. Gripping must be adequate to handle the imposed force.

When pulling cables in a horizontal run, the worker simply has to reduce the amount of operator pull in order to slow down the cable pull. Releasing all operator pull stops the pulling force entirely. However, when pulling vertical runs, the rope must be tied off after stopping the pull to keep the cable from reversing in the raceway.

The entire operation should be supervised from start to finish. Therefore, communications equipment must be utilized on both ends of the pull at all times.

During the pull, make sure that the pulling rope remains free and is not wrapped around any part of the body.

A typical cable-pulling setup is shown in *Figure 20*, including cable reels, reel stands, a power cable puller, conduit system, and other accessories necessary for the pull.

### 5.1.0 The Feeding End—Sheaves and Rollers

The feed-in setup should unreel the cable along its natural curvature, as shown in *Figure 21(A)*, as opposed to a reverse S curvature, as shown in *Figure 21(B)*. Feed-in setups are shown in *Figure 22*

Figure 20 Typical cable-pulling setup.

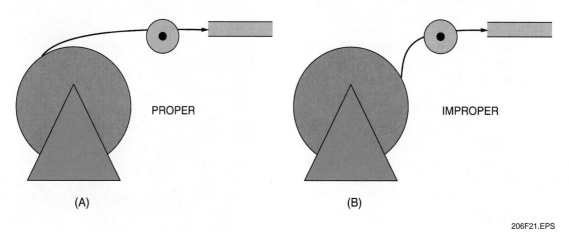

Figure 21 Unreel the cable along its natural curvature.

for manhole, underfloor duct, and overhead cable tray. Note the use of auxiliary equipment in some of these drawings, that is, cable reels, guide-in tubes, sheaves, and rollers.

Single sheaves, such as those shown in *Figure 23(A)*, may be used only for guiding cables. Multiple blocks should be arranged to hold the cable-bending radius wherever the cable is deflected. For pulling around bends, use a conveyor sheave assembly of the appropriate radius series, such as the one shown in *Figure 23(B)*.

Sheaves and pulleys must be positioned to ensure the effective curvature is smooth and deflected evenly at each pulley. Never allow a polygon curvature to occur, as shown in *Figure 24*.

## 6.0.0 SUPPORTING CONDUCTORS

Conductors in vertical raceways must be supported in accordance with *NEC Section 300.19* if the vertical rise exceeds the values in *Table 2*. In general, one conductor support must be provided at the top of the vertical raceway or as close to the top as practical. Intermediate supports must also be provided as necessary to limit supported conductor lengths to not greater than those values specified in *NEC Table 300.19(A)*.

The *NEC®* allows several different methods of supporting conductors in vertical raceways; the following are typical:

- Conductors may be supported by clamping devices constructed of or employing insulating wedges inserted in the ends of the conduit, as shown in *Figure 25*. Where clamping of insulation does not adequately support the cable, the conductor itself must also be clamped.
- Conductors may be supported by inserting boxes at the required intervals in which insulating supports are installed and secured in a satisfactory manner to withstand the weight of the conductors attached to them. The boxes must be provided with covers.

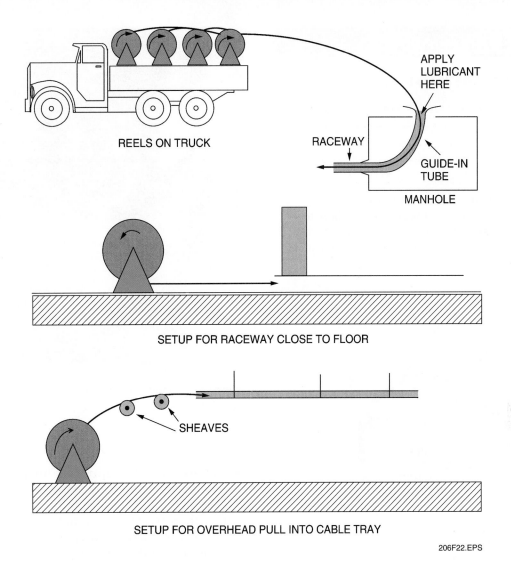

REELS ON TRUCK

APPLY LUBRICANT HERE

RACEWAY

GUIDE-IN TUBE

MANHOLE

SETUP FOR RACEWAY CLOSE TO FLOOR

SHEAVES

SETUP FOR OVERHEAD PULL INTO CABLE TRAY

206F22.EPS

*Figure 22*   Cable feed-in setups.

(A) SINGLE SHEAVES

(B) SHEAVE ASSEMBLY

206F23.EPS

*Figure 23*   Cable sheaves.

# Feed-In Setups

Why is an S-curve setup not desirable?

206F25.EPS

*Figure 25*  Conductors supported with wedges.

**NEVER ARRANGE PULLEYS LIKE THIS!**

206F24.EPS

*Figure 24*  Never allow a polygon curvature to occur in a cable-pulling operation.

**Table 2**  Spacings for Conductor Supports in Vertical Raceways [Data from *NEC Table 300.19(A)*]

| Conductor Size | Support (Aluminum or Copper-Clad Aluminum Conductor) | Support (Copper Conductor) |
|---|---|---|
| 18 AWG through 8 AWG | ≤ 100' | ≤ 100' |
| 6 AWG through 1/0 AWG | ≤ 200' | ≤ 100' |
| 2/0 AWG through 4/0 AWG | ≤ 180' | ≤ 80' |
| Over 4/0 through 350 kcmil | ≤ 135' | ≤ 60' |
| Over 350 kcmil through 500 kcmil | ≤ 120' | ≤ 50' |
| Over 500 kcmil through 750 kcmil | ≤ 95' | ≤ 40' |
| Over 750 kcmil | ≤ 85' | ≤ 35' |

- Cables may be supported in junction boxes by deflecting the cables not less than 90° and carrying them horizontally to a distance not less than twice the diameter of the cable. The cables are carried on two or more insulating supports and additionally secured to these supports by tie wires, if desired. Where this method is used, cables must be supported at intervals not greater than 20% of those mentioned in *NEC Table 300.19(A)*.
- Cables may be supported by a method of equal effectiveness.

A variety of supports are manufactured specifically for supporting cable in vertical conduit runs, and many ideas can be obtained from the manufacturers' catalogs. Therefore, make an effort to obtain such catalogs from electrical suppliers or manufacturer's representatives. In fact, manufacturers' catalogs of electrical tools and equipment are excellent study guides for any phase of the electrical industry. Manufacturers of electrical equipment want their equipment used, and they have found that one of the best ways to accomplish this is to provide easy-to-understand instructions and examples of practical applications. Most manufacturers go to great expense to provide this information, but it is usually free of charge for those working in the industry.

There are several precautionary measures that must be taken when pulling cables in vertical runs. The worst danger is runaways; that is, the weight of the cable combined with gravity exceeds the speed of the pull and falls at a rapid rate down the raceway run. Such a situation can cause injury to workers on both ends of the pull. Consequently, braking systems should be used on all long vertical cable pulls. To do otherwise is asking for trouble. Get specific instructions from your supervisor before beginning a vertical pull—either up or down.

## 7.0.0 PULLING CABLE IN CABLE TRAYS

When long lengths of cable are to be installed in raceways or cable trays, problems are frequently encountered, particularly when the cable has to be pulled directly into the tray and changes in direction of the tray sections are involved. An entire cable-pulling system has to be planned and set up so that the cable may be pulled into the trays without scuffing or cutting the sheathing and insulation, and also to avoid damaging the cable trays or the tray hangers. To accomplish a successful cable tray pull, a complete line of installation tools is available for pulling lengths of cable up to 1,000 feet or longer. These tools consist mainly of conveyor sheaves and cable rollers. *Figure 26* shows a partial cable tray system with a sheave and roller in place.

Short lengths of cable can be laid in place without tools or pulled with a basket grip. Long lengths of small cable (2" or less in diameter) can also be pulled with a basket grip. Larger cables, however, should be pulled by the conductor and the braid, sheath, or armor. This is done with a pulling eye applied at the cable factory or by tying the conductor to the eye of a basket grip and taping the tail end of the grip to the outside of the cable.

In general, the pull exerted on the cables pulled with a basket grip that is not attached to the conductor should not exceed 1,000 pounds. For heavier pulls, care should be taken not to stretch the insulation, jacket, or armor beyond the end of the conductor nor bend the ladder, trough, or channel out of shape.

The bending radius of the cable should not be less than the values recommended by the cable manufacturer, which range from four times the diameter for a rubber-insulated cable with a 1" maximum outside diameter without lead, shield, or armor, to eight times the diameter for interlocked armor cable. Cables or special construction such as wire armor and high-voltage cables require a larger radius bend.

Best results are obtained in installing long lengths of cable up to 1,000 feet with as many as a dozen bends by pulling the cable in one continuous operation at a speed of 20 to 25 feet per minute. It may be necessary to brake the reel to reduce sagging of the cables between rollers and sheaves.

The pulling line diameter and length will, of course, depend on the pull to be made and the tools and equipment available. Winch and power units must be of adequate size for the job and capable of developing the high pulling speeds required for the best and most economical results.

In general, single or multi-cable rollers are placed in the bottom of trays to protect the cable as it is pulled along. Sheaves are placed at each change of direction—either horizontally or vertically. The bottom rollers may be secured to the tray bottoms except at vertical changes in direction. Extra support is necessary at these locations to prevent damage or movement of the tray system.

Sheaves must be supported in the opposite direction of the pull. For example, all right-angle conveyor sheaves should be supported at two locations, as shown in *Figure 26*, to compensate for the pull of the cable.

SUPPORTS ARE PROVIDED IN THE OPPOSITE DIRECTION FROM THE FORCE EXERTED BY THE CABLE

CABLE EXERTS FORCE IN THESE DIRECTIONS DURING THE PULL

RIGHT-ANGLE CONVEYOR SHEAVE

CABLE ROLLER

206F26.EPS

*Figure 26*  Typical cable tray cable-pulling arrangement.

**NOTE**

Power cable pulls should not be stopped unless absolutely necessary. However, anyone associated with the pull—upon evidence of danger to either the cable or the workers—may stop a cable-pulling operation. Communication is the most important factor in these cases.

**CAUTION**

Workers feeding a cable pull must carefully inspect the cable as it is paid off the cable reel. Any visible defects in the cable at the feeding end warrants stopping the pull.

**WARNING!**

At the first sign of any type of malfunctioning equipment, broken sheaves, or other events that could present a danger to either the workers or the cable, the pull should be stopped. Make certain that all communications equipment is in proper order before starting a pull.

## 8.0.0 PHYSICAL LIMITATIONS OF CABLE

Consideration must be given to the physical limitations of a cable as it is being pulled into position. Pulling subjects cable to extreme stress and, if done improperly, can displace a cable's components. Thus, it is important that the following guidelines be observed:

- While reels are in storage—either before or after a pull—the conductor ends must be sealed to prevent moisture from entering or creeping into the cable ends.
- The minimum ambient working temperatures for cable-pulling operations depend on the cable jackets. In cold weather, cable reels should be stored in a warm area overnight so that the cable jackets will be at the proper temperature for pulling the next day.
- Calculate and stay within the cable's maximum pulling tension, maximum sidewall load, and minimum bending radii.
- Ensure that the raceway joints are aligned and that the wiring space is sufficient.
- Train the cable to avoid dragging on the edge of the raceway; also avoid laying or dragging cable on the ground.
- If using a basket grip, secure it to the cable with steel strapping and cut well behind the areas it covers once the cable is in place. The portion of the cable under the grip should be discarded.
- Ensure that the elongation of the pull rope minimizes jerking.
- Pull with a capstan and no faster than 40 feet per minute.

- Do not stop a pull unless absolutely necessary.
- Never pull the middle of the cable.
- Seal the ends with appropriate putty or silicone caulking and overwrap with tape until the conductors are terminated.

## 8.1.0 Maximum Pulling Tension

Maximum pulling tension should not exceed the smaller of the following values:

- Allowable tension on pulling device
- Allowable tension on conductor
- Allowable sidewall load

### 8.1.1 Allowable Tension on Pulling Device

Do not exceed the working load stated by the manufacturer of the pulling devices (pulling eyes, ropes, anchors, basket grips, etc.). If catalog information is not available, work at 10% of the rated braking tensile strength.

The allowable tension with a basket grip must not exceed the lbs/cmil value (as shown in *Table 3*) or 1,000 pounds, whichever is smaller. Exceptions to this rule, however, do occur, but seldom will this figure rise to over 1,250 pounds.

### 8.1.2 Allowable Tension on Conductors

The metallic phase conductors are the tensile members of the cables and should bear all of the pulling force. Never use shielding drain wires or braids for pulling. *Table 3* provides the allowable pulling tensions of various types of conductors. The listed values should never be exceeded.

Reduce the maximum pulling tension by 20% to 40% if several conductors are being pulled simultaneously since the tension is not always evenly distributed among the conductors.

**CAUTION**

When smaller conductors are pulled with large conductors, the smaller conductors may be damaged.

**Table 3** Physical Limitations of Cable

| Material | Cable Type | Temper | Lbs/Cmil |
|----------|------------|--------|----------|
| Copper | All | Soft | 0.008 |
| Aluminum | Power | Hard | 0.008 |
| Aluminum | Power | ¾ Hard | 0.006 |
| Aluminum | Power | AWM | 0.005 |
| Aluminum | URD (solid) | Soft (½ hard) | 0.003 |
| All | Thermocouple | — | 0.008 |

## 8.2.0 Calculating Pulling Tension

Normally, the maximum tension for a specific type of cable can be found using data from the cable manufacturer.

In general, the maximum tension for a single-conductor cable should not exceed 6,000 pounds. The maximum tension for two or more conductors should not exceed 10,000 pounds.

The maximum stress for leaded cables must not exceed 1,500 pounds per square inch of lead sheath area when pulled with a basket grip.

The maximum tension must not exceed 1,000 pounds per square inch of insulation area for non-leaded cables when pulled with a basket grip.

The maximum tension at a bend must not exceed 500 times the radius of curvature of the conduit or duct expressed in feet.

### 8.2.1 Tension in Horizontal Pulls

The pulling tension in a given horizontal raceway section may be calculated as follows:

For a straight section, the pulling tension is equal to the length of the duct run multiplied by the weight per foot of the cable and the coefficient of friction, which will vary depending on the type and amount of lubrication used. Therefore, the equation is as follows:

$$T = L \times w \times f$$

*Where:*

T = total pulling tension

L = length of raceway run in feet

w = weight of cable in pounds per foot

f = coefficient of friction

For ducts having curved sections, the following equation applies:

$$T_{OUT} = T_{in}\, e^{fa}$$

*Where:*

$T_{OUT}$ = tension of bend

$T_{in}$ = tension into bend

f = coefficient of friction

e = Napierian logarithm base 2.718

a = angle of bends in radians

To aid in solving the above equation, values of $e^{fa}$ for specific angles of bend and coefficients of friction are listed in *Table 4*. For more precise values, tables are available from cable manufacturers.

# Installation Aids

Other tools available to simplify the pulling of cables into cable tray systems include triple pulleys, bull wheels, and both wide and narrow rollers. Be sure to position these pulleys and rollers at the proper locations to prevent damage to the cables during the installation and also to help the installation proceed as quickly as possible. Various rollers are shown here.

**SNAP-IN SPINDLE AND CABLE ROLLER**

**CABLE ROLLER WITH ADJUSTABLE BRACKETS**

**ELBOW ROLLER**

206SA07.EPS

**Table 4**  Angle of Bend vs. Coefficients of Friction

| Angle of Bend (degrees/radians) | Values of $e^{fa}$ for Coefficients of Friction | | |
|---|---|---|---|
| | f = 0.75 | f = 0.50 | f = 0.35 |
| 15/0.2618 | 1.22 | 1.14 | 1.10 |
| 30/0.5236 | 1.48 | 1.30 | 1.20 |
| 45/0.7854 | 1.80 | 1.48 | 1.32 |
| 60/1.0472 | 2.19 | 1.68 | 1.44 |
| 90/1.5708 | 3.25 | 2.20 | 1.73 |

## 8.3.0 Sidewall Loading

Before calculating cable-pulling tension, sidewall loading or sidewall bearing pressure must be considered.

The sidewall load is the radial force exerted on a cable being pulled around a conduit bend or sheave. Excessive sidewall loading can crush a cable and is, therefore, one of the most restrictive factors in installations having bends or high tensions. *Figure 27* shows a section taken across a conduit run in a 90° bend. Note that pulling tension

## Software Programs

Software programs are now available to take the math out of calculating pulling tensions. These programs typically calculate pulling tension and sidewall pressures based on conductor size using their own data on friction coefficients. These programs also determine conduit fill, conductor configuration, jam ratios, and the amount of lubricant required for the pull.

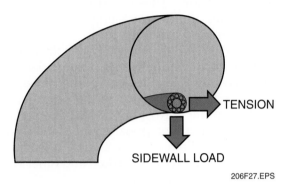

*Figure 27*   Sidewall loading.

is exerted parallel with the walls of the conduit. However, due to the 90° bend, pressure is also exerted downward against the wall of the conduit. Once again, this is known as the sidewall load. Sidewall loading is reduced by increasing the radius of bends.

In general, the sidewall load on any raceway run should not exceed 500 pounds/foot of bend radius. This pressure, however, must be reduced on some types of cables. For example, *Table 5* shows one manufacturer's recommendations for

**Table 5**   Maximum Sidewall Pressures for Various Types of Cable

| Cable Type | Sidewall Pressure in Pounds/Foot of Bend Radius |
|---|---|
| 600V nonshielded control | 300 |
| 600V and a kV nonshielded EP power | 500 |
| 5kV and 15kV EP power | 500 |
| 25kV and 35kV power | 300 |
| Interlocked armored cable (all voltages) | 300 |

the maximum sidewall pressures permitted for various types of cables. Always refer to the cable manufacturer's instructions for the type of cable being used.

### 8.4.0 Practical Applications

*Figure 28* shows a typical raceway system containing three 500 kcmil lead sheath copper conductors. Note the straight 300' run from A to B, a 45° kick, and then another 100' straight run from C to D.

To find the calculated pulling tension from D to A, refer to cable data in the manufacturer's catalogs. Suppose this cable weighs 8 pounds per foot and has a 0.141" lead sheath. The outside diameter of the three-conductor cable assembly is 3". Use 0.5 as the coefficient of friction and calculate the pulling tension from A to B (*Figure 28*) as follows:

*Step 1*   Find the tension between points A and B.

$$\text{Tension at B} = T_1 = L_1 \times w \times f$$

$L_1$ is the length between points A and B while w is the weight of the cable per foot, and f is the coefficient of friction, which is 0.5. Substituting the known values in the equation gives:

$$T_1 = 300 \times 8 \times 0.5 = 1{,}200 \text{ pounds}$$

*Step 2*   Find the tension between points B and C.

$$\text{Tension at C} = T_1\, e^{fa}$$

Since the distance between B and C involves an angle, refer to *Table 4* of this module. Looking in the left column of the table, we see that 45° equals 0.7854 radians; this figure multiplied by the coefficient of friction (0.5) equals 0.3927. Radians of angles may also be found with electronic pocket calculators if they have scientific functions. Follow the instructions in the manual accompanying the calculator. The exact key strokes will vary with the brand of calculator, but most require pressing

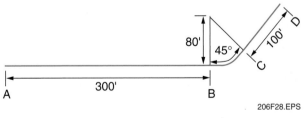

*Figure 28*   Sample conduit run.

the degree key, entering the numeral for degrees, then pressing the convert key, and finally pressing the radian or RAD key. The radians of the entered angle will be displayed.

Once again, refer to *Table 4*. Find the 45° angle of bend in the left column; read across the row to the column under 0.50—the coefficient of friction figure. The number is 1.48, the value of $e^{fa}$. Substituting these values in the equation gives the following:

$$A = 45° = 0.7854 \text{ radians}$$
$$fa = 0.7854 \times 0.5 = 0.3927$$
$$e^{fa} = 1.48$$
$$1,200 \times 1.48 = 1,776 \text{ pounds}$$

Therefore, the tension at C is 1,776 pounds.

*Step 3* Find the tension from points C to D.

$$\text{Tension from C to D} = T_2 = L_2 \times w \times f$$
$$T_2 = 100 \times 8 \times 0.5 = 400 \text{ pounds}$$

*Step 4* Find the total pulling tension by adding the figures obtained previously.

$$T = T_2 + T_1 e^{fa} = 400 + 1,776$$
$$= 2,176 \text{ pounds}$$

The maximum pulling force using a basket grip for this size cable should not exceed 1,900 pounds. Therefore, if the pull is made from point A to point D, a pulling eye will have to be used since the total pulling tension exceeds 1,900 pounds. However, if the pull is reversed—pulling from point D to point A—the total pulling tension will be reduced since the distance from point D to the 45° angle (point C) is ⅓ the distance from point A to the 45° angle (point B).

$$\text{Tension at C} = 400 \text{ pounds}$$
$$\text{Tension at B} = 400 \times 1.48 = 592 \text{ pounds}$$
$$\text{Total tension at A} = 1,200 + 592$$
$$= 1,792 \text{ pounds}$$

Therefore, if the cable is pulled from point D to point A, either a pulling eye or basket grip may be used.

> **NOTE**
>
> A lower tension is obtained by feeding the pull from the end nearest the bend.

## 9.0.0 CABLE-PULLING INSTRUMENTS

There are several instruments used in conjunction with cable-pulling operations. Since details of operation vary with the manufacturer, these instruments will only be briefly discussed in this module. Study the operation manuals for all instruments before using them.

- *Cable length meter* – Cable length meters (*Figure 29*) are available for direct reading in lengths from 2,000' to 20,000'. Most are calibrated for different wire sizes whereas the sizes are selected with a selector switch on the instrument. Controls may also be set for either copper or aluminum conductors. These instruments are ideal for determining the exact length of conductors on reels prior to making a pull.
- *Circuit tester/wire sorter* – This instrument is used to trace conductors on unenergized circuits. See *Figure 30*. One lead of the transmitter is attached to ground while the other lead is attached to the wire being traced. The receiver is then taken to the opposite end of the circuit, where it will show a strong signal on the traced conductor. That wire is marked and other wires are traced in the same way.

206F29.EPS

*Figure 29* Cable length meters.

### Case History

## Cable Pulling

At a large commercial job site recently, workers began a complex pull shortly before the end of the day. At 5 PM, they left for the day, having completed only a portion of the pulling operation. The next day, they resumed work promptly at 8 AM, but the pulling lubricant had already dried in the conduit. As a result of the excess friction on the pull, the rope broke, delaying the job and causing extensive rework.

**The Bottom Line:** Don't stop a pull in the middle of a job.

206F30.EPS

*Figure 30* Circuit tester.

# Cable Pulling

In June 2006, an electrician was tagging instrumentation cable on a work platform about 30 feet off the ground. At the same time, a crew was pulling an assembly in an adjacent cable tray but failed to install caution signs or barrier controls. The inexperienced crew did not monitor the tension on the pulling assembly and it failed during the pull, whipping back along the cable tray and picking up a loose cable roller, which propelled into the victim at high speed. It amputated his arm, broke his ribs and femur, and ruptured several internal organs. He died from his injuries.

The Bottom Line: Only qualified individuals may conduct a cable pull. All pulling equipment must be tested and certified for the expected load, and the area must be secured with caution/barrier controls during the pull.

# Complex Installations

Complex cable-pulling installations are best handled by highly experienced contractors. This is one small portion of an award-winning installation that took over 70,000 hours to complete.

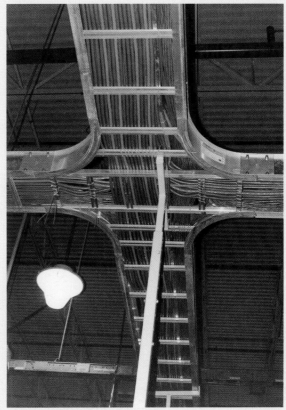

206SA08.EPS

- *Dynamometer (force gauge)* – This type of meter is designed to read dynamic friction (pulling force) during a cable pull. Many are designed for a specific cable-pulling tool and are shipped as an integral or optional part of the cable puller. For example, one common cable puller electronically displays both actual speed and actual force while running. It automatically shuts down when maximum preset force is reached to prevent damage to the cable being pulled. Others are portable units for use with any type of cable-pulling equipment. Such instruments are invaluable for use during high-force cable pulls to avoid damage to conductors during the pull. When the instrument indicates that the maximum pulling force has been reached, the pull can be stopped before damage occurs to the cable or conductors.

## SUMMARY

A cable-pulling operation involves careful planning. Furthermore, the proper use of appropriate equipment is crucial to a successful cable installation. Think of everything that can go wrong, and then plan accordingly.

Communications equipment is another crucial part of a cable installation. Workers on both ends must be in constant contact with each other in case something goes wrong. The workers feeding the cable must carefully inspect the cable as it is paid off the reels and stop the pull immediately if the cable appears damaged. In fact, anyone involved with the pull should be able to stop the pull whenever the safety of the cable or personnel is threatened. Again, good communication is crucial.

Safety precautions must be followed exactly during any cable installation. Use the correct size and type of equipment for the job. Never exceed the maximum force of the weakest component. Do not position yourself in a direct line with a cable pull.

1. Who is allowed to stop the pull at the first sign of danger to the cable, raceway system, or personnel?

   a. Only the project supervisor
   b. Only workers on the feeding end of the run
   c. Only workers on the pulling end of the run
   d. Anyone involved with the pull

2. After some length of conductor has been used from a cable reel, what should be done before placing the reel in storage?

   a. A stress test should be made to see if the cable has been damaged.
   b. An insulation test should be made using a megger.
   c. The ends of the conductors should be sealed.
   d. The conductor jacket should be warmed before storing.

3. Which of the following precautions should be taken when making a cable pull in extremely cold weather?

   a. The cable should be stored in a warm area overnight.
   b. The cable should be stored outside overnight.
   c. Antifreeze should be mixed with the pulling lubricant.
   d. Electric warming blankets should be used during the pull.

4. Which of the following pulling methods should be used when the pulling tension exceeds 1,000 pounds?

   a. A basket grip by itself
   b. A pulling eye
   c. Conductors bent around a hook and taped
   d. Conductors bent around a snake hook and left untaped

5. All the following are acceptable methods of transporting or storing cable reels *except* _____.

   a. reels should be stored in an upright position
   b. reels should be laid flat on their sides
   c. reels should be lifted by a rigged crane with a spindle through the reel
   d. reels should be carried with a forklift so that the forks do not touch the cable

6. Where is the best place to seek information about a wire-pulling lubricant that is compatible with a particular type of cable insulation?

   a. The lubricant manufacturer
   b. The cable manufacturer
   c. Both the cable and lubricant manufacturers
   d. The *NEC*®

7. How should the feed-in setup be handled during a cable installation?

   a. The cable should unreel with a reverse S curvature.
   b. It makes no difference.
   c. Only the pulling end is important when unreeling the cable.
   d. The cable should unreel along its natural curvature.

8. All of the following may be used to guide a cable during a pull *except* _____.

   a. a conveyor sheave
   b. a polygon setup
   c. a guide-in tube
   d. a cable reel

9. Which of the following is the most important consideration when pulling cables in vertical pulls?

   a. Calculate the cable weight before installation.
   b. Install the cable from the bottom up.
   c. Guard against runaways.
   d. Use less wire lubricant than in horizontal applications.

10. Which of the following best describes the location of sheave supports for pulling cable in cable trays?

    a. They should be supported in the opposite direction of the pull.
    b. They should be supported in the same direction as the pull.
    c. They should be supported only by the tray assembly itself.
    d. No support is necessary when sheaves are used in cable trays.

# Supplemental Exercises

1. Prior to installing conductors in a long conduit run, it is a good idea to use _____ to check for hidden obstructions.
   a. a go/no-go mandrel
   b. ¼" fish tape
   c. a flashlight
   d. a blow gun system

2. If a conduit presently contains energized cables and it must be fished, you should _____.
   a. use fiberglass fish tape
   b. tape the end of the metal fish tape
   c. use a vacuum system
   d. not fish it until the cables are de-energized

3. When installing large cables which require more pulling force, select a rope that is rated for at least _____ times the strength of the required pulling force.

4. True or False? If commercial wire lubricant is not available, any suitable petroleum-based product may be substituted.

5. When pulling vertical runs, be sure to _____ the rope after stopping the pull to keep the cable from reversing in the raceway.

6. When installing conductors in long vertical runs, you should install _____ to help prevent runaways.

7. True or False? When pulling long runs of cable, stopping on a regular basis will allow the cable to cool and will avoid stressing the cable.

8. True or False? Cable length meters may be used to determine the length of conductors on reels prior to making a pull.

9. When applying the capstan theory to a power puller, more wraps on the capstan will result in _____ pulling force.

10. What type of pulling device uses a flexible steel mesh grip on the ends of cable and attaches to the pulling rope?
    a. Mandrel
    b. Fish tape
    c. Basket grip
    d. Pulling eye

# Trade Terms Introduced in This Module

**Basket grip:** A flexible steel mesh grip that is used on the ends of cable and conductors for attaching the pulling rope. The more force exerted on the pull, the tighter the grip wraps around the cable.

**Cable grip:** A device used to secure ends of cables to a pulling rope during cable pulls.

**Capstan:** The turning drum of a cable puller on which the rope is wrapped and pulled. An increase in the number of wraps increases the pulling force of the cable puller.

**Clevis:** A device used in cable pulls to facilitate connecting the pulling rope to the cable grip.

**Conductor support:** The act of providing support in vertical conduit runs to support the cables or conductors. The *NEC®* gives several methods in which cables may be supported, including wedges in the tops of conduits, supports to change the direction of cable in pull boxes, etc.

**Conduit piston:** A cylinder of foam rubber that fits inside the conduit and is then propelled by compressed air or vacuumed through the conduit run to pull a line, rope, or measuring tape. Also called a mouse.

**Fish line:** Light cord used in conjunction with vacuum/blower power fishing systems that attaches to the conduit piston to be pushed or pulled through the conduit. Once through, a pulling rope is attached to one end and pulled back through the conduit for use in pulling conductors.

**Fish tape:** A flat iron wire or fiber cord used to pull conductors or a pulling rope through conduit.

**Setscrew grip:** A cable grip, usually with built-in clevis, in which the cable ends are inserted in holes and secured with one or more setscrews.

**Sheave:** A pulley-like device used in cable pulls in both conduit and cable tray systems.

**Soap:** Slang for wire-pulling lubricant.

# Additional Resources

This module presents thorough resources for task training. The following resource material is suggested for further study.

*Cable Installation Manual*, Latest Edition. New York: Cablec Corp.

*National Electrical Code® Handbook*, Latest Edition. Quincy, MA: National Fire Protection Association.

# Figure Credits

AGC of America, Module opener

Topaz Publications, Inc., Figure 1, SA01, Figures 12A, 12B, 14B, 25, and SA08

Greenlee Textron, Inc., a subsidiary of Textron, Inc., Figures 2, 3, SA02, Figures 8, 10, SA03, Figures 12C, 13, 14A, SA04, Figure 16, SA05, Figures 17–19, SA06, Figures 20, 23, SA07, Figures 29 and 30

National Fire Protection Association, Table 2 Reprinted with permission from *NFPA 70®*-2011, *National Electrical Code®*, Copyright © 2010, National Fire Protection Association, Quincy, MA. This reprinted material is not the complete and official position of the National Fire Protection Association on the referenced subject, which is represented only by the standard in its entirety. *NFPA 70®*, *National Electrical Code®*, and *NEC®* are registered trademarks of the National Fire Protection Association, Inc., Quincy, MA.

Tim Dean, Figure 26

# CONTREN® LEARNING SERIES — USER UPDATE

NCCER makes every effort to keep its textbooks up-to-date and free of technical errors. We appreciate your help in this process. If you find an error, a typographical mistake, or an in-accuracy in NCCER's Contren® materials, please fill out this form (or a photocopy), or complete the online form at www.nccer.org/olf. Be sure to include the exact module number, page number, a detailed description, and your recommended correction. Your input will be brought to the attention of the Authoring Team. Thank you for your assistance.

*Instructors* – If you have an idea for improving this textbook, or have found that additional materials were necessary to teach this module effectively, please let us know so that we may present your suggestions to the Authoring Team.

**NCCER Product Development and Revision**

3600 NW 43rd Street, Building G, Gainesville, FL 32606

**Fax:** 352-334-0932
**Email:** curriculum@nccer.org
**Online:** www.nccer.org/olf

☐ Trainee Guide  ☐ AIG  ☐ Exam  ☐ PowerPoints  Other _____

Craft / Level: _____  Copyright Date: _____

Module Number / Title: _____

Section Number(s): _____

Description: _____

_____

_____

_____

Recommended Correction: _____

_____

_____

_____

Your Name: _____

Address: _____

_____

Email: _____  Phone: _____

# Cable Tray

### Dallas/Fort Worth International Airport

Hensel Phelps Construction Co. won an ABC Excellence in Construction award in the mega-projects division for its work on the automated people mover stations at Dallas/Fort Worth International Airport.

## 26207-11

Trainees with successful module completions may be eligible for credentialing through NCCER's National Registry. To learn more, go to **www.nccer.org** or contact us at **1.888.622.3720.** Our website has information on the latest product releases and training, as well as online versions of our *Cornerstone* newsletter and Pearson's Contren® product catalog.

Your feedback is welcome. You may email your comments to **curriculum@nccer.org,** send general comments and inquiries to **info@nccer.org,** or use the User Update form at the back of this module.

 V.17/11

## Objectives

When you have completed this module, you will be able to do the following:

1. Describe the components that make up a cable tray assembly.
2. Explain the methods used to hang and secure cable tray.
3. Describe how cable enters and exits cable tray.
4. Select the proper cable tray fitting for the situation.
5. Explain the *National Electrical Code®* (*NEC®*) requirements for cable tray installations.
6. Select the required fittings to ensure equipment grounding continuity in cable tray systems.
7. Interpret electrical working drawings showing cable tray fittings.
8. Size cable tray for the number and type of conductors contained in the system.

## Performance Tasks

Under the supervision of the instructor, you should be able to do the following:

1. Generate a list of materials for a cable tray layout. List all the components required, including the fasteners required to complete the system.
2. Join two straight, ladder-type cable tray sections together.

## Trade Terms

Barrier strip
Cable pulley
Cross
Direct rod suspension
Dropout
Dropout plate
Elbow

Expansion joints
Fittings
Interlocked armor cable
Ladder tray
Pipe racks
Swivel plates
Tee

Trapeze mounting
Tray cover
Trough
Unistrut®
Wall mounting
Wye

## Required Trainee Materials

1. Paper and pencil
2. Appropriate personal protective equipment
3. Copy of the latest edition of the *National Electrical Code®*

Note: *NFPA 70®*, *National Electrical Code®*, and *NEC®* are registered trademarks of the National Fire Protection Association, Inc., Quincy, MA 02269. All *National Electrical Code®* and *NEC®* references in this module refer to the 2011 edition of the *National Electrical Code®*.

# Contents

*Topics to be presented in this module include:*

# Figures and Tables

# 1.0.0 INTRODUCTION

*NEC Section 392.2* defines a cable tray system as a unit or assembly of units or sections and associated *fittings* forming a structural system used to securely fasten or support cables and raceways. *NEC Article 392* covers cable tray installations, along with the types of conductors to be used in various systems. Whenever a question arises concerning cable tray installations, this is the *NEC*® article to use.

Cable trays are the usual means of supporting cable systems in industrial applications. The trays themselves are usually made up into a system of assembled, interconnected sections and associated fittings, all of which are made of metal or noncombustible units. The finished system forms into a continuous rigid assembly for supporting and carrying single, multiconductor, or other electrical cables and raceways from their origin to their point of termination, frequently over considerable distances.

Cable tray is fabricated from both aluminum and steel. Some manufacturers provide an aluminum cable tray that is coated with PVC for installation in caustic environments. Nonmetallic trays are also available; this type of tray is ideally suited for use in corrosive areas and in areas requiring voltage isolation. Cable tray is available in various forms, including ladder, *trough*, center rail, and solid bottom, and can be supported by either side mounts or center mounts.

*Ladder tray*, as the name implies, consists of two parallel channels connected by rungs, similar in appearance to a conventional straight or extension ladder. Trough types consist of two parallel channels (side rails) having a corrugated, ventilated bottom. The solid bottom cable tray is similar to the trough. All of these types are shown in *Figure 1*. Ladder, trough, and solid bottom trays are completely interchangeable; that is, all three types can be used in the same run when needed.

*Figure 1*  Typical cable tray system.

207F01.EPS

# Cable Tray Installation

Cable trays are widely used in many commercial and industrial installations.

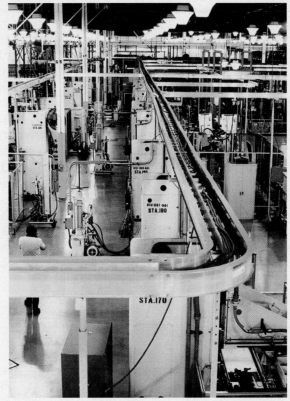

207SA01.EPS

Cable tray is manufactured in 12' and 24' lengths. Common widths range from 6" to 36". All sizes are provided in either 3", 4", 5", or 6" depths.

Cable tray sections are interconnected with various types of fittings. Fittings are also used to provide a means of changing the direction or dimension of the cable tray system. Some of the more common fittings include:

- Horizontal and vertical tees
- Wyes
- Horizontal and vertical bends
- Horizontal crosses
- Reducers
- Barrier strips
- Covers
- Splice plates
- Box connectors

The area of a cable tray cross section that is usable for cables is defined by width (W) × depth (D), as shown in *Figure 2*. The overall dimensions of a cable tray, however, are greater than W times D because of the side flanges and seams. Therefore, overall dimensions vary according to the tray design. Cables rest on the bottom of the tray and are held within the tray area by two longitudinal side rails, as shown in *Figure 3*.

A channel is used to carry one or more cables from the main tray system to the vicinity of the cable termination (see *Figure 4*). Conduit is then used to finish the run from the channel to the actual termination.

Certain *NEC*® regulations and National Electrical Manufacturers Association (NEMA) standards should be followed when designing or installing cable tray. Consequently, practically all projects of any great size will have detailed drawings and specifications for the workers to follow. Shop drawings may also be provided.

# Special-Application Cable Management Systems

This lightweight, flexible nonmetallic cable management system is used to route wiring to the large motors in a manufacturing facility.

207SA02.EPS

207F02.EPS

*Figure 2*  Cross section of cable tray comparing usable dimensions to overall dimensions.

CABLES

CABLE TRAY
SIDE RAIL

207F03.EPS

*Figure 3*  Cables rest on the bottom of the tray and are held in place by the longitudinal side rails.

MAIN TRAY SYSTEM

MAIN TRAY SYSTEM

CHANNEL

CONDUIT BRACKET
AND CLAMP

CHANNEL

CONDUIT

JUNCTION
BOX

MOTOR

MOTOR

FLEXIBLE METAL
CONDUIT

207F04.EPS

*Figure 4*  Two applications of cable tray channel.

## On Site

### Ladder Tray

Ladder tray is currently used for a majority of today's installations. It offers numerous advantages:

- Ladder tray has greater air circulation than other trays, which helps to dissipate heat.
- The ladder rungs are convenient for tying down conductors.
- Ladder tray does not collect moisture.
- Conductors can be dropped through the bottom of the tray at any point.

207SA03.EPS

# Basket Tray

Basket tray offers many of the same advantages as ladder tray. It is lightweight and easy to install using special clips that attach to threaded rods.

207SA04.EPS

Basket tray can be easily cut using bolt cutters or a basket tray cutter, such as that shown here.

207SA05.EPS

One additional advantage of basket tray is that it can be cut, bent, and connected to create tees, crosses, and elbows without the need for separate fittings.

SPLICE PLATE ———→

207SA06.EPS

## 2.0.0 CABLE TRAY LOADING

The *NEC®* requires that cable trays have sufficient strength to support all contained wiring, and that they be supported at the intervals provided in the manufacturer's installation instructions. The following sections describe some additional load considerations in the design of a cable tray system.

### 2.1.0 Load Factors

The load capacity of cable tray varies with each manufacturer and depends on the shape and thickness of the side rails, the shape and thickness of the bottom members, spacing of rungs (if any), material used, safety factor used, method used to determine the allowable load, method of supporting the tray, and volume capacity. Consequently, each manufacturer publishes load data.

### 2.2.0 Determining Fill

The density of fill can only be determined by personnel laying out the system. In doing so, however, be aware that cables packed closely together can impair each other's efficiency (see *Figure 5*).

### 2.3.0 Determining the Load on Supports

Each support should be capable of safely supporting approximately 1.25 times the full weight of the cable and tray on a typical span, as shown in *Figure 6*.

### 2.4.0 Deflection Under Load

Consider the case of a cable tray spanning only two supports (simple span). As the tray is loaded, the side rails take the deflected form. Simultaneously, in cross section, the side rails rotate inward and the tray bottom deflects. The amount of inward or outward movement of the side rails is a critical factor in the ability of the tray to carry a load.

### 2.5.0 Failure Under Load

There are two types of failures that can occur with a loaded cable tray. These are longitudinal (side rail) failures and transverse (rung) failures.

AVOID PACKING CABLES
CLOSE TOGETHER

PACK CABLES LOOSELY

207F05.EPS

*Figure 5*  Cables packed closely together can impair efficiency.

Transverse (rung) failures occur when the load applied causes the fibers on the tension side (bottom edge) of the rung to stretch and permanently deform. Simultaneously, fibers on the compression side (top edge) are permanently crushed. Longitudinal (side rail) failures occur either as bending failures when the tray is supported on a larger span, or, on longer spans, buckling failures may occur because the side rails of the tray have little resistance to inward or outward movement. As the tray deflects, the side rails rotate and the top (compression) flanges of the tray buckle. Bending failures occur on short spans because the side rails of the tray have greater resistance to rotation and remain reasonably upright. The tray does not fail until the load is such that it causes the fibers on the tension side (bottom edge) of the side rail to stretch and permanently deform. Simultaneously, fibers on the compression side (top edge) are permanently crushed. For any tray on an intermediate span, it is difficult to anticipate whether a bending or buckling failure would occur (see *Figure 7*).

### 2.6.0 Splicing Straight Sections

In *Figure 8*, the load on the cable tray creates bending moments along the spans. The stress in the side rails of the tray is directly related to the bending moments at all points along the tray. The magnitude of the bending moment at any point is determined by measuring the vertical height of the shaded portion. In any cable tray system, a splice is a point of weakness. Consequently, splices should be located at the points of least stress. Ideally, splices would be located at the points of zero bending moment, and the strength

---

**Think About It**

## Cable Fill

In what way can cables installed too closely together impair each other's efficiency?

---

NCCER — *Contren® Learning Series*  26207-11

$$\text{UNIFORM LOAD OF TRAY AND CABLE} = \frac{\text{FULL WEIGHT OF TRAY AND CABLE}}{\text{TOTAL LINEAR FEET}}$$

207F06.EPS

***Figure 6*** Each tray support should be capable of safely supporting 1.25 times the weight of the entire tray and cable assembly span.

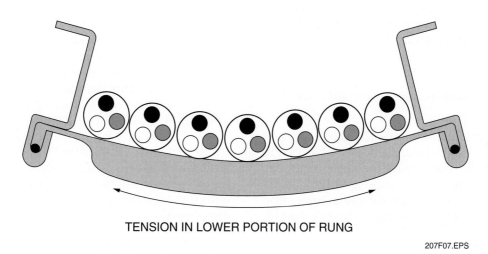

TENSION IN LOWER PORTION OF RUNG

207F07.EPS

***Figure 7*** Bending of loaded tray.

207F08.EPS

***Figure 8*** Load of cable creates bending moments along the span.

of the tray system would be at a maximum. In actual practice, if the splice is located within one-fourth of the span's distance from the support, the result will be close to ideal.

Locating the splice within the one-fourth points of the span requires extra labor on the part of the installer. When a splice occurs within the central length of a span between the one-fourth points, the tray will support the load for which it was designed, but the safety factor will be greatly reduced.

## 2.7.0 Cable Placement in Tray

Cables are placed in the tray either by being pulled along the tray or by being laid in over the side. *Figure 9* shows a cable pulley being used to facilitate a cable pull in a tray.

## 2.8.0 Cable Tray Cover

A cable tray cover is used primarily for two reasons:

- To protect the insulation of the cables against damage that might be caused if an object were to fall into the tray. Prime hazards are tools, discarded cigarettes, and weld splatter.
- To protect certain types of cable insulation against the damaging effects of direct sunlight.

### 2.8.1 Cover Selection

When maximum protection is desired, solid covers should be used. However, if accumulation of heat from the cables is expected, caution should be used. Ventilated covers should be used if some protection of the cable is desired and provisions

must be made to allow the escape of heat developed by the cables.

## 2.9.0 Cable Exit from Tray

Several different ways in which cables may exit from a cable tray are shown in *Figure 10*. While all of these methods are *NEC®*-approved and endorsed by most cable tray manufacturers, the engineering specifications on some projects may prohibit the use of some of these methods. Most notable are the dropout between rungs method and the dropout from the end of the tray method. The cable radius may be too short with either of these methods. Also, since no dropout plates are used, the cable or conductors are not protected. Although *NEC Section 392.100(B)* specifically requires that cable trays have smooth edges to ensure that cable will not be damaged, accidents do occur. For example, a tool might be dropped on a cable tray rung during the installation, which may cause a burr or other sharp edge on the rung. Then, after the cable is installed and the system is in use, vibrating machinery may cause this burr to cut into the cable insulation, resulting in a ground fault and possible power outage. Always review the project specifications carefully and/or check with the project supervisor before using either of these methods.

### 2.9.1 Dropout Plates

A dropout plate provides a curved surface for the cable to follow as it passes from the tray, as shown in *Figure 11(A)*. Without a dropout plate, cables can be bent sharply, causing damage to the insulation. See *Figure 11(B)*.

CABLE PULLEY

207F09.EPS

*Figure 9*  Cable pulley used to facilitate a cable pull in a tray.

SIMPLY UPWARD

DROPOUT FROM END OF
TRAY (NO DROPOUT PLATE)

DROPOUT FROM END OF
TRAY (WITH DROPOUT PLATE)

DROPOUT BETWEEN
RUNGS (NO PLATE)

OVER THE SIDE RAIL

DROPOUT BETWEEN RUNGS
(WITH DROPOUT PLATE)

CONDUIT BUSHING
DROPOUT

CONDUIT CLAMP

DROPOUT INTO CONDUIT
ATTACHED TO TRAY BOTTOM

DROPOUT INTO CONDUIT
CLAMPED TO THE SIDE RAIL

207F10.EPS

*Figure 10*   Several ways in which cables may exit from a cable tray.

## 2.10.0  Cable Supports in Vertical Trays

A cable hanger elbow is used to suspend cables in long vertical runs. Care should be taken to ensure that the weight of the suspended length of cable does not exceed the cable manufacturer's recommendation for the maximum allowable tension in the cable.

In short vertical runs, the weight of cables can be supported either by the outside vertical riser elbow or by the vertical straight section when the cables are tied to it. *Figure 12* shows a typical application of a cable hanger elbow in a vertical run.

## 2.11.0  Cable Edge Protection

The bottom of the solid bottom tray might be convex, concave, or flat. When two pieces of tray are butted together, the bottoms may be out of alignment. An alignment strip, also known as an H bar, is placed between the tray bottoms.

## 2.12.0  Splice Plates

There are several types of splice plates available, including vertical, horizontal, and expansion plates.

(A) WITH DROPOUT PLATE

(B) WITHOUT DROPOUT PLATE

207F11.EPS

*Figure 11*   A dropout plate provides a curved surface for the cable to follow as it leaves the tray.

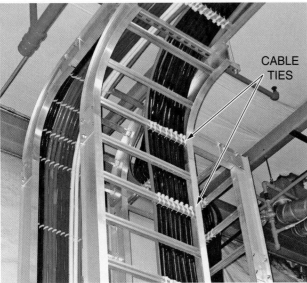

CABLE TIES

207F12.EPS

*Figure 12*   Typical application of supports in a vertical run.

### 2.12.1 Vertical Adjustment Splice Plates

Vertical adjustment splice plates are used to change the elevation in a run of cable tray. They should not be used when it is important to maintain a cable bending radius. Vertical adjustment splice plates are useful when the change in elevation is so slight or the angle so unusual that it would not be possible to install a standard outside vertical riser elbow and an inside vertical riser elbow in the space available.

In general, four swivel plates are used to build an offset in a cable tray system. Once the proper angles have been calculated, proceed as follows:

*Step 1* Bolt four swivel plates together at the proper angles, using the inner holes as the center or pivot hole (see *Figure 13*).

*Step 2* Using a flat surface such as a bench or concrete deck, space two swivel plates at the proper center-to-center distance apart (refer to *Figure 13*).

*Step 3* Measure and cut the amount of tray needed to complete the offset.

### 2.12.2 Horizontal Adjustment Splice Plates

These plates are sometimes used in place of horizontal elbows to change the direction in a run of cable tray. They are used primarily where there is insufficient space or an unusual angle that prevents the use of a standard elbow.

### 2.12.3 Expansion Splice Plates

These plates are used at intervals along a straight run of cable tray to allow space for thermal expansion or contraction of the tray to occur, or where offsets or expansion joints occur in the supporting structure.

To enable the expansion joint to function properly, the cable tray must be allowed to slide freely on its supports. Any cable tray hold-down device used in an installation subject to expansion or contraction must give clearance to the tray. An expansion joint and splice plates are shown in *Figure 14*.

### 2.13.0 Barrier Strips

Barrier (divider) strips are used to separate certain types of cable as a result of the nature of the installation, types of circuits used, type of equipment used, local codes, or the *NEC*®. Some reasons for using divider strips are to:

• Separate or isolate electrical circuits
• Separate or isolate cables of different voltages
• Separate cable or wire runs from each other to prevent fire or ground fault damage from spreading to other cables or wires in the same tray
• Aid neatness in the arrangement of the cables
• Warn electricians of the difference between cables on each side of the divider strip

### 2.13.1 Barrier Strip Cable Protectors

Barrier strip cable protectors are used to bind any raw metallic edge over which a cable is to pass. Their purpose is to protect the cable insulation against damage.

**Figure 13** Fabricating a cable tray offset with swivel plates.

**Figure 14** Expansion joint and splice plates.

## 3.0.0 CABLE TRAY SUPPORT

Proper supports for cable tray installations are very important in obtaining a good overall layout. Cable is usually supported in one or more of the following ways:

- Trapeze mounting
- Direct rod suspension
- Wall mounting
- Center hung support
- Pipe rack mounting

### 3.1.0 Trapeze Mounting

When trapeze mounting is used, a structural member—usually a steel channel or Unistrut®—is connected to the vertical supports to provide an appearance similar to a swing or trapeze (*Figure 15*). The cable tray is mounted to the structural member using bolts, anchor clips, or J-clamps. The underside of the channel or Unistrut® may also be used to support conduit.

### 3.2.0 Direct Rod Suspension

The direct rod suspension method of supporting cable tray uses threaded rods and hanger clamps. The threaded rod is connected to the ceiling or other overhead structure and is connected to the hanger clamps that are attached to the cable tray side rails, as shown in *Figure 16(A)*.

### 3.3.0 Wall Mounting

Wall mounting is accomplished by supporting the tray with structural members attached to the wall, as shown in *Figure 16(B)*. This method of support is often used in tunnels (mining operations) and other underground or sheltered installations where large numbers of conductors interconnect equipment that is separated by long distances. When using this or any other method of supporting cable tray, always examine the structure to which the hangers are attached, and make absolutely certain that the structure is of adequate strength to support the tray system.

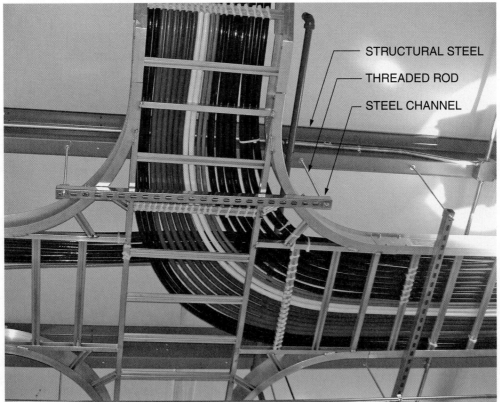

STRUCTURAL STEEL

THREADED ROD

STEEL CHANNEL

207F15.EPS

*Figure 15*   Channel support.

### 3.4.0 Trapeze Mounting and Center Hung Support

Trapeze mounting of cable tray is similar to direct rod suspension mounting. The difference is in the method of attaching the cable tray to the threaded rods. A structural member, usually a steel channel or strut, is connected to the vertical supports to provide an appearance similar to a swing or trapeze. The cable tray is mounted to the structural member. Often, the underside of the channel or strut is used to support conduit. A trapeze mounting assembly is shown in *Figure 16(C)*.

A method that is similar to trapeze mounting is a center hung tray support, as shown in *Figure 16(D)*. In this case, only one rod is used and

it is centered between the cable tray side rails. A bushing sleeve, such as a short piece of small-diameter PVC, may be used over the center rod to protect the conductors.

### 3.5.0 Pipe Rack Mounting

Pipe racks are structural frames used to support the piping that interconnects equipment in outdoor industrial facilities. Usually, some space on the rack is reserved for conduit and cable tray. Pipe rack mounting of cable tray is often seen in petrochemical plants where power distribution and instrumentation wiring is routed over a large area and for long distances.

**(A) DIRECT ROD SUSPENSION**

**(B) WALL MOUNTING**

**(C) TRAPEZE MOUNTING**

**(D) CENTER HUNG SUPPORT**

207F16.EPS

*Figure 16*   Alternate ways to hang cable tray.

## Center Hung Tray

With center hung tray, care must be taken to balance the weight of the cable on either side of the tray to prevent tipping. In addition, extra support may be required for a long drop, such as that shown here.

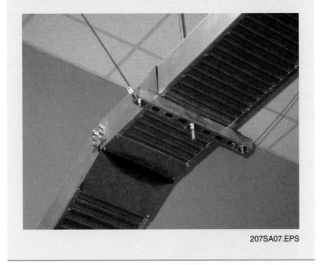

207SA07.EPS

### 4.0.0 CENTER RAIL CABLE TRAY SYSTEMS

Center rail or monorail cable tray systems (*Figure 17*) are light, easy to install, and provide open sides for ready access when changing or adding cables. Center rail cable tray is used in light-duty applications such as sound, telephone, and other communications systems. In addition to its weight limitations, it cannot be used with dividers so its use is restricted to systems where dividers are not necessary.

207F17.EPS

*Figure 17*   Center rail cable tray.

## Cable Tray in Raised-Floor Systems

Cable tray can also be located beneath raised floors as shown here. This provides design flexibility because outlets can be located anywhere on the raised floor and conductors routed to the outlet. These systems are used in many office areas, libraries, and data rooms.

207SA08.EPS

## 5.0.0 *NEC®* REQUIREMENTS

*NEC Article 392* deals with cable tray systems along with related wire, cable, and raceway installations.

*Figures 18, 19*, and *20* summarize the *NEC®* requirements for cable tray installations. For in-depth coverage, always refer to the *NEC®*. Cable tray manufacturers also have some excellent reference material available that details the installation of their products. Consult your local dealer for more information.

Welding cable may only be installed in dedicated cable tray per *NEC Section 630.42.*

Insulated conductors and jacketed cable must be identified as sunlight-resistant if exposed to direct sunlight. *NEC Section 392.10*

Where conductors No. 1/0 through 4/0 are run in ladder-type trays, the maximum rung spacing must not exceed 9". *NEC Section 392.10(B)(1)(a)*

Single conductors must be No. 1/0 or larger. *NEC Section 392.10(B)(1)(a)*

SUNLIGHT-RESISTANT

9" MAX.

RUNGS

Single conductor cable must be marked for use in cable tray. *NEC Section 392.10(B)(1)(a)*

Nonmetallic cable tray is permitted for use in corrosive areas and in areas requiring voltage isolation. *NEC Section 392.10(D)*

Cable tray systems must not be used in hoistways or where subject to severe physical damage nor in ducts or plenums used for environmental air, but may be used in other environmental air spaces provided that the tray has a solid metal bottom and a solid metal cover. *NEC Sections 392.12 and 300.22(C)(2)*

Electrical conductors must not be installed in the same raceways or cable tray with steam, water, gas, air, or drainage pipes. *NEC Section 300.8*

ELECTRICAL WIRING

GAS PIPE

STEAM PIPE

207F18.EPS

*Figure 18*   *NEC®* regulations governing the use of cable tray.

**NEC Section 392.100,** Construction Specifications

Cable tray must include fittings or other suitable means for changes in direction and elevation of runs.
*NEC Section 392.100(E)*

Cable tray must have side rails or equivalent structural members.
*NEC Section 392.100(D)*

Nonmetallic cable tray must be made of flame-retardant materials.
*NEC Section 392.100(F)*

**NEC Article 392, Part II,** Installation

Cable tray must be installed as a complete system, and each run must be completed before installing cable.
*NEC Section 392.18(A) and (B)*

Sufficient space must be provided and maintained around cable tray to permit adequate access for installing and maintaining the cables.
*NEC Section 392.18(F)*

Cable tray must have suitable strength and rigidity to provide adequate support for all contained wiring.
*NEC Section 392.100(A)*

Tray edges must be smooth, with no sharp edges, burrs, or projections.
*NEC Section 392.100(B)*

If ferrous metal tray is used, it must be protected against corrosion.
*NEC Section 392.100(C)*

*NEC Section 392.18(D)* states that cable tray may be extended through partitions and walls or vertically through platforms and floors if the conditions in *NEC Section 300.21* are met.

In portions of the cable tray run where additional protection is required, tray covers are normally employed; they must be of a material compatible with the tray system.
*NEC Section 392.18(C)*

Cable tray may be used as support for raceways under certain conditions.
*NEC Section 392.18(G)*

Cables 600 volts or less on this side of barrier, cables over 600 volts on opposite side.

Where multiconductor cables rated over 600 volts are installed in the same cable tray with conductors rated 600 volts or less, the two types of conductors must be separated by a solid fixed barrier of material compatible with the cable tray unless the cables over 600 volts are Type MC.
*NEC Section 392.20(B)(1) and (2)*

THREADED ROD

CABLE TRAY CONTAINING SINGLE AND/OR MULTI-CONDUCTOR CABLES

ANCHOR CLIP

CHANNEL FRAMING (UNISTRUT®)

CONDUIT SECURED TO BOTTOM OF TRAPEZE HANGER

207F19.EPS

*Figure 19* NEC® regulations governing cable tray construction and installation.

*NEC Section 392.60,* Grounding and Bonding

BUILDING STEEL

GROUNDING STRAP

Steel or aluminum cable tray systems are permitted to be used as equipment grounding conductors if they meet all of the requirements in *NEC Section 392.60(B).*

Proper grounding lessens hazards due to ground faults. Therefore, the *NEC®* requires all metallic cable tray to be grounded as required for conductor enclosures in accordance with *NEC Section 250.96.*

Per *NEC Section 392.60(B)(2),* the minimum cross-sectional area of cable tray must conform to the requirements in *NEC Table 392.60(A).*

FLEXIBLE BONDING JUMPER

BOLT, NUT, AND WASHER

Where supervised by qualified personnel, grounded metallic cable tray may also be used as an equipment grounding conductor. *NEC Section 392.60(A).*

Cable tray sections and fittings must be bonded in accordance with *NEC Section 250.96.*

207F20.EPS

*Figure 20*   *NEC®* regulations governing cable tray grounding.

## 6.0.0 CABLE INSTALLATION

*NEC Section 392.20* covers the general installation requirements for all conductors used in cable tray systems; that is, splicing, securing, and running conductors in parallel. For example, cable splices are permitted in cable trays provided they are made and insulated by *NEC®*-approved methods. Furthermore, any splices must be readily accessible and must not project above the side rails of the tray.

In most horizontal runs, the cables may be laid in the tray without securing them in place. However, on vertical runs or any runs other than horizontal, the cables must be secured to transverse members of the cable tray.

Cables may enter and leave a cable tray system in a number of different ways, as discussed

previously. In general, no junction box is required where such cables are installed in bushed conduit or tubing. Where conduit or tubing is used, it must be secured to the tray with the proper fittings. Further precautions must be taken to ensure that the cable is not bent sharply as it enters or leaves the conduit or tubing.

### 6.1.0 Conductors Connected in Parallel

Where single-conductor cables comprising each phase, neutral, or grounded conductor of a circuit are connected in parallel as permitted in *NEC Section 310.10(H),* the conductors must be installed in groups consisting of not more than one conductor per phase, neutral, or grounded conductor to prevent a current imbalance in the paralleled conductors due to inductive reactance. This also prevents excessive movement due to fault current magnetic forces.

## 6.2.0 Number of Cables Allowed in Cable Tray (2,000V or Less)

The number of multiconductor cables rated at 2,000V or less that are permitted in a single cable tray must not exceed the requirements of *NEC Section 392.22(A)*. This section applies to both copper and aluminum conductors.

### 6.2.1 All Conductors Size 4/0 or Larger

Per *NEC Section 392.22(A)(1)(a)*, where all of the cables installed in ladder or ventilated trough tray are 4/0 or larger, the sum of the diameters of all cables shall not exceed the cable tray width, and the cables must be installed in a single layer. For example, if a cable tray installation is to contain three 4/0 multiconductor cables (1.5" in diameter), two 250 kcmil multiconductor cables (1.85"), and two 350 kcmil multiconductor cables (2.5"), the minimum width of the cable tray is determined as follows:

$$3(1.5) + 2(1.85) + 2(2.5) = 13.2"$$

The closest standard cable tray size that meets or exceeds 13.2" is 18". Therefore, this is the size to use.

### 6.2.2 All Conductors Smaller Than 4/0

Per *NEC Section 392.22(A)(1)(b)*, where all of the cables are smaller than 4/0, the sum of the cross-sectional area of all cables smaller than 4/0 must not exceed the maximum allowable cable fill area as specified in Column 1 of *NEC Table 392.22(A)*; this gives the appropriate cable tray width. To use this table, however, you must have the manufacturer's data for the cables being used. This will give the cross-sectional area of the cables.

The steps involved in determining the size of cable tray for multiconductors smaller than 4/0 AWG are as follows:

*Step 1* Calculate the total cross-sectional area of all cables used in the tray. Obtain the area of each from the manufacturer's data.

*Step 2* Look in Column 1 of *NEC Table 392.22(A)* and find the smallest number that is at least as large as the calculated number.

*Step 3* Look at the number to the left of the row selected in Step 2. This is the minimum width of cable tray that may be used.

For example, determine the minimum cable tray width required for the following multiple conductor cables—all less than 4/0 AWG:

- Four at 1.5" diameter
- Five at 1.75" diameter
- Three at 2.15" diameter

*Step 1* Determine the cross-sectional area of the cables from the equation:

$$A = \frac{\pi \times D^2}{4}$$

*Where:*

A = area

D = diameter

The area of a 1.5"-diameter cable is:

$$\frac{(3.14159)(1.5^2)}{4} = 1.7671 \text{ square inches}$$

The area of the four 1.5" cables is:

$$4 \times 1.7662 \text{ square inches}$$
$$= 7.0648 \text{ square inches}$$

The area of 1.75"-diameter cable is:

$$\frac{(3.14159)(1.75^2)}{4} = 2.4053 \text{ square inches}$$

The area of the five 1.75" cables is:

$$5 \times 2.4041 \text{ square inches}$$
$$= 12.0205 \text{ square inches}$$

The area of the three 2.15" cables is:

$$\frac{(3.14159)(2.15^2)}{4} = 3.6305 \text{ square inches}$$

Therefore, the total area of the three cables is:

$$3 \times 3.6287 \text{ square inches}$$
$$= 10.8861 \text{ square inches}$$

The total cross-sectional area is found by adding the above three totals to obtain:

$$7.0648 + 12.0205 + 10.8861$$
$$= 29.9714 \text{ square inches}$$

*Step 2* Look in Column 1 of *NEC Table 392.22(A)* and find the smallest number that is at least as large as 29.9714 square inches. The number is 35.

*Step 3* Look to the left of 35 and you will see the inside tray width of 30". Therefore, the minimum tray width that can be used for the given group of conductors is 30".

# Low-Voltage Cable

Increasingly, cable trays are being used to carry many low-voltage conductors in communication centers. In a large commercial building, thousands of telecommunications cables are distributed in bundles from the equipment room to the telecommunications closets on each floor. Cable trays offer a convenient means of running these cables through the building as well as a much easier method of allowing for system expansion. Instead of having to access conduit buried within the building walls, a new communication cable can simply be added to the cable tray system, which is normally accessible through a dropped ceiling.

207SA09.EPS

207SA10.EPS

### 6.2.3 Combination Cables

Per *NEC Section 392.22(A)(1)(c)*, where 4/0 or larger cables are installed in the same ladder or ventilated trough cable tray with cables smaller than 4/0, the sum of the cross-sectional area of all cables smaller than 4/0 must not exceed the maximum allowable fill area from Column 2 of *NEC Table 392.22(A)* for the appropriate cable tray width. The 4/0 and larger cables must be installed in a single layer, and no other cables can be placed on them.

To determine the tray size for a combination of cables as discussed in the above paragraph, proceed as follows:

*Step 1* Repeat the steps from the procedure used previously to determine the minimum tray width required for the multiconductor cables having conductors sized 4/0 and larger.

*Step 2* Repeat the steps from the procedure used previously to determine the cross-sectional area of all multiconductor cables having conductors smaller than 4/0 AWG.

*Step 3* Multiply the result of Step 1 by the constant 1.2 and add this product to the result of Step 2. Call this sum A. Search

Column 2 of *NEC Table 392.22(A)* for the smallest number that is at least as large as A. Look to the left in that row to determine the minimum size cable tray required.

To illustrate these steps, assume that you need to find the minimum cable tray width of two multiconductor cables, each with a diameter of 2.54" (conductors size 4/0 or larger); three cables with a diameter of 3.30" (conductors size 4/0 or larger), plus eight cables with a diameter of 1.92" (conductors less than 4/0).

*Step 1* The sum of all the diameters of cable having conductors 4/0 or larger is:

$$2(2.54) + 3(3.30) = 14.98"$$

*Step 2* The sum of the cross-sectional areas of all cables having conductors smaller than 4/0 is:

$$\frac{(8)(3.14159)(1.92^2)}{4} = 23.1623 \text{ square inches}$$

*Step 3* Multiply the result of Step 1 by 1.2 and add this product to the result of Step 2:

$$(1.2 \times 14.98) + 23.1506$$
$$= 41.1266 \text{ square inches}$$

The smallest number in Column 2 of *NEC Table 392.22(A)* that is not larger than 41.1383 is 42. The tray width that corresponds to 42 is 36". Therefore, select a cable tray width of 36".

### 6.2.4 Solid Bottom Tray

Per *NEC Section 392.22(A)(3)*, where solid bottom cable trays contain multiconductor power or lighting cables, or any mixture of multiconductor power, lighting, control, and signal cables, the maximum number of cables must conform to the following:

- Where all of the cables are 4/0 or larger, the sum of the diameters of all cables must not exceed 90% of the cable tray width, and the cables must be installed in a single layer.
- Where all of the cables are smaller than 4/0, the sum of the cross-sectional areas of all cables must not exceed the maximum allowable cable fill area in Column 3 of *NEC Table 392.22(A)* for the appropriate cable tray width.
- Where 4/0 or larger cables are installed in the same cable tray with cables smaller than 4/0, the sum of the cross-sectional areas of all of the smaller cables must not exceed the maximum allowable fill area resulting from the computation in Column 4, *NEC Table 392.22(A)* for the appropriate cable tray width. The 4/0 and larger cables must be installed in a single layer, and no other cables can be placed on them.

Where a solid bottom cable tray with a usable inside depth of 6" or less contains multiconductor control and/or signal cables only, the sum of the cross-sectional areas of all cables at any cross section must not exceed 40% of the interior cross-sectional area of the cable tray. A depth of 6" must be used to compute the allowable interior cross-sectional area of any cable tray that has a usable inside depth of more than 6".

In a previous example, it was determined that the minimum tray size for multiconductor cables with all conductors size 4/0 or larger was 18". The sum of all cable diameters for this example was 13.2". To see if an 18" solid bottom tray can be used, multiply the tray width by 0.90 (90%):

$$18" \times 0.9 = 16.2"$$

Therefore, 16.2" is the minimum width allowed for solid bottom tray. Since we are using 18" tray, this meets the requirements of *NEC Section 392.22(A)(3)(a)*.

When dealing with solid bottom trays and using *NEC Table 392.22(A)*, use Columns 3 and 4 instead of Columns 1 and 2, as used for ladder and trough-type cable trays.

### 6.2.5 Single-Conductor Cables

Calculating cable tray widths for single-conductor cables (2,000V or under) is similar to the calculations used for multiconductor cables, with the following exceptions:

- Conductors that are 1,000 kcmil and larger are treated the same as multiconductor cables having conductors size 4/0 or larger.
- Conductors that are smaller than 1,000 kcmil are treated the same as multiconductor cables having conductors smaller than size 4/0.

*NEC Section 392.22(B)* covers the details of installing single-conductor cables with rated voltages of less than 2,000V in cable tray systems.

### 6.3.0 Ampacity of Cable Tray Conductors

*NEC Section 392.80* gives the requirements for cables used in tray systems with rated voltages of 2,000 volts or less. Cables in cable tray use the same ampacity tables as other cable (see *NEC Article 310*), with additional derating applied for the following conditions:

- Cable tray construction
  - Open or covered
  - Solid or ventilated/ladder
- Type(s) of cable
  - Ampacity (less than 2,000V or over 2,001V)
  - Single conductor or multiconductor
- Number of cables and cable configuration in the tray

*NEC Section 392.80(B)* covers the installation and ampacity of cables with voltages of 2,001 and over.

## *What's wrong with this picture?*

207SA11.EPS

## 7.0.0 CABLE TRAY DRAWINGS

For an economical and satisfactory installation, working out the details of supports and hangers for a cable tray system is usually done beforehand by the engineering department or project engineer and is seldom left to the judgment of a field force that is not acquainted with the loads and forces to be encountered. As a result, drawings and specifications will usually be furnished to the work crew to provide details about the cable tray system. All workers involved with the installation should know how to interpret these drawings.

The exact method of showing cable tray systems on working drawings will vary, so always consult the symbol list or legend before beginning the installation. Also, study any shop or detail drawings that might accompany the construction documents.

If space permits, many engineers prefer to draw the cable tray system as close to scale as possible,

using various symbols to show the different types of cable tray to be installed.

Look at the floor plan drawing in *Figure 21*. The cable tray system in this project originates at several power panels and motor control centers to feed and control motors in other parts of the building. The trays run from the motor control centers, are offset to miss beams and other runs of cable tray, and then branch off to various parts of the building.

Although experienced workers in the industrial electrical trade will have little trouble reading the information in this drawing, new electricians may have some difficulty in visualizing the system. However, if a supplemental drawing were provided with the floor plan drawing in *Figure 21*, it would provide a clearer picture of the system and leave little doubt as to how the cable tray system is to be installed. Even new workers in the

SEE CABLE TRAY SCHEDULE IN THE WRITTEN
SPECIFICATIONS FOR CABLE STRUT CATALOG
NUMBERS OF EACH TRAY TYPE SPECIFIED

207F21.EPS

*Figure 21*   Sample floor plan of a cable tray system.

trade would be able to see how the system should be installed, and this would take some of the load off experienced workers and give them more time to accomplish other tasks.

In actual practice, however, consulting engineering firms seldom furnish the isometric drawing; they merely show the layout in plan view. Consequently, plan views involving the construction details of cable tray systems must be studied carefully during the planning stage—before the work is begun.

## 8.0.0 PULLING CABLE IN TRAY SYSTEMS

When installing cables in tray systems, proper precautions must be taken to avoid damaging the cables. A complete line of installation tools is available for pulling long lengths of cable up to 1,000' or longer. These tools save considerable installation time.

Short lengths of cable can be laid in place without power pulling tools, or the cable can be pulled manually using a basket grip and pulling rope. Long lengths of small cable, 2" or less in diameter, can also be pulled with a basket grip and pulling rope. Larger cables, however, should be pulled by the conductor and the braid, sheath, or armor. This is usually done with a pulling eye applied to the cable at the factory, or by tying the conductor to the eye of a basket grip and taping the tail end of the grip to the outside of the cable.

### Case History

## Pulling Cable in a Tray

At a clothing manufacturing plant in New York, an electrician attempted to pull cable in a tray already partly filled with energized cables. He used plenty of cable-pulling lubricant, but did not use sheaves and rollers to isolate the new cables from the existing ones. During the pull, the new cables wedged against the live ones, yanking one hot wire apart at a splice. The wire grounded itself on the tray and arced for 30' down the tray, fatally shocking the electrician.

The Bottom Line: Use sheaves and rollers when pulling cables, especially when pulling them between existing live cables.

### On Site

## Big Pulls

For very long pulls, some contractors use an auxiliary winch. Power winches may exceed their preset tension limits and abort the pull, so a second winch is installed in a high-tension area of the pull in a straight section of tray. The second winch is used to pull a loop of slack cable and reduce the tension on the main winch, which now pulls only the length of slack cable between itself and the auxiliary winch.

In general, the pull exerted on the cables pulled with a basket grip that is not attached to the conductor should not exceed 1,000 pounds. For heavier pulls, care should be taken not to stretch the insulation, jacket, or armor beyond the end of the conductor nor bend the ladder, trough, or channel out of shape.

The bending radius of the cable should not be less than the values recommended by the cable manufacturer, which range from four times the diameter for a rubber-insulated cable that has a 1" maximum outside diameter without lead, shield, or armor, to eight times the diameter for interlocked armor cable. Cables of special construction such as wire armor and high-voltage cables require a larger radius bend.

When installing long lengths of cable up to 1,000' with as many as a dozen bends, best results are obtained by pulling the cable in one continuous operation at a speed of 20' to 25' per minute. The pulling line diameter and length will depend on the pull to be made and construction equipment available. The winch and power unit must be of an adequate size for the job and capable of developing the high pulling speed required for the best and most economical results. A complete description of pulling equipment was covered in the *Conductor Installations* module.

## 9.0.0 SAFETY

Installing cable tray means working at heights above the floor. Consequently, workers must take the necessary precautions. In general, workers installing cable tray will use ladders, scaffolds, lifts, work from the tray assembly itself, or a combination of all these.

Keep the tray assembly uncluttered during installation. Tools, tray fittings, and the like are ideal obstacles for tripping workers. They may also fall off the tray and injure workers below.

To help prevent the latter, set up work barriers beneath the section of tray assembly being worked on. If you are working on the ground, never penetrate or move these barriers until the work above is complete.

Learn to secure your lifeline properly, and also make sure your full body harness is a proper fit. A harness that is too large may slip, resulting in a serious injury or even death.

**Think About It**

## Putting It All Together

If possible, examine the conduit system at your workplace or school. Did the installers make use of cable tray? If not, is the system large enough to warrant the use of it?

## SUMMARY

A cable tray system includes the assembly of units or sections and associated fittings that form a rigid structural system used to support cables and raceways. Cable tray systems are commonly used in industrial applications and are normally constructed of either aluminum or steel, although PVC-coated and all-nonmetallic trays are also available.

Cable tray is available in various forms, including ladder, trough, center rail, and solid bottom, and can be supported by either side mounts or center mounts. *NEC Article 392* covers cable tray installations, along with the types of conductors to be used in various cable tray systems.

1. When a cable tray section branches off from a main section in two 90° turns, the section is called a(n) _____.

    a. wye
    b. tee
    c. divider strip
    d. ell

2. A section of cable tray that makes a single horizontal 90° turn is known as a(n) _____.

    a. tee
    b. wye
    c. elbow
    d. cross

3. Which of the following best describes the type of fitting that will be used with a solid cover cable tray?

    a. An Ericson coupling
    b. Vertical splices
    c. A horizontal cross
    d. A cover clamp

4. Which of the following branches in four different directions?

    a. A horizontal wye
    b. A horizontal cross
    c. An inside vertical riser
    d. A dropout

5. Which of the following best describes how closely packed cables will be affected when energized?

    a. Their efficiency will be increased.
    b. Their efficiency will be decreased.
    c. No change will be encountered.
    d. The operating temperature of each cable will be lower.

6. When a cable tray system must be protected from dropping objects that may damage the cable, a _____.

    a. solid bottom tray is used
    b. trough-type tray is used
    c. tray cover is used
    d. blind end is used

7. When cable exits downward from a tray without a dropout plate, which of the following could occur?

    a. Current surges
    b. Reduced amperage
    c. Voltage surges
    d. Insulation damage

8. The main purpose of vertical adjustment splice plates is to _____.

    a. increase the strength of a cable tray section
    b. support the cable tray system
    c. change the elevation in a run of cable tray
    d. secure the cable within the tray

9. A barrier strip is a _____.

    a. division strip installed in a raceway to separate certain types of cables
    b. strip used to provide the tray with added support
    c. glass strip used as a cable tray insulator
    d. metal strip used solely for grounding purposes

10. Which of the following is a violation of *NEC Section 300.8?*

    a. Placing an instrumentation air line in the same tray with electrical conductors
    b. Spacing tray rungs 9" or less apart
    c. Running single conductors in trays that are larger than size 1/0
    d. Using nonmetallic tray systems in corrosive areas

1. Cable supports should be capable of supporting _____ times the full weight of the cable and tray.

2. Where solid bottom cable trays contain multiconductor power or lighting cables, and all cables are larger than 4/0, the sum of the diameters of all cables must not exceed _____ of the cable tray width.

3. The two types of cable tray failure under load are _____ and _____.

4. Cable tray is commonly manufactured in _____ and _____ lengths.

5. List the five reasons for using divider strip in cable:

   _____

   _____

   _____

   _____

   _____

6. Locate splices within _____ of the span's distance from the nearest support.

7. True or False? When installing long lengths of cable, it is best to pull the cable in one continuous operation.

8. List five types of cable tray supports:

   _____

   _____

   _____

   _____

   _____

9. True or False? Long runs of instrumentation wiring are typically supported by pipe rack mounting.

10. Cable tray is covered in _____.
    a. *NEC Article 300*
    b. *NEC Article 392*
    c. *NEC Article 517*
    d. *NEC Article 550*

11. Cable tray must be bonded in accordance with _____.

    a. *NEC Section 250.96*
    b. *NEC Section 300.96*
    c. *NEC Section 300.21*
    d. *NEC Section 392.6(J)*

12. True or False? Water piping may be run in cable tray with electrical conductors.

13. When conductors 1/0 through 4/0 are run in ladder-type trays, the maximum rung spacing must not exceed _____.

    a. 9"
    b. 8"
    c. 7"
    d. 6"

14. All of the following are basic forms of cable tray *except* _____.

    a. ladder
    b. ratchet
    c. trough
    d. solid bottom

15. Cable in vertical trays should be supported _____.

    a. per the manufacturer's recommendations
    b. every 6" or 8"
    c. every 12"
    d. every 24"

# Trade Terms Introduced in This Module

**Barrier strip:** A metal strip constructed to divide a section of cable tray so that certain kinds of cable may be separated from each other.

**Cable pulley:** A device used to facilitate pulling conductor in cable tray where the tray changes direction. Several types are available (single, triple, etc.) to accommodate almost all pulling situations.

**Cross:** A four-way section of cable tray used when the tray assembly must branch off in four different directions.

**Direct rod suspension:** A method used to support cable tray by means of threaded rods and hanger clamps. One end of the threaded rod is secured to an overhead structure, while the other end is connected to hanger clamps that are attached to the cable tray side rails.

**Dropout:** Cable leaving the tray assembly and travelling directly downward; that is, the cable is not routed into a conduit or channel.

**Dropout plate:** A metal plate used at the end of a cable tray section to ensure a greater cable bending radius as the cable leaves the tray assembly.

**Elbow:** A section of cable tray used to change the direction of the tray assembly a full quarter turn (90°). Both vertical and horizontal elbows are common.

**Expansion joints:** Plates used at intervals along a straight run of cable tray to allow space for thermal expansion or contraction of the tray.

**Fittings:** Devices used to assemble and/or change the direction of cable tray systems.

**Interlocked armor cable:** Mechanically protected cable; usually a helical winding of metal tape formed so that each convolution locks mechanically upon the previous one (armor interlock).

**Ladder tray:** A type of cable tray that consists of two parallel channels connected by rungs, similar in appearance to the common straight ladder.

**Pipe racks:** Structural frames used to support the piping that interconnects equipment in outdoor industrial facilities.

**Swivel plates:** Devices used to make vertical offsets in cable tray.

**Tee:** A section of cable tray that branches off the main section in two other directions.

**Trapeze mounting:** A method of supporting cable tray using metal channel, such as Unistrut®, Kindorf®, etc., supported by two threaded rods, and giving the appearance of a swing or trapeze.

**Tray cover:** A flat piece of metal, fiberglass, or plastic designed to provide a solid covering that is needed in some locations where conductors in the tray system may be damaged.

**Trough:** A type of cable tray consisting of two parallel channels (side rails) having a corrugated, ventilated bottom or a corrugated, solid bottom.

**Unistrut®:** A brand of metal channel used as the bottom bracket for hanging cable trays. Double Unistrut® adds strength and stability to the trays and also provides a means of securing future runs of conduit.

**Wall mounting:** A method of supporting cable tray systems using supports secured directly to the wall.

**Wye:** A section of cable tray that branches off the main section in one direction.

## METRIC CONVERSION CHART

# METRIC CONVERSION CHART

| INCHES | | METRIC | INCHES | | METRIC | INCHES | | METRIC |
|--------|---------|--------|--------|---------|--------|--------|---------|--------|
| Fractional | Decimal | mm | Fractional | Decimal | mm | Fractional | Decimal | mm |
| . | 0.0039 | 0.1000 | . | 0.5512 | 14.0000 | . | 1.8898 | 48.0000 |
| . | 0.0079 | 0.2000 | 9/16 | 0.5625 | 14.2875 | . | 1.9291 | 49.0000 |
| . | 0.0118 | 0.3000 | . | 0.5709 | 14.5000 | . | 1.9685 | 50.0000 |
| 1/64 | 0.0156 | 0.3969 | 37/64 | 0.5781 | 14.6844 | 2 | 2.0000 | 50.8000 |
| . | 0.0157 | 0.4000 | . | 0.5906 | 15.0000 | . | 2.0079 | 51.0000 |
| . | 0.0197 | 0.5000 | 19/32 | 0.5938 | 15.0813 | . | 2.0472 | 52.0000 |
| . | 0.0236 | 0.6000 | 39/64 | 0.6094 | 15.4781 | . | 2.0866 | 53.0000 |
| . | 0.0276 | 0.7000 | . | 0.6102 | 15.5000 | . | 2.1260 | 54.0000 |
| 1/32 | 0.0313 | 0.7938 | 5/8 | 0.6250 | 15.8750 | . | 2.1654 | 55.0000 |
| . | 0.0315 | 0.8000 | . | 0.6299 | 16.0000 | . | 2.2047 | 56.0000 |
| . | 0.0354 | 0.9000 | 41/64 | 0.6406 | 16.2719 | . | 2.2441 | 57.0000 |
| . | 0.0394 | 1.0000 | . | 0.6496 | 16.5000 | 2 1/4 | 2.2500 | 57.1500 |
| . | 0.0433 | 1.1000 | 21/32 | 0.6563 | 16.6688 | . | 2.2835 | 58.0000 |
| 3/64 | 0.0469 | 1.1906 | . | 0.6693 | 17.0000 | . | 2.3228 | 59.0000 |
| . | 0.0472 | 1.2000 | 43/64 | 0.6719 | 17.0656 | . | 2.3622 | 60.0000 |
| . | 0.0512 | 1.3000 | 11/16 | 0.6875 | 17.4625 | . | 2.4016 | 61.0000 |
| . | 0.0551 | 1.4000 | . | 0.6890 | 17.5000 | . | 2.4409 | 62.0000 |
| . | 0.0591 | 1.5000 | 45/64 | 0.7031 | 17.8594 | . | 2.4803 | 63.0000 |
| 1/16 | 0.0625 | 1.5875 | . | 0.7087 | 18.0000 | 2 1/2 | 2.5000 | 63.5000 |
| . | 0.0630 | 1.6000 | 23/32 | 0.7188 | 18.2563 | . | 2.5197 | 64.0000 |
| . | 0.0669 | 1.7000 | . | 0.7283 | 18.5000 | . | 2.5591 | 65.0000 |
| . | 0.0709 | 1.8000 | 47/64 | 0.7344 | 18.6531 | . | 2.5984 | 66.0000 |
| . | 0.0748 | 1.9000 | . | 0.7480 | 19.0000 | . | 2.6378 | 67.0000 |
| 5/64 | 0.0781 | 1.9844 | 3/4 | 0.7500 | 19.0500 | . | 2.6772 | 68.0000 |
| . | 0.0787 | 2.0000 | 49/64 | 0.7656 | 19.4469 | . | 2.7165 | 69.0000 |
| . | 0.0827 | 2.1000 | . | 0.7677 | 19.5000 | 2 3/4 | 2.7500 | 69.8500 |
| . | 0.0866 | 2.2000 | 25/32 | 0.7813 | 19.8438 | . | 2.7559 | 70.0000 |
| . | 0.0906 | 2.3000 | . | 0.7874 | 20.0000 | . | 2.7953 | 71.0000 |
| 3/32 | 0.0938 | 2.3813 | 51/64 | 0.7969 | 20.2406 | . | 2.8346 | 72.0000 |
| . | 0.0945 | 2.4000 | . | 0.8071 | 20.5000 | . | 2.8740 | 73.0000 |
| . | 0.0984 | 2.5000 | 13/16 | 0.8125 | 20.6375 | . | 2.9134 | 74.0000 |
| 7/64 | 0.1094 | 2.7781 | . | 0.8268 | 21.0000 | . | 2.9528 | 75.0000 |
| . | 0.1181 | 3.0000 | 53/64 | 0.8281 | 21.0344 | . | 2.9921 | 76.0000 |
| 1/8 | 0.1250 | 3.1750 | 27/32 | 0.8438 | 21.4313 | 3 | 3.0000 | 76.2000 |
| . | 0.1378 | 3.5000 | . | 0.8465 | 21.5000 | . | 3.0315 | 77.0000 |
| 9/64 | 0.1406 | 3.5719 | 55/64 | 0.8594 | 21.8281 | . | 3.0709 | 78.0000 |
| 5/32 | 0.1563 | 3.9688 | . | 0.8661 | 22.0000 | . | 3.1102 | 79.0000 |
| . | 0.1575 | 4.0000 | 7/8 | 0.8750 | 22.2250 | . | 3.1496 | 80.0000 |
| 11/64 | 0.1719 | 4.3656 | . | .8858 | 22.5000 | . | 3.1890 | 81.0000 |
| . | 0.1772 | 4.5000 | 57/64 | .89063 | 22.6219 | . | 3.2283 | 82.0000 |
| 3/16 | 0.1875 | 4.7625 | . | .9055 | 23.0000 | . | 3.2677 | 83.0000 |
| . | 0.1969 | 5.0000 | 29/32 | .90625 | 23.0188 | . | 3.3071 | 84.0000 |
| 13/64 | 0.2031 | 5.1594 | 59/64 | .92188 | 23.4156 | . | 3.3465 | 85.0000 |
| . | 0.2165 | 5.5000 | . | .9252 | 23.5000 | . | 3.3858 | 86.0000 |
| 7/32 | 0.2188 | 5.5563 | 15/16 | .93750 | 23.8125 | . | 3.4252 | 87.0000 |
| 15/64 | 0.2344 | 5.9531 | . | .9449 | 24.0000 | . | 3.4646 | 88.0000 |
| . | 0.2362 | 6.0000 | 61/64 | .95313 | 24.2094 | 3 1/2 | 3.5000 | 88.9000 |
| 1/4 | 0.2500 | 6.3500 | . | .9646 | 24.5000 | . | 3.5039 | 89.0000 |
| . | 0.2559 | 6.5000 | 31/32 | .96875 | 24.6063 | . | 3.5433 | 90.0000 |
| 17/64 | 0.2656 | 6.7469 | . | .9843 | 25.0000 | . | 3.5827 | 91.0000 |
| . | 0.2756 | 7.0000 | 63/64 | .98438 | 25.0031 | . | 3.6220 | 92.0000 |
| 9/32 | 0.2813 | 7.1438 | 1 | 1.000 | 25.40 | . | 3.6614 | 93.0000 |
| . | 0.2953 | 7.5000 | . | 1.0039 | 25.5000 | . | 3.7008 | 94.0000 |
| 19/64 | 0.2969 | 7.5406 | . | 1.0236 | 26.0000 | . | 3.7402 | 95.0000 |
| 5/16 | 0.3125 | 7.9375 | . | 1.0433 | 26.5000 | . | 3.7795 | 96.0000 |
| . | 0.3150 | 8.0000 | . | 1.0630 | 27.0000 | . | 3.8189 | 97.0000 |
| 21/64 | 0.3281 | 8.3344 | . | 1.0827 | 27.5000 | . | 3.8583 | 98.0000 |
| . | 0.3346 | 8.5000 | . | 1.1024 | 28.0000 | . | 3.8976 | 99.0000 |
| 11/32 | 0.3438 | 8.7313 | . | 1.1220 | 28.5000 | . | 3.9370 | 100.0000 |
| . | 0.3543 | 9.0000 | . | 1.1417 | 29.0000 | 4 | 4.0000 | 101.6000 |
| 23/64 | 0.3594 | 9.1281 | . | 1.1614 | 29.5000 | . | 4.3307 | 110.0000 |
| . | 0.3740 | 9.5000 | . | 1.1811 | 30.0000 | 4 1/2 | 4.5000 | 114.3000 |
| 3/8 | 0.3750 | 9.5250 | . | 1.2205 | 31.0000 | . | 4.7244 | 120.0000 |
| 25/64 | 0.3906 | 9.9219 | 1 1/4 | 1.2500 | 31.7500 | 5 | 5.0000 | 127.0000 |
| . | 0.3937 | 10.0000 | . | 1.2598 | 32.0000 | . | 5.1181 | 130.0000 |
| 13/32 | 0.4063 | 10.3188 | . | 1.2992 | 33.0000 | . | 5.5118 | 140.0000 |
| . | 0.4134 | 10.5000 | . | 1.3386 | 34.0000 | . | 5.9055 | 150.0000 |
| 27/64 | 0.4219 | 10.7156 | . | 1.3780 | 35.0000 | 6 | 6.0000 | 152.4000 |
| . | 0.4331 | 11.0000 | . | 1.4173 | 36.0000 | . | 6.2992 | 160.0000 |
| 7/16 | 0.4375 | 11.1125 | . | 1.4567 | 37.0000 | . | 6.6929 | 170.0000 |
| . | 0.4528 | 11.5000 | . | 1.4961 | 38.0000 | . | 7.0866 | 180.0000 |
| 29/64 | 0.4531 | 11.5094 | 1 1/2 | 1.5000 | 38.1000 | . | 7.4803 | 190.0000 |
| 15/32 | 0.4688 | 11.9063 | . | 1.5354 | 39.0000 | . | 7.8740 | 200.0000 |
| . | 0.4724 | 12.0000 | . | 1.5748 | 40.0000 | 8 | 8.0000 | 203.2000 |
| 31/64 | 0.4844 | 12.3031 | . | 1.6142 | 41.0000 | . | 9.8425 | 250.0000 |
| . | 0.4921 | 12.5000 | . | 1.6535 | 42.0000 | 10 | 10.0000 | 254.0000 |
| 1/2 | 0.5000 | 12.7000 | . | 1.6929 | 43.0000 | 20 | 20.0000 | 508.0000 |
| . | 0.5118 | 13.0000 | . | 1.7323 | 44.0000 | 30 | 30.0000 | 762.0000 |
| 33/64 | 0.5156 | 13.0969 | 1 3/4 | 1.7500 | 44.4500 | 40 | 40.0000 | 1016.000 |
| 17/32 | 0.5313 | 13.4938 | . | 1.7717 | 45.0000 | 60 | 60.0000 | 1524.000 |
| . | 0.5315 | 13.5000 | . | 1.8110 | 46.0000 | 80 | 80.0000 | 2032.000 |
| 35/64 | 0.5469 | 13.8906 | . | 1.8504 | 47.0000 | 100 | 100.0000 | 2540.000 |

TO CONVERT TO MILLIMETERS; MULTIPLY INCHES X 25.4
TO CONVERT TO INCHES; MULTIPLY MILLIMETERS X 0.03937*
*FOR SLIGHTLY GREATER ACCURACY WHEN CONVERTING TO INCHES; DIVIDE MILLIMETERS BY 25.4

207A01.EPS

 Cable Tray

# Additional Resources

This module presents thorough resources for task training. The following resource material is suggested for further study.

*National Electrical Code® Handbook*, Latest Edition. Quincy, MA: National Fire Protection Association.

# Figure Credits

Associated Builders and Contractors, Inc., Module opener

Topaz Publications, Inc., SA01, SA02, SA09, and SA10

John Traister, Figures 1–3 and 18

Tim Dean, SA03, SA04, Figures 9, 12, 15, SA07, and SA08

Greenlee Textron, Inc., a subsidiary of Textron, Inc., SA05

Jim Mitchem, SA06

Tim Ely, SA11

# CONTREN® LEARNING SERIES — USER UPDATE

NCCER makes every effort to keep its textbooks up-to-date and free of technical errors. We appreciate your help in this process. If you find an error, a typographical mistake, or an inaccuracy in NCCER's Contren® materials, please fill out this form (or a photocopy), or complete the online form at www.nccer.org/olf. Be sure to include the exact module number, page number, a detailed description, and your recommended correction. Your input will be brought to the attention of the Authoring Team. Thank you for your assistance.

*Instructors* – If you have an idea for improving this textbook, or have found that additional materials were necessary to teach this module effectively, please let us know so that we may present your suggestions to the Authoring Team.

### NCCER Product Development and Revision
3600 NW 43rd Street, Building G, Gainesville, FL 32606

**Fax:** 352-334-0932
**Email:** curriculum@nccer.org
**Online:** www.nccer.org/olf

☐ Trainee Guide    ☐ AIG    ☐ Exam    ☐ PowerPoints    Other _____

Craft / Level: _____ Copyright Date: _____

Module Number / Title: _____

Section Number(s): _____

Description: _____

_____

_____

_____

_____

Recommended Correction: _____

_____

_____

_____

_____

Your Name: _____

Address: _____

_____

Email: _____ Phone: _____

# Conductor Terminations and Splices

**Daytona International Speedway Infield**

QuestCom won an ABC Excellence in Construction award for its work on updating Daytona International Speedway's 45-year-old facilities with state-of-the-art telecommunications, fire alarm systems, and concert-quality sound.

## 26208-11

Trainees with successful module completions may be eligible for credentialing through NCCER's National Registry. To learn more, go to **www.nccer.org** or contact us at **1.888.622.3720.** Our website has information on the latest product releases and training, as well as online versions of our *Cornerstone* newsletter and Pearson's Contren® product catalog.

Your feedback is welcome. You may email your comments to **curriculum@nccer.org,** send general comments and inquiries to **info@nccer.org,** or use the User Update form at the back of this module.

 V.1 7/11

# Conductor Terminations and Splices

## Objectives

When you have completed this module, you will be able to do the following:

1. Describe how to make a good conductor termination.
2. Prepare cable ends for terminations and splices and connect using lugs or connectors.
3. Train cable at termination points.
4. Understand the *National Electrical Code®* (*NEC®*) requirements for making cable terminations and splices.
5. Demonstrate crimping techniques.
6. Select the proper lug or connector for the job.

## Performance Tasks

Under the supervision of the instructor, you should be able to do the following:

1. Terminate conductors using selected crimp-type and mechanical-type terminals and connectors.
2. Terminate conductors on a terminal strip.
3. Insulate selected types of wire splices and/or install a motor connection kit.

## Trade Terms

| | | |
|---|---|---|
| AL-CU | Insulating tape | Splice |
| Amperage capacity | Lug | Strand |
| Connection | Mechanical advantage (MA) | Terminal |
| Connector | Pressure connector | Termination |
| Drain wire | Reducing connector | |
| Grooming | Shielding | |

## Required Trainee Materials

1. Pencil and paper
2. Copy of the latest edition of the *National Electrical Code®*
3. Appropriate personal protective equipment

Note: *NFPA 70®*, *National Electrical Code®*, and *NEC®* are registered trademarks of the National Fire Protection Association, Inc., Quincy, MA 02269. All *National Electrical Code®* and *NEC®* references in this module refer to the 2011 edition of the *National Electrical Code®*.

# Contents

*Topics to be presented in this module include:*

# Figures and Tables

# 1.0.0 INTRODUCTION

Anyone involved with electrical systems of any type must be familiar with wire **connectors** and splicing, as they are both necessary to make the numerous electrical joints required during the course of an electrical installation. A properly made **splice** and **connection** should last as long as the insulation on the wire itself, while a poorly made connection will always be a source of trouble; that is, the joints will overheat under load and eventually fail with the potential for starting a fire.

The basic requirements for a good electrical connection include the following:

- It must be mechanically and electrically secure.
- It must be insulated as well as or better than the existing insulation on the conductors.
- These characteristics should last as long as the conductor is in service.

> **NOTE**
>
> Every splice is a point of potential failure. Therefore, splicing should not be used as a solution to providing shorter conductor pulls. In fact, some applications prohibit the use of splices in critical areas.

There are many different types of electrical joints, and the selection of the proper type for a given application often depends on how and where the splice or connection is used. Electrical joints are normally made with a solderless **pressure connector** or **lug** to save time.

# 2.0.0 STRIPPING AND CLEANING CONDUCTORS

Before any connection or splice can be made, the ends of the conductors must be properly cleaned and stripped. To ensure a low-resistance connection and avoid contaminating the termination, clean the areas of the cable where it is to be cut and stripped. Remove any pulling compound, dirt, oil, grease, or water to avoid contaminating the exposed conductor.

Stripping is the removal of insulation from the end of the conductor or at the location of the splice. Conductors should only be stripped using the appropriate stripping tool. This will help to prevent cuts and nicks in the wire, which can reduce the conductor area and weaken the conductor.

Poorly stripped conductors can result in nicks, scrapes, or burnishes. Any of these can lead to a stress concentration in the damaged area. Heat, rapid temperature changes, mechanical vibration, and oscillatory motion can aggravate the damage, causing faults in the circuitry or even total failure. Lost **strands** are a problem in splice or crimp-type **terminals**, while exposed strands might be a safety hazard.

Faulty stripping can pierce, scuff, or split the insulation. This can cause changes in dielectric strength and lower the conductor's resistance to moisture and abrasion. Insulation particles often get trapped in solder and crimp joints. These form the basis for a defective **termination**. A variety of factors determine how precisely a conductor can be stripped, including the wire size and type of insulation.

It is a common misconception that a certain gauge of stranded conductor has the same diameter as a solid conductor. This is a very important consideration in selecting the proper blades for strippers. *Table 1* shows the nominal sizes referenced for the different wire gauges.

To eliminate nicking, cutting, and fraying, wires should only be stripped using the appropriate stripping tool. The specific tool used depends on the size and type of wire being stripped.

## 2.1.0 Stripping Small Conductors

There are many kinds of wire strippers available. *Figure 1* shows a common type of wire stripper for small conductors. It can be used to cut, strip, and crimp wires from No. 22 through No. 10 AWG. To use this tool, insert the conductor into the proper size knife groove, then squeeze the tool handles. The tool cuts the conductor insulation, allowing the conductor to be easily removed

**Table 1** Dimensions of Common Wire Sizes

| Size (AWG/kcmil) | Area (Circular Mils) | Overall Diameter in Inches | |
| --- | --- | --- | --- |
| | | Solid | Stranded |
| 18 | 1,620 | 0.040 | 0.046 |
| 16 | 2,580 | 0.051 | 0.058 |
| 14 | 4,130 | 0.064 | 0.073 |
| 12 | 6,530 | 0.081 | 0.092 |
| 10 | 10,380 | 0.102 | 0.116 |
| 8 | 16,510 | 0.128 | 0.146 |
| 6 | 26,240 | — | 0.184 |
| 4 | 41,740 | — | 0.232 |
| 3 | 52,620 | — | 0.260 |
| 2 | 66,360 | — | 0.292 |
| 1 | 83,690 | — | 0.332 |
| 1/0 | 105,600 | — | 0.373 |
| 2/0 | 133,100 | — | 0.419 |
| 3/0 | 167,800 | — | 0.470 |
| 4/0 | 211,600 | — | 0.528 |
| 250 | — | — | 0.575 |
| 300 | — | — | 0.630 |
| 350 | — | — | 0.681 |
| 400 | — | — | 0.728 |
| 500 | — | — | 0.813 |
| 600 | — | — | 0.893 |
| 700 | — | — | 0.964 |
| 750 | — | — | 0.998 |
| 800 | — | — | 1.03 |
| 900 | — | — | 1.09 |
| 1,000 | — | — | 1.15 |
| 1,250 | — | — | 1.29 |
| 1,500 | — | — | 1.41 |
| 1,750 | — | — | 1.52 |
| 2,000 | — | — | 1.63 |

(A)

(B)

208F02.EPS

*Figure 2* Wire strippers.

208F01.EPS

*Figure 1* Wire stripper/crimper.

without crushing its stripped end. The length of the strip is regulated by the amount of wire extending beyond the blades when it is inserted in the knife groove.

*Figure 2* shows production-grade stripping tools. *Figure 2(A)* can be used to strip conductors from No. 20 to No. 10 AWG, while *Figure 2(B)* strips wires from No. 18 to No. 6 AWG. Note that this tool has front entry jaws for use in tight spaces.

## 2.2.0 Stripping Power Cables and Large Conductors

Larger conductors can be cut using a ratchet-type cable cutter. The cable cutter shown in *Figure 3* can be used to cut cables up to 1,000 kcmil. Heavy-duty strippers are used to strip large power cables. *Figure 4* shows a heavy-duty stripper used to strip power cables with outside diameters ranging from 1/0 through 1,000 kcmil. Strippers can be used to remove insulation from the end of a cable or to make window cuts (*Figure 5*). All stripping tools should be operated according to the manufacturer's instructions. The procedures for using the tool shown in *Figure 4* to strip insulation from the end of a cable and to make a window cut are described here.

208F03.EPS

*Figure 3*  Ratchet-type cable cutter.

END
TERMINATION

WINDOW
CUT

SPIRAL
CUT

CIRCUMFERENTIAL (SQUARE) CUT
LONGITUDINAL CUT

CIRCUMFERENTIAL AND
LONGITUDINAL CUTS

208F05.EPS

*Figure 5*  Types of cable stripping.

CAP ASSEMBLY

BLADE

LOCKING KNOB

208F04.EPS

*Figure 4*  Heavy-duty cable stripper.

**On Site**

# Cable Strippers

Open-blade (knife) stripping can damage the conductors and is prohibited on many job sites. Adjustable strippers such as the tool shown here are used instead.

208SA01.EPS

To strip insulation from the end of a cable, proceed as follows:

**Step 1** Loosen the locking knob to open the tool to the maximum position. Place the cable in the V-groove and close the tool firmly around the cable. Tighten the locking knob.

**Step 2** Turn the cap assembly until the blade reaches the required depth.

**Step 3** Rotate the tool around the cable, advancing to the required strip length.

**Step 4** Rotate the tool in the reverse direction to produce a square end cut.

**Step 5** Loosen the locking knob to release the tool and remove it from the cable.

**Step 6** Peel off the insulation.

To make a window cut, proceed as follows:

**Step 1** With the tool opened to the maximum position, place the cable in the V-groove and close the tool firmly around the cable. Tighten the locking knob.

**Step 2** Turn the cap assembly until the blade reaches the required depth.

**Step 3** Rotate the tool to produce the first square cut.

**Step 4** Rotate the tool in the reverse direction to cut the required window strip length.

**Step 5** Rotate the tool in the original direction to produce the second square cut.

**Step 6** Loosen the locking knob assembly to release the tool and remove it from the cable.

**Step 7** Peel off the insulation.

*Figure 6* shows a round cable slitting and ringing tool that can be used to strip single- or multi-conductor cables. This tool can be used to cut around the cable (square cut) or slit the length of the cable jacket (longitudinal cut) for easy removal. The tool blade is adjustable to accommodate different jacket thicknesses.

### 2.3.0 Stripping Control and Signal Cable/Conductors

A scissor-action type of cutting tool is preferable to cutting tools with jaws that butt against each other. These tools have a tendency to produce a flattened chisel end on the conductor, especially when the cutting edges become dull, as shown in *Figure 7*. This makes it difficult to insert the conductor into the barrel of a terminal.

Observe the following points when stripping conductors:

• Remove the cable jacket using strippers with an adjustable blade or a die designed for the particular wire size (*Figure 8*). The terminal manufacturer will recommend a stripping length. Be careful to avoid nicking or stretching the conductor or insulation in a multi-conductor cable. If this type of damage occurs, it is likely that the wrong groove of the stripping tool was used, the tool was improperly set or damaged, the wrong type of stripping blade was used, or the tool was used incorrectly.

208F06.EPS

*Figure 6*   Round cable slitting and ringing tool.

DULL EDGE LEAVES
CHISEL POINT ON WIRE

208F07.EPS

*Figure 7*   Chisel point on a conductor.

SEMI-AUTOMATIC
DIE STRIPPER

SCISSOR
STRIPPER/
DIMPLE
CRIMPER

208F08.EPS

*Figure 8*   Cable and wire stripping tools.

## Cutting Large Power Cable

When cutting large power cables prior to stripping, you can make the job easier by using a battery-powered cable cutter such as the one shown here. Electric and hydraulic cable cutters are also available.

208SA02.EPS

- Cut the insulation so that no frayed pieces or threads extend past the point of cutoff. Frayed pieces or threads of insulation indicate the use of an improper tool or dull cutter blades.
- If possible, do not twist, spread, or disturb the wire strands from their original position in the cable. Retwisting or tightening the twist of the strands will eventually result in damage.
- Terminate stripped wire as soon as possible. The exposed strands will invariably become bent and spread, making termination difficult. A minimum amount of handling and storage after stripping will result in better terminations.

*Figure 9* shows the positioning of the wire in the crimp barrel when stripped to the proper length. The conductor insulation must be in the belled mouth of the terminal. This relieves stress on the

**STRIPPING THAT IS TOO LONG WILL INTERFERE WITH THE TERMINAL SCREW**

**STRIPPING THAT IS TOO SHORT DOES NOT PROVIDE ENOUGH CONTACT SURFACE**

**PROPER STRIPPING LENGTH WITH INSULATION INSIDE THE TERMINAL**

208F09.EPS

*Figure 9*  Proper stripping length.

---

**Think About It**

# Mechanical Compression Connectors

Can you name a use for a compression connector with a two-hole tongue?

---

strands or wire and increases the strength of the connection. Allowing conductor strands to protrude out of the inspection hole more than $\frac{1}{32}$" will interfere with the terminal screw. Cutting the strands too short will reduce the contact surface area.

## 3.0.0 WIRE CONNECTIONS UNDER 600 VOLTS

*NEC Section 110.14* governs electrical connections, including terminations and splices. Wire connections are used to connect a wire or cable to electrically operated devices such as fan coil units, duct heaters, oil burners, motors, pumps, and control circuits of all types.

A variety of wire connectors for stranded wire are shown in *Figure 10*. These connectors are available in various sizes to accommodate wire sizes No. 22 AWG and larger. They can be installed with crimping tools having a single or double indenter. The range is normally stamped on the tongue of each terminal.

Mechanical compression-type terminators are also available to accommodate wires from No. 8 AWG through 1,000 kcmil. One-hole lugs, two-hole lugs, split-bolt connectors, and other types are shown in *Figure 11*.

208F10.EPS

*Figure 10*  Crimp-on wire lugs.

---

NCCER — *Contren® Learning Series*  26208-11

ONE BARREL, OFFSET TONGUE
ONE HOLE
NO. 14 AWG THROUGH 1,000 KCMIL

ONE BARREL, STRAIGHT TONGUE
TWO HOLE
NO. 14 AWG THROUGH 1,000 KCMIL

ONE BARREL, FIXED TONGUE
ONE HOLE
NO. 14 AWG THROUGH 500 KCMIL

SINGLE HOLE
NO. 14 AWG THROUGH 4/0

ONE BARREL, STRAIGHT TONGUE
ONE HOLE
NO. 14 AWG THROUGH 1,000 KCMIL

TWO HOLE, PANELBOARD CONNECTOR
NO. 2 AWG THROUGH 750 KCMIL

ONE BARREL, OFFSET TONGUE
TWO HOLE
NO. 14 AWG THROUGH 1,000 KCMIL

TWO BARRELS, OFFSET TONGUE
ONE HOLE
NO. 6 AWG THROUGH 500 KCMIL

PARALLEL-TAP CONNECTOR
WITH INSULATED COVER
(VARIOUS WIRE SIZE COMBINATIONS)

SPLIT BOLT CONNECTOR
(2) NO. 14 AWG THROUGH (2) 1,000 KCMIL
RUN AND TAP COMBINATIONS

208F11.EPS

*Figure 11*  Various mechanical compression connectors.

Crimp-type connectors for wires smaller than No. 8 AWG are normally made to accept at least two wire sizes and are often color coded. For example, one manufacturer's color code is red for No. 18 or No. 20 wire, blue for No. 16 or No. 14 wire, and yellow for No. 12 or No. 10 wire. Crimp-type connectors for wire sizes No. 8 and larger, commonly called lugs, are made to accept one specific conductor size. Crimp-type reducing connectors are used to connect two different size wires. Mechanical compression-type connectors and lugs are made to accommodate a range of different wire sizes.

The parallel-tap connector with an insulated cover shown in *Figure 11* is one example of a pre-insulated, molded mechanical compression connector. There are several kinds available. They come in setscrew/pressure plate and insulation-piercing configurations made for use in a variety of feeder tap and splice applications. Because they are equipped with an insulating cover, the requirement for taping the joint is eliminated.

## Tightening Compression Connector Screws and Bolts

Mechanical compression connectors must be tightened to a specified torque using a torque screwdriver or torque wrench. Overtightening can cut the wires or break the fitting, while undertightening may lead to loose connections, resulting in overheating and failure.

208SA03.EPS

### 3.1.0 Aluminum Connections

Aluminum has certain properties that are different from copper and special precautions must be taken to ensure reliable connections. These properties are: cold flow (aluminum does not spring back in the same way as copper and has a tendency to deform), coefficient of thermal expansion, susceptibility to galvanic corrosion, and the formation of oxide film on the surface.

Because of the thermal expansion and cold flow of aluminum, standard copper connectors cannot be safely used on aluminum wire. Most manufacturers design their aluminum connectors with greater contact area to counteract this problem. Tongues and barrels of all aluminum connectors are larger or deeper than comparable copper connectors.

The electrolytic action between aluminum and copper can be controlled by plating the aluminum with a neutral metal (usually tin). The plating prevents electrolysis from taking place, and the joint remains tight. Connectors should also be tin-plated and prefilled with an oxide-inhibiting compound.

The insulating aluminum oxide film must be removed or penetrated before a reliable aluminum joint can be made. Aluminum connectors are

designed to bite through this film as they are applied to conductors. The conductor should also be wire brushed and coated with a joint compound to ensure a reliable joint.

> **CAUTION**
> Never use connectors designed strictly for use on copper conductors on aluminum conductors. Connectors listed for use on both metals will be marked AL-CU. All connectors must be applied and installed in the manner for which they are listed and labeled.

*NEC Section 110.14* prohibits conductors made of dissimilar metals (copper and aluminum, copper and copper-clad aluminum, or aluminum and copper-clad aluminum) from being intermixed in a terminal or splicing connector unless the device is identified for the purpose. As a general rule:

- Connectors marked with only the wire size should only be used with copper conductors.
- Connectors marked with AL and the wire size should only be used with aluminum conductors.
- Connectors marked with AL-CU and the wire size may be safely used with either copper or aluminum.

### 3.2.0 Heat-Shrink Insulators

Heat-shrink insulators for small connectors provide skintight insulation protection and are fast and easy to use. They are designed to slip over wires, taper pins, connectors, terminals, and splices. When heat is applied, the insulation becomes semi-rigid and provides positive strain relief at the flex point of the conductor. A vaporproof band seals and protects the conductor against abrasion, chemicals, dust, gasoline, oil, and moisture. Extreme temperatures, both hot and cold, will not affect the performance of these insulators. The source of heat can be any number of types, but most manufacturers of these insulators also produce a heat gun especially designed for use on heat-shrink insulators. It is similar to a conventional hair dryer, as shown in *Figure 12*.

In general, a heat-shrink insulator may be thought of as tubing with a memory. After it is initially manufactured, it is heated and expanded to a predetermined diameter and then cooled. Upon a second application of heat, the tubing compound "remembers" its original size and shrinks to that smaller diameter. This property enables it to conform to the contours of any object. Various types of heat-shrink tubing are available, some of which include the following:

- *PVC* – This is a general-purpose, economical tubing widely used in the electronics industry. It provides good electrical and mechanical protection and resists cracking and splitting.
- *Polyolefin* – Polyolefin tubing has a wide range of uses for wire bundling, harnessing, strain relief, and other applications where cables and components require additional insulation. It is flame-retardant, flexible, and comes in a variety of colors.
- *Double wall* – This tubing is designed for outstanding protective characteristics. It is a semi-rigid tubing with an inner wall that melts and an outer wall that shrinks to conform to the melted area.
- *Teflon®* – This type is considered by many users to be the best overall heat-shrink tubing—physically, electrically, and chemically. Its high temperature rating of 250°C resists brittleness from extended exposure to heat and will not support combustion.
- *Neoprene* – This is a highly durable and flexible tubing that provides superior protection against abrasion.
- *Kynar®* – This is a thin-wall, semi-rigid tubing with outstanding resistance to abrasion. This transparent tubing enables easy inspection of components.

Tubing is available in a wide variety of colors and configurations. The manufacturer's tubing selector guide can help in the selection of the best tubing for any given application. A typical tubing selector guide appears in *Table 2*.

HEAT GUN

SLIP INSULATOR OVER OBJECT TO BE INSULATED, THEN APPLY HEAT FOR A FEW SECONDS

WHEN FINISHED, IT PROVIDES PERMANENT INSULATION PROTECTION

208F12.EPS

*Figure 12*   Method of installing heat-shrink insulators.

**Table 2**  Tubing Selector Guide

| Type | Material | Temp. Range (°C) | Shrink Ratio | Max. Long. Shrinkage (%) | Tensile Strength (psi) | Colors | Dielectric Strength (V/mil) |
|---|---|---|---|---|---|---|---|
| Nonshrinkable | PVC | +105 | — | — | 2,700 | White, red, clear, black | 800 |
| Shrinkable | PVC | −35 to +105 | 2:1 | 10 | 2,700 | Clear, black | 750 |
| Nonshrinkable | Teflon® | −65 to +260 | — | — | 2,700 | Clear | 1,400 |
| Shrinkable | Flexible polyolefin | −55 to +135 | 2:1 | 5 | 2,500 | Black, white, red, yellow, blue, clear | 1,300 |
| Nonshrinkable | Teflon® | −65 to +260 | — | — | 7,500 | Clear | 1,400 |
| Shrinkable | Polyolefin double wall | −55 to +110 | 6:1 | 5 | 2,500 | Black | 1,100 |
| Shrinkable | Kynar® | −55 to +175 | 2:1 | 10 | 8,000 | Clear | 1,500 |
| Shrinkable | Teflon® | +250 | 1.2:2 | 10 | 6,000 | Clear | 1,500 |
| Shrinkable | Teflon® | +250 | 1½:1 | 10 | 6,000 | Clear | 1,500 |
| Shrinkable | Neoprene | +120 | 2:1 | 10 | 1,500 | Black | 300 |

# 4.0.0 Control and Signal Cable

For the most part, electricians install all the associated power cables and many of the signal cables for the instrumentation systems in industrial facilities. It is necessary to select the proper type and rating of control and signal cables to meet *National Electrical Code*® (*NEC*®) requirements, local codes, and equipment manufacturer requirements for the system being installed. With the exception of optical fiber or communication cables that enter a structure and are terminated in an enclosure within 50' of the entrance, the *NEC*® requires that control and signal cables be listed and marked with an appropriate classification code.

## 4.1.0 *NEC*® Classifications and Ratings

Control and signal cables are type-classified, rated, and listed for use in various areas of a structure in accordance with the following:

- *NEC Article 725*, *Remote Control, Signaling, and Power-Limited Circuits (Class 1, 2, and 3 Circuits)*
- *NEC Article 727*, *Instrumentation Tray Cable*
- *NEC Article 760*, *Fire Alarm Systems*
- *NEC Article 770*, *Optical Fiber Cables and Raceways*
- *NEC Article 800*, *Communications Circuits*

All cable conforming to the requirements of the *NEC*® is normally marked by the cable manufacturer with the appropriate classification code. For more detail, refer to the latest edition of the *NEC*®.

## 4.2.0 Fire Alarm, Instrumentation, and Control Cable

Fire alarm, instrumentation, and control cable carry low-level signals that require less current than power cable. Conductor sizes range from No. 12 to No. 24 AWG, depending on the circuit requirements. Conductors can be tinned or bare, solid or stranded, and twisted or parallel. Shielding, if used, is usually an aluminum-coated Mylar® film with one or more drain wires (ground wires) inside a jacket. The jacket and conductor insulation are rated for the applicable usage. *Figure 13* shows typical fire alarm and instrumentation cable.

To be effective, the shielding drain wire(s) must be grounded. The installation loop diagrams or instructions for the system equipment will indicate how and where the shields are connected and grounded. Typically, only one end is grounded, and the drain wire at the other end is isolated by folding it back and taping over it. This prevents the electrical noise collected by the shield from recirculating through the system and interfering with the control signals.

FIRE ALARM CABLE

INSTRUMENTATION CABLE

208F13.EPS

*Figure 13*  Fire alarm and instrumentation cable.

# 5.0.0 LOW-VOLTAGE CONNECTORS AND TERMINALS

Low-voltage control circuits typically use compression-type crimp connectors. These connectors are often color coded by wire size.

## 5.1.0 Crimp Connectors for Screw Terminals

Compression-type connectors for connecting conductors to screw terminals for low-voltage circuits include those in which hand tools indent or crimp tube-like sleeves that hold a conductor. Proper crimping action changes the size and shape of the connector and deforms the conductor strands enough to provide good electrical conductivity and mechanical strength.

*Figure 14* shows the basic structure of a crimp connector. The crimp barrel receives the wire and is crimped to secure it in place. The V's or dimples inside the barrel improve the wire-to-terminal conductivity and also increase the termination tensile strength. Most crimp connectors have nylon or vinyl insulation covering the barrel to reduce the possibility of shorting to adjacent terminals. The insulation is color coded according to the connector's wire range to reduce the problem of wire-to-connector mismatch. An inspection hole is provided at the end of the barrel to allow visual inspection of the wire position. For the smaller wire sizes, a sleeve is crimped over the conductor insulation in the process of crimping the barrel, providing strain relief for the conductor.

The barrel is connected to the terminal tongue, which physically connects the wire to the termination point, such as a terminal screw. Information about the connector size and conductor range is usually stamped on the tongue by the manufacturer. Tongue styles vary depending on termination requirements. *Figure 15* shows standard tongue styles. The styles most frequently used are the ring tongue and flanged or locking fork. These types are preferred because the terminals will not slip off the terminal screw as the screw is tightened. They are also compatible with most vendor-supplied termination points.

## 5.2.0 Color Codes

Most manufacturers color code the barrel insulation to provide quick identification for installation and as an aid to inspection. Different colors, or a combination of colors, have special meanings. Although manufacturers vary, common or standard colors have been accepted. *Table 3* lists typical color codes.

*Figure 14*   Basic crimp connector structure.

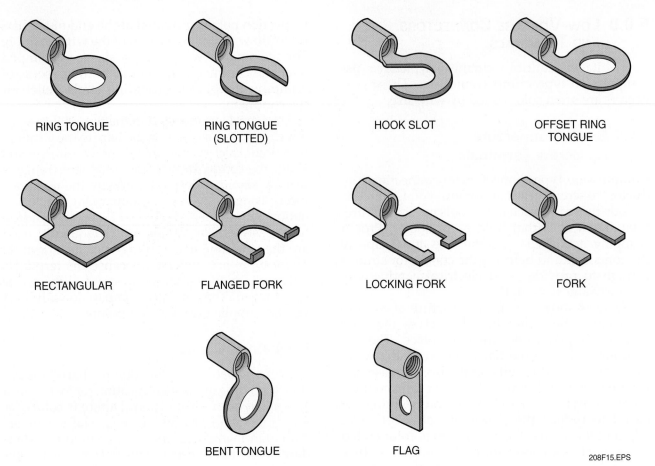

RING TONGUE     RING TONGUE (SLOTTED)     HOOK SLOT     OFFSET RING TONGUE

RECTANGULAR     FLANGED FORK     LOCKING FORK     FORK

BENT TONGUE     FLAG

208F15.EPS

*Figure 15*   Standard tongue styles of crimped connectors.

**Table 3**  Typical Color Codes

| AWG Wire Size | Color Code |
|---|---|
| 22–16 | Red |
| 16–14 | Blue |
| 12–10 | Yellow |

Color combinations are sometimes varied to indicate the class or grade rating of an individual lug or splice. Some manufacturers use a clear plastic or other suitable insulation on the crimp barrel with a colored line to indicate wire size range.

## 6.0.0 GUIDELINES FOR INSTALLING CONNECTORS

Before beginning work on incoming line connections, refer to all drawings and specifications dealing with the project at hand. Details of terminations are usually furnished on larger installations. Guidelines for installing the different types of connectors covered in this module are given in the following sections.

## 6.1.0 Installing Compression Connectors

The task of fastening a compression connector to a wire requires the use of the proper connector, crimping tool, and installation procedure. Always use the correct tool for the connector and follow the manufacturer's instructions.

### 6.1.1 Crimping Tools

With a compression connector, an electrical connection between a wire and a terminal can be made by tightly compressing the crimp barrel with an ordinary pair of pliers. However, such a connection would not necessarily be made to the required pressure or in the correct location to ensure a good connection. A crimping tool is required to produce consistently good connections.

*Figure 16* shows the relationship between the amount of crimping force and the mechanical and electrical performance. The maximum mechanical strength (A) occurs at a lower crimping force than the maximum electrical performance (B). The point of intersection (C) represents the ideal crimping force. Using a crimping die that is too

**Figure 16** Mechanical strength versus electrical performance of a crimped connector.

ELECTRICAL ‑ ‑ ‑
MECHANICAL ▬▬▬

208F16.EPS

large results in poor electrical performance, and using a die that is too small produces a weak mechanical connection.

A simple pliers-type crimping tool was the earliest crimping tool developed and continues to be used for repair operations or where only a few installations are to be made. These tools are similar in construction to ordinary mechanics' pliers except the jaws are specially shaped, as shown in *Figure 17*.

It takes a tremendous amount of force at the lug to make a good crimp. Ratchet tools provide a means of increasing the human output force. The typical force capability of a hand for repetitive operations is 75 pounds for adult men and 50 pounds for adult women. The amount the tool multiplies the hand force is termed the **mechanical advantage (MA)** of the tool.

Simple pliers are basically constant-MA tools; the MA is the same whether the crimp is being started or finished. By adjusting linkage or cam mechanisms connecting the handles to the crimp dies, the MA can be varied so that it is low at the start and high at the finish of the crimp stroke. Thus, when the handles start to close from the open position and little or no crimping is being done, the MA is low. As the handles are closed farther, the crimp dies begin to compress the terminal, and the MA increases. In this manner, the MA is patterned to the crimp force requirements and distributed so that a high MA is achieved over the portion of the cycle where required.

*Figure 18* shows a high MA type of tool equipped with a ratchet control. The ratchet mechanism prevents opening the tool and removing the crimped terminal before the handle has been closed all the way and the crimp completed. These provide a consistent, reliable crimp that meets the terminal manufacturer's requirements.

The crimp dies of this tool are interchangeable and may contain two or three positions for crimping different size terminals. These dies may be color coded to be used with a color-coded terminal lug for easy identification and to ensure proper crimping force. The crimp die of the tool determines the completed crimp configuration. There are a variety of configurations in use such as the simple nest and indenter type of die or the more complicated four-indent die.

*Figure 19* shows hand-operated and hydraulic crimping tools typical of those used to crimp connectors for stranded wires ranging from No. 8 AWG to 750 kcmil. These tools normally develop about 12 tons of compression force at 10,000 pounds per square inch (psi). Guidelines for the use of these tools are described here. Many other tools operate in a similar manner.

208F17.EPS

**Figure 17** Hand crimpers.

208F18.EPS

**Figure 18** Leveraged crimping tool.

**DIE SET**

**HAND-OPERATED**

**HYDRAULIC**

208F19.EPS

*Figure 19*  Crimping tools used to crimp large connectors.

**NOTE**

Some manufacturers design a crimping tool that must be used with their connectors. Use the crimping tool recommended by the manufacturer.

Several different configurations may work equally well for some applications, while for others, a certain shape is superior. Many considerations affect the determination of crimp die configuration, including the type of terminal (size, shape, material, and function), as well as the type and size of wires to be accommodated.

To use a hand-operated crimping tool, proceed as follows:

**Step 1**  Select the proper die for use with the connector to be crimped. Do not operate the tool without the die.

**Step 2**  Push the die release button on the C-head and slide one of the die halves into position until the retainer snaps. Insert the other die half in the piston body by pushing the die release button and sliding the die in until the retainer snaps.

**Step 3**  Place the tool C-head in position over the connector to be crimped. Pump the handle until compression is complete, as indicated by the dies touching at their flat surfaces nearest the throat of the C-head.

**Step 4**  Retract the ram and remove the connector after completion of the crimp. This is done by raising the pump handle slightly, rotating it clockwise until it stops, and then pushing the handle down in a pumping motion until the pressure release snaps.

**WARNING!**

Always read and follow the manufacturer's instructions when using power tools.

To operate a hydraulic crimping tool, proceed as follows:

**Step 1**  Connect the hydraulic pump to the crimping tool using a suitable hydraulic hose.

**Step 2**  Select the proper dies for use with the connector to be crimped. Do not operate the tool without the dies.

**Step 3**  Push the die release button on the C-head and slide one of the die halves into position until the retainer pin snaps. In a similar manner, install the other die half in the piston body.

**Step 4**  Place the tool C-head in position over the connector to be crimped. Operate the remote pump until compression is complete, as indicated by the dies touching on the frame side.

**Step 5**  Release the pressure at the hydraulic pump to retract the lower die half, and then remove the connector from the tool.

In addition to hand-operated and hydraulic crimpers, battery-operated and corded crimping tools are also available (*Figures 20* and *21*). Battery-operated tools offer the advantage of freedom of movement, while corded tools allow extensive use without worrying about battery changes. Universal crimping tools are also available (*Figure 22*). They operate in the same way as other crimpers but do not require separate dies.

208F20.EPS

*Figure 20*  Battery-operated crimping tool.

208F21.EPS

*Figure 21*  Corded crimping tool.

208F22.EPS

*Figure 22*  Universal crimping tool.

### 6.1.2 General Compression (Crimp) Connector Installation Procedure

To make a crimp connection, proceed as follows:

*Step 1*  Select a crimp connector of proper size and appropriate material for the wire size you are using. Copper connectors should be used with copper wires and aluminum connectors with aluminum wires. Dual-rated connectors may be used with both copper and aluminum wires.

*Step 2*  Use a suitable wire stripper to remove the insulation from the end of the wire. Be careful not to nick the wire. Strip the insulation back far enough so that the bare conductor will go fully into the connector. Make sure not to strip off too much insulation; it should fit close to the connector when the wire is fully inserted into the connector.

*Step 3*  Clean the stripped portion of the wire. Use a wire brush for large wire sizes. Also clean the related unplated terminal pad and the surface to which the connector will be attached.

*Step 4*  Obtain the crimping tool and dies made for the type and size of connector to be crimped.

*Step 5*  Insert the stripped end of the wire completely into the connector. Position the crimping tool in place over the connector, and then operate the tool to fully crimp the connector, as directed in the tool manufacturer's instructions. Make sure that the crimping tool jaws are fully closed, indicating that a full compression crimp has been made. If multiple crimps are required (*Figure 23*), crimp from the lug back to the barrel base, rotating the crimper as necessary to avoid deforming the connector.

*Step 6*  Using a bolt or screw and washers (if required), secure the crimped connector and attached wire to the correct terminal in the equipment. Tighten the terminal bolt and torque to the level specified by the equipment manufacturer. Too little or too much torque can adversely affect the performance of the connection.

*Table 4* lists some torque values typical of those used for tightening common sizes of steel and aluminum terminal bolts.

**SPRING-LOADED TERMINAL BLOCK**

**SCREW-TYPE TERMINAL BLOCK**

CRIMPED ON CONDUCTOR AND INSERTED HERE

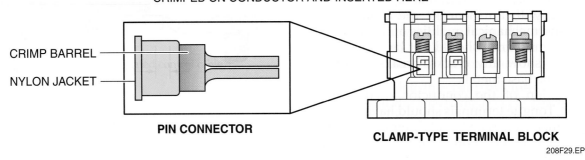

CRIMP BARREL

NYLON JACKET

**PIN CONNECTOR**

**CLAMP-TYPE TERMINAL BLOCK**

208F29.EPS

*Figure 29*  Terminal blocks.

- *Spring-type terminal block* – To install stripped wires in a spring-loaded terminal block, release the spring contact using a flat-blade screwdriver and insert the wire into the hole on top of the block. After the wire is inserted, remove the screwdriver. The wire is clamped in place by the spring contact.
- *Clamp-type terminal block* – To install stripped wires in a clamp-type terminal block, insert the wires into the boxed terminal and tighten the screw to clamp the wire in place.
- *Screw-type terminal block* – To install stripped wires in a common screw-type terminal block, curve the wires around the screw or fit them with a terminal lug that is inserted under the screw. Once the wire or terminal lug is under the head of the screw, tighten the screw to secure the wire.

> **NOTE**
>
> When installing wires in a common screw-type terminal block, only two terminals are permitted at any one terminal point. Also, only one flanged fork terminal may be used. When two types of terminals are located at one point, the bottom terminal should be installed upside down. This will provide easier installation and a neater appearance.

Care should be taken to strip the cable jacket to a point as close as possible to the first termination of the cable, but not to interfere with other terminations originating from that cable, as shown in *Figure 30*. Place the cable identification tag at that point on the cable jacket and make sure it can be read easily.

When multiple cables are installed, tie them neatly to a support without blocking access to the lower terminal blocks or interfering with the connection or disconnection of other wires.

Individual or multi-conductor cables should be routed parallel or at right angles to the frame or wireways. Take care to ensure that wires do not come in contact with sharp edges. Position shielded and coaxial cables on the outer perimeter of a cable bundle whenever possible and keep wire crossover to a minimum.

### 6.4.4 Cable/Conductor Routing and Inspection Considerations

Cable/conductor routing and inspection considerations include the following:

- *Wire bends* – For all wire and cable routing, use a minimum bend radius of three times the outside diameter of the wire or cable unless otherwise specified by the cable manufacturer.

**Figure 30** Routing cabling.

- *Neatness of routing* – Provide adequate clearance for movement of mechanical parts. Allow enough slack in the cable for servicing plugs and terminal boards. Support all cables to prevent strain on the conductors or terminals. Strain can cause breaks in the terminations and render the system inoperable. Ensure that cables do not cross maintenance openings or otherwise interfere with normal operation and replacement of components.
- *Protection of wires or cable* – Provide clearance between the conductors and any heat-radiating components to prevent heat-related deterioration of the insulation. Protect all wires against abrasion. Do not route wire over sharp screws, lugs, terminals, or openings. Use grommets, chase nipples, or similar means to protect the insulation when passing wire through a metal partition.
- *Service loops* – Allow sufficient slack at the termination of each wire to permit one repair (such as cutting off a terminal and re-terminating it).
- *Jumper wires* – Bare jumper wires are not permitted. Use only insulated wire for jumpers.

- *Terminal bending limits* – Do not bend terminals more than 30 degrees above or below the termination point, as shown in *Figure 31*.

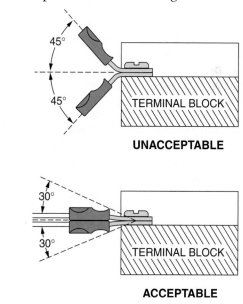

**Figure 31** Terminal bend radius.

## 7.0.0 BENDING CABLE AND TRAINING CONDUCTORS

Training is the positioning of cable so that it is not under tension. Bending is the positioning of cable that is under tension. When installing cable or any large conductors, the object is to limit the tension so that the cable's physical and electrical characteristics are maintained for the expected service life. Training conductors, rather than bending them, also reduces the tension on lugs and connectors, extending their service life considerably.

All bends made in cable must comply with the *NEC®*. The minimum bending radius is determined by the cable diameter and, in some instances, by the construction of the cable. For example, *NEC Section 330.24* states that bends in Type MC cable must be made so that the cable is not damaged, and the radius of the curve of the inner edge of any bend shall not be less than the following:

- *Smooth sheath* – Ten times the external diameter of the metallic sheath for cable not more than ¾" in external diameter, twelve times the external diameter of the metallic sheath for cable more than ¾" but not more than 1½" in external diameter, and fifteen times the external diameter of the metallic sheath for cable more than 1½" in external diameter.
- *Interlocked-type armor or corrugated sheath* – Seven times the external diameter of the metal sheath.
- *Shielded conductors* – Twelve times the overall diameter of the individual conductors or seven times the overall diameter of the multiconductor cable, whichever is greater.

Two common types of cable bending tools are the ratchet bender and the hydraulic bender. The ratchet cable bender in *Figure 32* bends 600V copper or aluminum conductors up to 500 kcmil, and the hydraulic bender in *Figure 33* is designed for cables from 350 kcmil through 1,000 kcmil. In addition, the hydraulic bender is capable of one-shot bends up to 90° and automatically unloading the cable when the bend is finished. Either type simplifies and speeds cable installation.

Conductors at terminals or conductors entering or leaving cabinets or cutout boxes and the like must comply with certain *NEC®* requirements, many of which are covered in *NEC Article 312*. The bending radii for various sizes of conductors that do not enter or leave an enclosure through the wall opposite its terminal are shown in *Table 5*. When using this table, the bending space at terminals must be measured in a straight line from the end of the lug or wire connector (in the direction

208F32.EPS

*Figure 32*   Ratchet bender.

208F33.EPS

*Figure 33*   Hydraulic bender.

that the wire leaves the terminal) to the wall, barrier, or obstruction, as shown in *Figure 34*.

An unshielded cable can tolerate a sharper bend than a shielded cable. This is especially true of cables having helical metal tapes, which, when bent too sharply, can separate or buckle and cut into the insulation. This causes increased electrical stress. The problem is compounded by the fact that most tapes are under jackets that conceal such damage. Damaged cable may initially pass acceptance testing, but often fails prematurely at the shield/insulation interface.

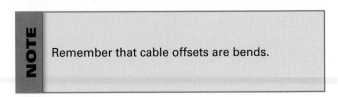

**NOTE**

Remember that cable offsets are bends.

## Table 5 Minimum Wire Bending Space for Conductors Not Entering or Leaving Opposite Wall [Data from *NEC Table 312.6(A)*]

| AWG or Circular-Mil Size of Wire | Wires per Terminal | | | | |
| --- | --- | --- | --- | --- | --- |
| | 1 | 2 | 3 | 4 | 5 |
| 14–10 | Not Specified | — | — | — | — |
| 8–6 | 1½ | — | — | — | — |
| 4–3 | 2 | — | — | — | — |
| 2 | 2½ | — | — | — | — |
| 1 | 3 | — | — | — | — |
| 1/0–2/0 | 3½ | 5 | 7 | — | — |
| 3/0–4/0 | 4 | 6 | 8 | — | — |
| 250 kcmil | 4½ | 6 | 8 | 10 | — |
| 300–350 kcmil | 5 | 8 | 10 | 12 | — |
| 400–500 kcmil | 6 | 8 | 10 | 12 | 14 |
| 600–700 kcmil | 8 | 10 | 12 | 14 | 16 |
| 750–900 kcmil | 8 | 12 | 14 | 16 | 18 |
| 1,000–1,250 kcmil | 10 | — | — | — | — |
| 1,500–2,000 kcmil | 12 | — | — | — | — |

Reprinted with permission from *NFPA 70®*, the *National Electrical Code®*. Copyright © 2010, National Fire Protection Association, Quincy, MA 02269. This reprinted material is not the complete and official position of the National Fire Protection Association on the referenced subject, which is represented only by the standard in its entirety.

## Table 6 Minimum Wire Bending Space for Conductors Entering or Leaving Opposite Wall [Data from *NEC Table 312.6(B)*]

| AWG or Circular-Mil Size of Wire | Wires per Terminal | | | |
| --- | --- | --- | --- | --- |
| | 1 | 2 | 3 | 4 or More |
| 14–10 | Not Specified | — | — | — |
| 8 | 1½ | — | — | — |
| 6 | 2 | — | — | — |
| 4 | 3 | — | — | — |
| 3 | 3 | — | — | — |
| 2 | 3½ | — | — | — |
| 1 | 4½ | — | — | — |
| 1/0 | 5½ | 5½ | 7 | — |
| 2/0 | 6 | 6 | 7½ | — |
| 3/0 | 6½ (½) | 6½ (½) | 8 | — |
| 4/0 | 7 (1) | 7½ (1½) | 8½ (½) | — |
| 250 | 8½ (2) | 8½ (2) | 9 (1) | 10 |
| 300 | 10 (3) | 10 (2) | 11 (1) | 12 |
| 350 | 12 (3) | 12 (3) | 13 (3) | 14 (2) |
| 400 | 13 (3) | 13 (3) | 14 (3) | 15 (3) |
| 500 | 14 (3) | 14 (3) | 15 (3) | 16 (3) |
| 600 | 15 (3) | 16 (3) | 18 (3) | 19 (3) |
| 700 | 16 (3) | 18 (3) | 20 (3) | 22 (3) |
| 750 | 17 (3) | 19 (3) | 22 (3) | 24 (3) |
| 800 | 18 | 20 | 22 | 24 |
| 900 | 19 | 22 | 24 | 24 |
| 1,000 | 20 | — | — | — |
| 1,250 | 22 | — | — | — |
| 1,500–2,000 | 24 | — | — | — |

Reprinted with permission from *NFPA 70®*, the *National Electrical Code®*. Copyright © 2010, National Fire Protection Association, Quincy, MA 02269. This reprinted material is not the complete and official position of the National Fire Protection Association on the referenced subject, which is represented only by the standard in its entirety.

When using *NEC Table 312.6(A)*, bending space at terminals must be measured in a straight line from the end of the lug or wire connector (in the direction that the wire leaves the terminals) to the wall, barrier, or obstruction.

208F34.EPS

*Figure 34*    Bending space at terminals is measured in a straight line.

When conductors enter or leave an enclosure through the wall opposite its terminals (*Figure 35*), *NEC Table 312.6(B)* applies. See *Table 6*. When using this table, the bending space at terminals must be measured in a straight line from the end of the lug or wire connector

Bending space at terminals must be measured in a straight line from the end of the lug or wire connector in a direction perpendicular to the enclosure wall. Use the values in *NEC Table 312.6(B)*.

208F35.EPS

*Figure 35*    Conductors entering an enclosure opposite the conductor terminals.

Conductor Terminations and Splices

in a direction perpendicular to the enclosure wall. For removable and lay-in wire terminals intended for only one wire, the bending space in the table may be reduced by the number of inches shown in parentheses.

## 8.0.0 *NEC*® TERMINATION REQUIREMENTS

There are many *NEC*® requirements governing the termination of conductors as well as the installation of enclosures containing conductors. *NEC Sections 110.14 and 312.6* cover most installations and terminations. However, other sections, such as *NEC Sections 300.15 and 430.10*, also apply for specific applications.

### 8.1.0 Overcurrent Protection

In general, all ungrounded conductors from a transformer secondary require overcurrent protection to comply with *NEC Section 240.21*. The conductors from the transformer secondary are the feeder to the service. The secondary conductors are protected by the transformer primary protection (assuming tap rules and other requirements are met) until the first overcurrent device (point of service) in the secondary circuit.

#### 8.1.1 Overcurrent Protection from Upstream Devices

Motor control centers that are fed from protective devices in a switchboard or other switchgear are not required to have a main breaker or disconnect switch in the motor control center (MCC). *Figure 36* shows feeders from a 480V switchboard terminating in the main lugs only (MLO) incoming line compartment of an MCC. This connection is made directly to the horizontal bus system distributing power to the vertical bus in each vertical section of the MCC.

#### 8.1.2 Overcurrent Protection within Equipment

*Figure 37* shows a circuit interrupter and fuses serving as the main disconnect and overcurrent protection for an MCC fed directly from a transformer secondary. It also illustrates the *NEC*® requirements for conductors entering enclosures

208F36.EPS

***Figure 36***   Incoming feeders connected to horizontal bus.

and the use of enclosures for routing or tapping conductors. The load side of the overcurrent protective device is connected to the MCC busbar system. Where the main disconnect is rated at 400A or less, the load connection may be made with stab connections to vertical bus sections connecting to the main horizontal bus.

#### 8.1.3 Short Circuit Bracing

All incoming lines to either incoming line lugs or main disconnects must be braced to withstand the mechanical force created by a high fault current. If the cables are not anchored sufficiently or the lugs are not tightened correctly, the connections become the weakest part of a panelboard or motor control center when a fault develops. In most cases, each incoming line compartment is equipped with a two-piece spreader bar located at a certain distance from the conduit entry. This spreader bar should be used along with appropriate lacing material to tie cables together where they can be bundled and to hold them apart where they must be separated. In other words, the incoming line cables should first be positioned and then anchored in place.

Manufacturers of electrical panelboards and motor control centers normally furnish detailed information on recommended methods of short circuit bracing; follow this information exactly.

Conductors must not be deflected within a cabinet or cutout box unless a gutter having a width in accordance with *NEC Table 312.6(A)* is provided.

Where ungrounded conductors of No. 4 or larger enter a raceway in a cabinet or other enclosure, the conductors must be protected by a substantial fitting providing a smoothly rounded insulated surface, unless the conductors are separated from the raceway fitting by substantial insulating material securely fastened in place.
*NEC Sections 312.6(C) and 300.4(G)*

CIRCUIT BREAKER

VERTICAL WIREWAY

Enclosures for switches or overcurrent devices must not be used as junction boxes, auxiliary gutters, or raceways for conductors feeding through or tapping off to other switches or overcurrent devices, unless the conditions of *NEC Section 312.8* are met.

FUSES

Cabinets and cutout boxes must have sufficient space to accommodate all conductors installed in them without crowding. *NEC Section 312.7*

STAB CONNECTORS

208F37.EPS

*Figure 37* MCC fed directly from a transformer secondary.

## 9.0.0 TAPING ELECTRICAL JOINTS

When it is not practical to protect a spliced joint by some other means, electrical tape may be used to insulate the joint. Joints must be taped carefully to provide the same quality of insulation over the splice as over the rest of the conductors.

Various types of electrical insulating tapes are available for use in specific applications. Some common types of electrical tape include vinyl plastic tape, linerless rubber tape, high-temperature silicone rubber tape, and glass cloth tape. Electrical tapes made of vinyl plastic are widely used as primary insulation on joints made with thermoplastic-insulated wires. They are used for splices up to 600V and for fixture and wire splices up to 1,000V. Depending on the product, they are made for indoor use, outdoor use, or both.

Linerless rubber splicing tape provides for a tight, void-free, moisture-resistant insulation without loss of electrical characteristics. It is typically used as primary insulation with all solid dielectric cables through 70kV. Other applications include jacketing on high-voltage splices and terminals, moisture-sealing electrical connections, busbar insulations, and end sealing high-voltage cables.

High-temperature silicone rubber tapes are used as a protective overwrap for terminating high-voltage cables. Glass cloth electrical tapes provide a heat-stable insulation for hot-spot applications such as furnace and oven controls, motor leads, and switches. They are also used to reinforce insulation where heavy loads cause high heat and breakdown of insulation, such as in motor control exciter feeds, etc. All-metal braid tapes are also available. These are used to continue electrostatic shielding across a splice. When taping a splice, begin by selecting the correct tape for the job. Always follow the tape manufacturer's recommendations.

A general procedure describing one method of taping a splice or joint, such as encountered when connecting motor lugs, is shown in *Figure 38*. A method for taping a split-bolt connector is shown in *Figure 39*. Prior to taping, make sure the

1
Make sure hardware
is fastened.

2
Place pieces of filler tape
over lugs and hardware.

3
Start by taping over
lugs and hardware.

4
With the joined lugs covered, go
beyond and wrap around each leg,
up to and over the insulation.

5
Cover both legs completely.

6
Finish off the winds at the tip, going
well beyond the lugs. Bend back the
tip and tape it back on the splice,
sealing it with additional winds of tape.

208F38.EPS

*Figure 38*   Typical method for taping motor lug connections.

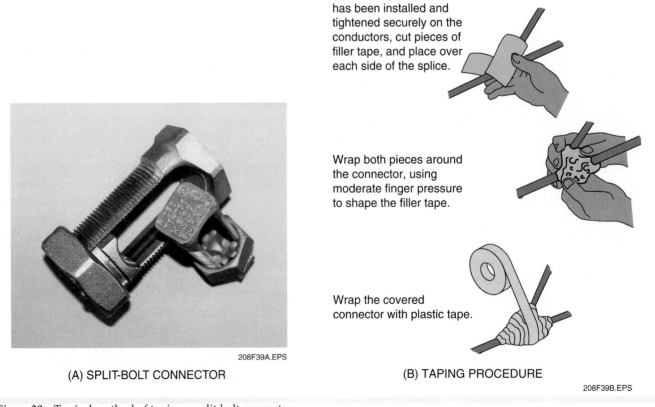

Once the split-bolt connector
has been installed and
tightened securely on the
conductors, cut pieces of
filler tape, and place over
each side of the splice.

Wrap both pieces around
the connector, using
moderate finger pressure
to shape the filler tape.

Wrap the covered
connector with plastic tape.

208F39A.EPS

(A) SPLIT-BOLT CONNECTOR

(B) TAPING PROCEDURE

208F39B.EPS

*Figure 39*   Typical method of taping a split-bolt connector.

## Electrical Tape

Most non-electricians think of electrical tape as only the simple black vinyl variety found in nearly every home toolbox. Electrical tape actually comes in a wide range of colors to be used for labeling various conductors when making terminations.

208SA04.EPS

joined lugs are securely fastened together with the appropriate hardware. Use pieces of a suitable filler tape or putty to fill any voids around the lugs and hardware and eliminate any sharp edges. This also helps to provide a smooth, even surface that makes taping easier.

> **NOTE**
>
> For all splices and joints where it is likely that the tape will have to be removed at some future date to perform work on the joint, apply an upside down (that is, adhesive side up) wrap of tape to the joint before applying the final layers of insulating tape to the joint in the usual manner. This will keep the area free of tape residue and facilitate the removal of the tape later on, if necessary.

### 10.0.0 MOTOR CONNECTION KITS

Motor connection kits are available to insulate bolted splice connections, such as those in motor terminal boxes. These kits eliminate the need for

taping and the use of filler tape or putty. To aid in joint reentry during rework, the insulator strips off easily, leaving a clean bolt area and thus eliminating the need to remove old tape and putty. Motor connection kits are available for use with stub (butt splice) connections (*Figure 40*)

STUB

IN-LINE

208F40.EPS

*Figure 40* Stub and in-line splice connections.

where there is insufficient room to make in-line connections. They are also made to insulate in-line splice connections where space permits. These insulating kits use a high-voltage mastic that seals the splice against moisture, dirt, and other contaminants.

One type of motor connection kit insulator is heat-shrinkable. It installs easily using heat from a propane torch to shrink the insulator in a manner similar to that of heat-shrink tubing (*Figure 41*). Another type of kit used for insulating stub connections comes in the form of an elastomeric insulating cap that is cold-applied by rolling it over the stub splice. Always follow the kit manufacturer's recommendations when selecting a kit to use for a particular application. Typically, the kit is selected based on the size of the motor feeder cable.

STUB CONNECTION

IN-LINE CONNECTION

STUB ROLL-ON INSULATING
CAP CONNECTION

208F41.EPS

*Figure 41*   Motor connection kits installed on splices.

# SUMMARY

A system is only as good as the weakest link. Poor quality terminations and splices weaken a system and can result in failure or fire. Good terminations and splices are mechanically and electrically secure, well-insulated, and should last as long as the conductor is in service.

Before a connection is made, the conductor must be stripped using the appropriate tool. Even minor damage to the conductor or insulation can affect the strength of the system.

Crimp connectors or terminals are used for many types of wire and cable. They are convenient for terminating conductors at terminal boards, control wiring terminals, and the like.

Lugs are provided for the larger wire sizes on panelboards and motor control centers. It is very important to tighten these lugs properly to provide a sound electrical connection as well as to provide short circuit bracing.

There are a variety of insulating tapes available to seal spliced joints. Always match the tape with the application.

# Review Questions

1. A poorly made splice is acceptable as long as it provides a path for current flow.
   a. True
   b. False

2. Which of the following best describes the term *stripping* as it applies to conductor splicing?
   a. Removal of the insulation from conductors
   b. Removal of the packing material from the carton
   c. Removal of any excess strands of wire from a splice
   d. Removal of the pulling ring from lead sheath conductors

3. What is the diameter, in inches, of 2,000 kcmil wire?
   a. 0.89
   b. 0.99
   c. 1.29
   d. 1.63

4. All of the following are types of stripping *except* _____.
   a. end terminations
   b. window cuts
   c. spiral cuts
   d. indent cuts

5. Which of the following best describes the purpose of a reducing connector?
   a. To temporarily connect a circuit
   b. To join two different size conductors
   c. To join conductors of the same size
   d. To make a 90° bend in parallel conductors

6. Information about connector size and conductor range is _____.
   a. stamped on the connector tongue
   b. color coded
   c. printed on the insulation
   d. stamped on the connector barrel

7. The most frequently used tongue styles are _____.
   a. ringed tongue and locking fork
   b. slotted and flanged fork
   c. hook shot and rectangular
   d. fork and flag

8. When using a crimping tool, use _____.
   a. the maximum crimping force possible
   b. a lower mechanical force to achieve maximum electrical performance
   c. the highest force needed for the maximum mechanical strength
   d. an amount of force between optimum mechanical and electrical performance

9. Using a crimp die that is too large will _____.
   a. destroy the crimp barrel
   b. weaken the conductor
   c. provide a weak mechanical connection
   d. provide poor electrical performance

10. The crimp indent is properly placed on the _____.
    a. right side of the barrel
    b. bottom of the barrel
    c. seam of the barrel
    d. left side of the barrel

11. Which of the following tools is best for positioning cable?
    a. Split-bolt connectors
    b. A heat gun
    c. A power connector indenter
    d. Wire-bending tools (either ratchet or hydraulic)

12. The minimum wire bending space for a single 3/0 conductor *not* entering or leaving the wall opposite its terminal is _____.
    a. 2 inches
    b. 3 inches
    c. 4 inches
    d. 5 inches

13. The minimum wire bending space for two 1/0 conductors *not* entering or leaving the wall opposite the terminal is _____.
    a. 2 inches
    b. 3 inches
    c. 4 inches
    d. 5 inches

14. The minimum wire bending space for three 3/0 conductors entering or leaving the wall opposite the terminal is _____.

    a. 7 inches
    b. 8 inches
    c. 9 inches
    d. 10 inches

15. Which of the following types of electrical tape is best for taping most joints for voltages up to 600V?

    a. High-temperature silicone rubber
    b. Rubber
    c. Glass cloth
    d. Vinyl plastic

1. Heavy-duty strippers are used to strip power cables with outside diameters ranging from _____ to _____.

2. Properly made splices and connections should last _____ _____.

3. Aluminum has certain properties that are different than copper. These properties are:

   _____

   _____

   _____.

4. The electrolytic action between aluminum and copper can be controlled by _____ _____.

5. Connectors marked AL should only be used with _____ wire, while connectors marked AL-CU can be used with either wire.

6. True or False? Poorly stripped conductors can result in changes in the dielectric strength.

7. Nearly all cable installations require the cable to be bent at terminations and other points along the cable route (these bends must comply with the *NEC®*). Minimum bending radii are determined by

   _____

   _____.

8. The bending radii for various sizes of conductors entering or leaving cabinets or cutout boxes must comply with certain *NEC®* requirements, many of which are covered in _____

   _____.

9. All incoming lines to either incoming line lugs or main disconnects must be braced to withstand the mechanical force created by a high fault current. Detailed information on the recommended methods of short circuit bracing is normally furnished by _____

   _____.

10. All wire joints not protected by some other means should be taped carefully to _____

    _____

    _____.

# Trade Terms Introduced in This Module

**AL-CU:** An abbreviation for aluminum and copper, commonly marked on terminals, lugs, and other electrical connectors to indicate that the device is suitable for use with either aluminum conductors or copper conductors.

**Amperage capacity:** The maximum amount of current that a lug can safely handle at its rated voltage.

**Connection:** That part of a circuit that has negligible impedance and joins components or devices.

**Connector:** A device used to physically and electrically connect two or more conductors.

**Drain wire:** A wire that is attached to a coaxial connector to allow a path to ground from the outer shield.

**Grooming:** The act of separating the braid in a coaxial conductor.

**Insulating tape:** Adhesive tape that has been manufactured from a nonconductive material and is used for covering wire joints and exposed parts.

**Lug:** A device for terminating a conductor to facilitate the mechanical connection.

**Mechanical advantage (MA):** The force factor of a crimping tool that is multiplied by the hand force to give the total force.

**Pressure connector:** A connector applied using pressure to form a cold weld between the conductor and the connector.

**Reducing connector:** A connector used to join two different sized conductors.

**Shielding:** The metal covering of a cable that reduces the effects of electromagnetic noise.

**Splice:** The electrical and mechanical connection between two pieces of cable.

**Strand:** A group of wires, usually stranded or braided.

**Terminal:** A device used for connecting cables.

**Termination:** The connection of a cable.

## Additional Resources

This module presents thorough resources for task training. The following resource material is suggested for further study.

*National Electrical Code® Handbook*, Latest Edition. Quincy, MA: National Fire Protection Association.

## Figure Credits

# CONTREN® LEARNING SERIES — USER UPDATE

NCCER makes every effort to keep its textbooks up-to-date and free of technical errors. We appreciate your help in this process. If you find an error, a typographical mistake, or an inaccuracy in NCCER's Contren® materials, please fill out this form (or a photocopy), or complete the online form at www.nccer.org/olf. Be sure to include the exact module number, page number, a detailed description, and your recommended correction. Your input will be brought to the attention of the Authoring Team. Thank you for your assistance.

*Instructors* – If you have an idea for improving this textbook, or have found that additional materials were necessary to teach this module effectively, please let us know so that we may present your suggestions to the Authoring Team.

### NCCER Product Development and Revision
3600 NW 43rd Street, Building G, Gainesville, FL 32606

**Fax:** 352-334-0932
**Email:** curriculum@nccer.org
**Online:** www.nccer.org/olf

☐ Trainee Guide ☐ AIG ☐ Exam ☐ PowerPoints Other _____

Craft / Level: _____ Copyright Date: _____

Module Number / Title: _____

Section Number(s): _____

Description: _____

_____

_____

_____

_____

Recommended Correction: _____

_____

_____

_____

_____

Your Name: _____

Address: _____

_____

Email: _____ Phone: _____

# Grounding and Bonding

**1600 Glenarm Place, Denver Colorado**

Greiner Electric won an ABC Pyramid Award for electrical contract work on this high-end downtown apartment building in Denver, Colorado.

**26209-11**

Trainees with successful module completions may be eligible for credentialing through NCCER's National Registry. To learn more, go to **www.nccer.org** or contact us at **1.888.622.3720.** Our website has information on the latest product releases and training, as well as online versions of our *Cornerstone* newsletter and Pearson's Contren® product catalog.

   Your feedback is welcome. You may email your comments to **curriculum@nccer.org,** send general comments and inquiries to **info@nccer.org,** or use the User Update form at the back of this module.

 V.1 7/11

# GROUNDING AND BONDING

## Objectives

When you have completed this module, you will be able to do the following:

1. Explain the purpose of grounding and bonding and the scope of *NEC Article 250*.
2. Distinguish between a short circuit and a ground fault.
3. Define the *National Electrical Code®* (*NEC®*) requirements relating to bonding and grounding.
4. Distinguish between grounded systems and equipment grounding.
5. Use *NEC Table 250.66* to size the grounding electrode conductor for various AC systems.
6. Explain the function of the grounding electrode system and determine the grounding electrodes to be used.
7. Define electrodes and explain the resistance requirements for electrodes using *NEC Section 250.53*.
8. Use *NEC Table 250.122* to size the equipment grounding conductor for raceways and equipment.
9. Explain the function of the main and system bonding jumpers in the grounding system and size the main and system bonding jumpers for various applications.
10. Size the main bonding jumper for a service utilizing multiple service disconnecting means.
11. Explain the importance of bonding equipment in clearing ground faults in a system.
12. Explain the purposes of the grounded conductor (neutral) in the operation of overcurrent devices.

## Performance Tasks

Under the supervision of the instructor, you should be able to do the following:

1. Using the proper fittings, connect one end of a No. 4 AWG bare copper grounding wire to a length of ¾" galvanized water pipe and the other end to the correct terminal in a main panelboard.
2. Install two lengths of Type NM cable in a switch box using Type NM cable clamps:
   - Strip the ends of the cable to conform with *National Electrical Code®* requirements.
   - Secure the cable in the switch box and tighten the cable clamps.
   - Connect and secure the equipment grounding conductors according to *NEC®* requirements, and secure to the switch box with either a ground clip or a grounding screw.
3. Size the minimum required grounding electrode conductor for a 200A service fed by 3/0 copper.
4. Size the minimum required equipment grounding conductor in each conduit for a 400A feeder gap using two parallel runs of 3/0 copper.
5. Size the minimum required bonding jumper for a copper water pipe near a separately derived system (transformer) where the secondary conductors are 500 kcmil copper.

## Trade Terms

| | | |
|---|---|---|
| Auxiliary electrodes | Ground resistance | Main bonding jumper |
| Bonding | Ground rod | Neutral |
| Effective ground fault path | Grounded | Resistivity |
| Equipment bonding jumper | Grounded conductor | Separately derived system |
| Equipment grounding conductor | Grounding clip | Short circuit |
| Ground | Grounding conductor | Step voltage |
| Ground current | Grounding connections | System grounding |
| Ground grids | Grounding electrode | Touch voltage |
| Ground mats | Grounding electrode conductor | Ungrounded conductors |

Note: *NFPA 70®*, *National Electrical Code®*, and *NEC®* are registered trademarks of the National Fire Protection Association, Inc., Quincy, MA 02269. All *National Electrical Code®* and *NEC®* references in this module refer to the 2011 edition of the *National Electrical Code®*.

## Required Trainee Materials

1. Pencil and paper
2. Appropriate personal protective equipment
3. Copy of the latest edition of the *National Electrical Code*®

## Contents

*Topics to be presented in this module include:*

## Figures and Tables

## 1.0.0 INTRODUCTION

The grounding system is a major part of the electrical system. Its purpose is to protect life and equipment against the various electrical faults that can occur. It is sometimes possible for higher-than-normal voltages to appear at certain points in an electrical system or in the electrical equipment connected to the system. Proper grounding ensures that the electrical charges that cause these high voltages are channeled to earth or ground before damaging equipment or causing danger to human life. Therefore, a circuit is grounded to limit the voltage on the circuit and improve overall operation of the electrical system and continuity of service.

When we refer to ground, we are talking about being connected to the earth. If a conductor is connected to the earth or to some conducting body that extends the ground connection, such as a driven ground rod (electrode) or cold-water pipe, the conductor is said to be grounded. The neutral in a three- or four-wire service, for example, is intentionally grounded and therefore becomes a grounded conductor. However, a wire used to connect this neutral conductor to a grounding electrode or electrodes is referred to as a grounding electrode conductor. Note the difference in the two meanings; one is grounded, while the other is grounding.

The equipment grounding conductor is the conductive path used to connect the noncurrent-carrying metal parts of equipment, raceways, and other enclosures to the system grounded conductor, the grounding electrode conductor, or both, at the service equipment or at the source of a separately derived system. See *Figure 1*.

*NEC Article 100, Informational Note No. 1* under the definition for Grounding Conductor, Equipment (ECG), recognizes that the equipment grounding conductor also performs bonding.

This module is designed to explain what a ground is and to teach proper grounding and bonding techniques. *NEC*® regulations and the testing of grounding systems will be thoroughly covered. Upon completion of this module, you should have a solid foundation in the principles of grounding and bonding, one of the most important aspects of an electrical system for the protection of life and property.

SERVICE-ENTRANCE CABLE

UNGROUNDED CONDUCTORS SERVICE 120/240-VOLT HOT WIRES

MAIN DISCONNECT SWITCH

CABLES TO HOUSE CIRCUITS

THREE-WIRE CABLE TO 240-VOLT APPLIANCES (CONNECT TO DOUBLE BREAKER)

NEUTRAL (GROUNDED) CONDUCTOR

PANELBOARD

NEUTRAL/GROUND BUSBAR

GROUNDING ELECTRODE CONDUCTOR

MAIN BONDING JUMPER

CABLES TO HOUSE CIRCUITS

THREE-WIRE CABLE TO 240-VOLT APPLIANCES (CONNECT TO DOUBLE BREAKER)

CIRCUIT BREAKERS

209F01.EPS

*Figure 1*   Panelboard showing grounded and grounding conductors.

## Neutral Conductor Size

The grounded neutral conductor is permitted to be smaller than the hot ungrounded phase conductors only when the neutral conductor is properly and adequately sized to carry the maximum load imbalance as computed according to *NEC Sections 215.2, 220.61, and 230.42*.

It must be pointed out that per *NEC Section 250.24(C)(1)* the grounded conductor must not be smaller than the size of the grounding electrode conductor. The purpose of the grounded conductor is not only to carry the unbalanced load, but also to provide a low-impedance path back to the source (utility) to complete a circuit and allow enough current to flow to open the overcurrent devices under ground fault conditions. This is important when you have three-phase services with little or no neutral loads to use as a basis for sizing the grounded conductor.

*NEC Section 215.2* (feeders) and *NEC Section 230.42* (service-entrance conductors) reference *NEC Article 220* for minimum rating and size computation requirements. *NEC Section 220.61* states that the feeder neutral load must be the maximum imbalance of the load, which is the maximum net computed load between the neutral and any one ungrounded conductor.

In summary, both the neutral current, as determined in *NEC Articles 215, 220, and 230*, and the minimum grounding electrode size, as determined in *NEC Article 250*, must be used to determine the actual neutral wire size.

## 2.0.0 PURPOSE OF GROUNDING AND BONDING

Systems are solidly grounded to limit the voltage to ground during normal operation and to prevent excessive voltages due to lightning and line surges. They are also grounded to stabilize the voltage to ground during normal operation. Conductive materials enclosing electrical conductors or equipment, or that form a part of the equipment, are grounded to limit the voltage to ground on these materials.

The conductive materials are bonded (connected together) to establish electrical continuity and conductivity. This facilitates the operation of overcurrent devices under ground fault conditions.

## 3.0.0 *NEC®* REQUIREMENTS

*NEC Article 250* is the primary governing article for the proper use and installation of grounding and bonding. This article covers general requirements for grounding and bonding of electrical installations. It presents the following specific requirements:

- Systems, circuits, and equipment that are required, permitted, or not permitted to be grounded
- Circuit conductor to be grounded on grounded systems
- Location of grounding connections
- Types and sizes of grounding and bonding conductors and electrodes
- Methods of grounding and bonding
- Conditions under which guards, isolation, or insulation may be substituted for grounding

*NEC Section 250.4(A)(5)* requires that the path to ground from circuits, equipment, and metal enclosures shall:

- Be permanent and continuous
- Have the capacity to safely conduct any fault current likely to be imposed on it
- Have sufficiently low impedance to facilitate the operation of the protective devices in the circuit or ground detector for high-impedance grounded systems
- Be capable of safely carrying the maximum ground fault likely to be imposed
- Not rely on the earth as an effective ground fault current path

Permanent, reliable, and continuous grounding systems are vital to the safety of electrical systems. Intermittent connections are likely to be unpredictable and may result in hazardous situations.

The minimum size grounding electrode conductor, equipment bonding jumpers, and equipment

### What's wrong with this picture?

209SA01.EPS

grounding conductor is necessary to ensure that the proper capacity to protect the system is in place. Methods to determine the proper size of these conductors will be discussed later in this module.

A low-impedance conductor is necessary because the higher the impedance, the more resistance there is to current flow. The current should flow with the least amount of resistance to ensure that personnel and equipment are protected when a fault does occur.

## 4.0.0 SHORT CIRCUIT VERSUS GROUND FAULT

Short circuits and ground faults are commonly misunderstood. Both faults stem from a failure of insulation resistance. It is important to have a common language for better understanding of these conditions.

### 4.1.0 Short Circuit

A short circuit is a conducting connection, whether intentional or accidental, between any of the conductors of an electrical system, whether it is from line-to-line or from line-to-ground. See *Figure 2*.

The failure might occur from one phase conductor to another phase conductor or from one phase conductor to the grounded conductor or neutral. The maximum value of fault current is dependent on the available capacity the system can deliver to the point of fault. The maximum value of short circuit current from line-to-ground will vary depending upon the distance from the source to the fault. The available short circuit current is further limited by the impedance of the arc where one is established, plus the impedance of the conductors to the point of short circuit.

### 4.2.0 Ground Fault

A ground fault, as defined in *NEC Article 100*, is an unintentional, electrically conducting connection between an ungrounded conductor of an electrical circuit and the normally noncurrent-carrying conductors, metallic enclosures, metallic raceways, metallic equipment, or the earth. While not specified in the *NEC®*, an unintentional electrically conducting connection between a grounded conductor of an electrical circuit and the normally noncurrent-carrying conductors, metallic enclosures, metallic raceways, metallic equipment, or earth would also be considered a ground fault. While the first instance may result in a large amount of fault current to flow in a properly installed grounded system, the latter may result in parallel paths being formed between the grounded conductor and the grounding system. These parallel paths would result in an unwanted current flow on the grounding system. See *Figure 3*.

## 5.0.0 TYPES OF GROUNDING SYSTEMS

The two general classifications of protective grounding are system and equipment grounding.

System grounding and bonding relates to the service-entrance equipment and its interrelated and bonded components. That is, one system conductor is grounded to limit voltages due to lightning, line surges, or unintentional contact with higher voltage lines, as well as to stabilize the voltage to ground during normal operation.

Equipment grounding conductors are used to connect the noncurrent-carrying metal parts of equipment, conduit, outlet boxes, and other enclosures to the system grounded conductor, the grounding electrode conductor, or both, at the service equipment or at the source of a separately derived system. Equipment grounding conductors are bonded to the system grounded conductor to provide a low impedance path for fault current that will facilitate the operation of overcurrent devices under ground fault conditions. *NEC Article 250* covers general requirements for grounding and bonding.

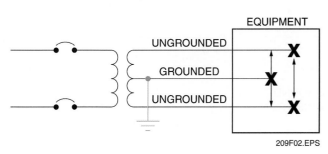

209F02.EPS

*Figure 2*   Short circuit.

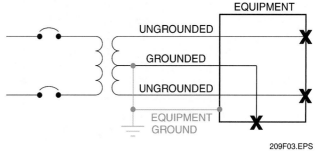

209F03.EPS

*Figure 3*   Ground fault.

## 5.1.0 Single-Phase Systems

To better understand a complete grounding system, we will examine a conventional residential or small commercial system beginning at the power company's high-voltage lines and transformer, as shown in *Figure 4*. The pole-mounted transformer is fed with a two-wire 7,200V system that is transformed and stepped down to a three-wire, 120/240V, single-phase electric service suitable for residential use. A wiring diagram of the transformer connections is shown in *Figure 5*. Note that the voltage between Leg A and Leg B is 240V.

However, by connecting a third wire (neutral) on the secondary winding of the transformer between the other two, the 240V splits in half, giving 120V between either Leg A or Leg B and the neutral conductor. Consequently, 240V is

209F04.EPS

*Figure 4*   Pole-mounted transformer.

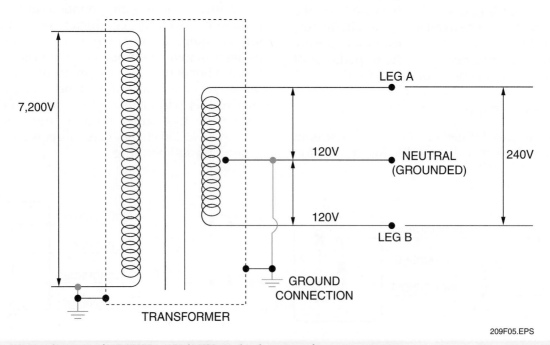

209F05.EPS

*Figure 5*   Wiring diagram of a 7,200V to 120/240V single-phase transformer connection.

available for household appliances such as ranges, hot water heaters, clothes dryers, and the like, while 120V is available for lights, small appliances, televisions, and similar appliances.

Referring again to the diagram in *Figure 5*, conductors A and B are ungrounded conductors, while the neutral is a grounded conductor. If only 240V loads were connected, the neutral or grounded conductor would carry no current. However, since 120V loads are present, the neutral will carry the unbalanced load and become a current-carrying conductor. For example, if Leg A carries 60A and Leg B carries 50A, the neutral conductor would carry only 10A (60 − 50 = 10). This is why the *NEC*® sometimes allows the neutral conductor in an electric service to be smaller than the ungrounded conductors.

The typical pole-mounted service entrance is normally routed by a grounded (neutral) messenger cable from a point on the pole to a point on the building being served, terminating at the service drop. Service-entrance conductors are routed between the meter housing and the main service switch or panelboard. This is the point where most systems are grounded—the neutral bus in the main panelboard. See *Figure 6*.

**WARNING!**

Exercise extreme caution when disconnecting a ground. Never grab a disconnected ground wire with one hand and the grounding electrode with the other. Your body will act as a conductor for any fault current; the results could be fatal.

### 5.2.0 Grounding Requirements

*NEC Sections 250.20 and 250.21* provide the grounding requirements for AC systems. The *NEC*® should always be referenced when working with grounded and ungrounded systems. There are various exceptions to the requirements that will not be covered in detail in this module.

*Figure 6*   Typical service entrance and related service equipment.

209F06.EPS

### 5.2.1 Systems Less Than 50 Volts

Grounding is required:

- Where supplied by transformers if the supply voltage to the transformer exceeds 150V to ground
- Where supplied by transformers if the transformer supply system is ungrounded
- Where installed as overhead conductors outside of the building
- In other circuits and systems provided they comply with the provisions of *NEC Article 250*

*Figure 7* shows examples of these requirements.

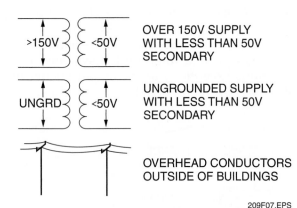

OVER 150V SUPPLY WITH LESS THAN 50V SECONDARY

UNGROUNDED SUPPLY WITH LESS THAN 50V SECONDARY

OVERHEAD CONDUCTORS OUTSIDE OF BUILDINGS

209F07.EPS

*Figure 7*  Systems less than 50V that must be grounded.

### 5.2.2 50-Volt to 1,000-Volt Systems

AC systems of 50V to 1,000V supplying premises wiring and premises wiring systems shall be grounded under any of the following conditions:

- Where the system can be grounded so that the maximum voltage to ground on the ungrounded conductors does not exceed 150 volts
- Where the system is three-phase, four-wire, wye-connected and the neutral is used as a circuit conductor
- Where the system is three-phase, four-wire, delta-connected and the midpoint of one phase winding is used as a circuit conductor
- In other circuits and systems provided they comply with the provisions of *NEC Article 250*

*Figure 8* shows examples of 50V to 1,000V grounded applications.

### 5.2.3 AC Systems 1kV and Over

AC systems over 1,000V must be grounded if they supply mobile or portable equipment as covered in *NEC Section 250.188*. Other AC systems over 1,000V are permitted (but not required) to be grounded.

Other circuits and systems shall be permitted to be grounded. If such systems are grounded, they shall comply with the provisions of *NEC Article 250*.

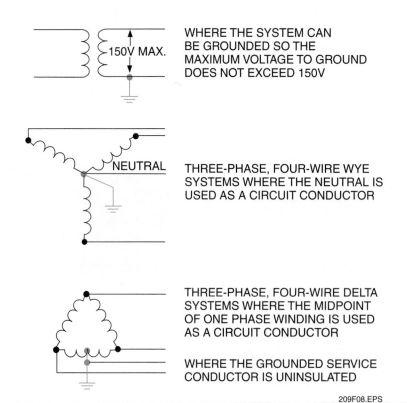

150V MAX.

WHERE THE SYSTEM CAN BE GROUNDED SO THE MAXIMUM VOLTAGE TO GROUND DOES NOT EXCEED 150V

NEUTRAL

THREE-PHASE, FOUR-WIRE WYE SYSTEMS WHERE THE NEUTRAL IS USED AS A CIRCUIT CONDUCTOR

THREE-PHASE, FOUR-WIRE DELTA SYSTEMS WHERE THE MIDPOINT OF ONE PHASE WINDING IS USED AS A CIRCUIT CONDUCTOR

WHERE THE GROUNDED SERVICE CONDUCTOR IS UNINSULATED

209F08.EPS

*Figure 8*  Systems that must be grounded.

### 5.2.4 Separately Derived Systems

A separately derived system is a premises wiring system whose power is derived from a source of electrical energy other than a service. These electrical systems must be grounded if the system that is derived complies with the conditions previously described. The separately derived system must then be grounded as specified in *NEC Section 250.30*.

Examples of separately derived systems include, but are not limited to:

- Transformers with no direct electrical connection between the primary and secondary
- Generator systems that supply power where the neutral is not connected to the utility system, such as for carnivals, rock crushers, or batch plants
- Generator systems used for emergency, required standby, or optional standby power that have all conductors, including a neutral, isolated from the neutral or grounded conductor of another system (usually via a transfer switch)
- AC or DC systems derived from inverters or rectifiers

### 5.2.5 Circuits That Cannot Be Grounded

According to *NEC Section 250.22*, only five types of circuits are not permitted to be grounded:

- Circuits for cranes that operate over combustible fibers in Class III locations, as provided in *NEC Section 503.155*
- For healthcare facilities, those isolated power circuits in hazardous (classified) locations as provided in *NEC Sections 517.61 and 517.160*
- Circuits for electrolytic cells as provided in *NEC Article 668*
- Secondary circuits of lighting systems as provided in *NEC Section 411.5(A)*
- Secondary circuits on lighting systems as provided in *NEC Section 668.3(C)(3)*

**NOTE**

Equipment located or used within the electrolytic cell line working zone or associated with the cell line DC power circuits are not required to comply with *NEC Article 250*.

*Figure 9* shows some examples of these circuits.

CRANES THAT OPERATE OVER COMBUSTIBLE FIBERS IN CLASS III LOCATIONS

ISOLATED POWER SYSTEMS IN HEALTHCARE FACILITIES FOR HAZARDOUS INHALATION ANESTHETIZING AND WET LOCATIONS

CIRCUITS FOR ELECTROLYTIC CELLS

SECONDARY CIRCUITS ON LIGHTING SYSTEMS

209F09.EPS

*Figure 9*   Circuits that are not permitted to be grounded.

# 6.0.0 *NEC®* REQUIREMENTS

The grounding equipment requirements established by Underwriters Laboratories, Inc. (UL) have served as the basis for approval for grounding in the *NEC®*. The *NEC®*, in turn, provides the grounding standard for the Occupational Safety and Health Administration (OSHA).

All electrical systems must be grounded and bonded in a manner prescribed by the *NEC®* to protect personnel and equipment. A grounded system must limit the voltage on the electrical system and protect it from:

- Lightning
- Voltage surges higher than that for which the circuit is designed
- An increase in maximum potential to ground due to abnormal voltages

## 6.1.0 Grounding Methods

The requirements for grounding of services are found in *NEC Sections 250.24(A), (B), (C), and (D)*. Methods of grounding an electric service are covered in *NEC Article 250, Part III*. In general, all of the following electrodes that are present must be bonded together to form the grounding electrode system:

- An underground metal water pipe in direct contact with the earth for no less than 10'
- The metal frame of a building as described in *NEC Section 250.52(A)(2)*
- An electrode encased by at least 2" of concrete, located within and near the bottom of a concrete foundation or footing that is in direct contact with the earth

> **NOTE**
> This electrode must be at least 20' long and must be made of electrically conductive coated steel reinforcing bars or rods of not less than ½" in diameter, or consisting of at least 20' of bare copper conductor not smaller than No. 4 AWG wire size.

- A ground ring encircling the building or structure, in direct contact with the earth at a depth not less than 2.5' below grade
- Rod, pipe, or plate electrodes
- Other local metal underground systems or structures

> **NOTE**
> This ring must consist of at least 20' of bare copper conductor not smaller than No. 2 AWG wire size. See *Figure 10*.

Grounding electrode systems used in industrial buildings will frequently use all of the methods shown in *Figure 10*, and the methods used will often surpass the *NEC®*, depending upon the manufacturing process and the calculated requirements made by plant engineers.

*Figure 11* shows a floor plan of a typical industrial grounding system.

The building in *Figure 12* has a metal underground water pipe that is in direct contact with the earth for more than 10', so this is one valid grounding source. The building also has a metal underground gas piping system, but this may not be used as a grounding electrode [*NEC Section 250.52(B)(1)*].

*NEC Section 250.53(D)(2)* further states that the underground water pipe must be supplemented by an additional electrode of a type specified in *NEC Sections 250.52(A)(2) through 250.52(A)(8)*. Since a grounded metal building frame, concrete-encased electrode, or ground ring is not available in this application, *NEC Section 250.50* must be used in determining the supplemental electrode. In most cases, this supplemental electrode will consist of either a driven rod (*Figure 13*) or pipe electrode, the specifications for which are as follows:

- Withstand and dissipate repeated surge circuits
- Provide corrosion resistance to various soil chemistries to ensure continuous performance for the life of the equipment being protected
- Provide rugged mechanical properties for easy driving with minimum effort and rod damage

An alternative to the pipe or rod method is a plate electrode. Each plate electrode must expose not less than two square feet of surface to the surrounding earth. Plates made of iron or steel must be at least ¼" thick, while plates of nonferrous metal like copper need only be 0.06" thick.

Rod, pipe, and plate electrodes must have a resistance to ground of 25 ohms (Ω) or less. If not, they must be augmented by an additional electrode spaced not less than 6' from the first. Many locations require two electrodes regardless of the resistance to ground. This is not an *NEC®*

ALL GROUNDING ELECTRODES THAT ARE PRESENT ON THE PREMISES MUST BE BONDED TOGETHER TO FORM THE GROUNDING SYSTEM
*NEC Section 250.50*

PANELBOARD

NEUTRAL BUS

MAIN BONDING JUMPER
*NEC Section 250.28*

GROUNDING ELECTRODE CONDUCTOR

EFFECTIVELY GROUNDED METAL FRAME OF BUILDING
*NEC Section 250.52(A)(2)*

METAL WATER PIPE
*NEC Section 250.52(A)(1)*

CONCRETE-ENCASED REBARS
*NEC Section 250.52(A)(3)*

RODS OR PIPES OF IRON OR STEEL
*NEC Section 250.52(A)(5)*

REBARS MUST BE AT LEAST 20' LONG
*NEC Section 250.52(A)(3)*

PLATE ELECTRODE MUST EXPOSE NOT LESS THAN 2 SQUARE FEET OF SURFACE TO EXTERIOR SOIL, MUST BE BARE OR CONDUCTIVELY COATED IRON OR STEEL, AND AT LEAST ¼" THICK
*NEC Section 250.52(A)(7)*

GROUND RING MADE OF AT LEAST 20' OF NO. 2 OR LARGER BARE COPPER BURIED NO LESS THAN 2.5' BELOW GRADE
*NEC Section 250.52(A)(4)*

209F10.EPS

*Figure 10*   *NEC®*-approved grounding electrodes.

requirement, but is required by some power companies and local ordinances in some cities and counties. Always check with the local inspection authority for such rules that may go beyond the requirements of the *NEC®*.

### 6.2.0 Grounding Electrode Conductor

The grounding electrode conductor is the sole connection between the grounding electrode and the grounded system conductor for a grounded system, or the sole connection between the grounding electrode and the service equipment enclosure for an ungrounded system. See *NEC Section 250.64*, which describes grounding electrode conductor installation.

A common grounding electrode conductor is required to ground both the circuit grounded conductor and the equipment grounding conductor. *Figure 14* shows the common connection for this system.

### 6.2.1 Sizing

The grounding electrode conductor must be sized in accordance with *NEC Table 250.66*. The size of this conductor is based on the size of the service-entrance conductor.

The size of the service-entrance conductor is determined from the ampacity of that system. *NEC Table 310.15(B)(16)* is used to determine the size of the service conductor from the ampacity.

COLD WATER PIPE

METAL FRAME OF BUILDING BONDED TO GROUND RING

ELECTRIC PANEL

GROUND RING OF BARE COPPER NOT SMALLER THAN NO. 2 AWG

FLOOR PLAN OF BUILDING

ELECTRODE ENCASED IN AT LEAST 2" OF CONCRETE AT THE BOTTOM OF THE FOOTING CONSISTING OF AT LEAST 20' OF ONE OR MORE ½" STEEL BARS, OR BARE COPPER AT LEAST NO. 2 AWG

IF MORE THAN ONE METHOD EXISTS IN ONE BUILDING, THEY MUST ALL BE BONDED TOGETHER

209F11.EPS

*Figure 11*    Floor plan of the grounding system for an industrial building.

MAIN PANELBOARD

NEUTRAL BUS

GROUNDING ELECTRODE CONDUCTOR

UPPER END OF ROD MUST BE BELOW PERMANENT MOISTURE LEVEL IF PRACTICABLE
*NEC Section 250.53(A)*

GROUNDING CLAMP (CONNECTION MUST BE MADE WITHIN 5 FEET FROM THE POINT WHERE THE PIPE ENTERS THE BUILDING)

METAL COLD WATER PIPE

**NOTE:** GROUNDING ELECTRODES TO BE SPACED AT LEAST 6' APART
*NEC Section 250.53(B)*

METAL UNDERGROUND GAS PIPING SYSTEM MAY NOT BE USED AS A GROUNDING ELECTRODE
*NEC Section 250.52(B)(1)*

209F12.EPS

*Figure 12*    Grounding requirements for non-industrial buildings.

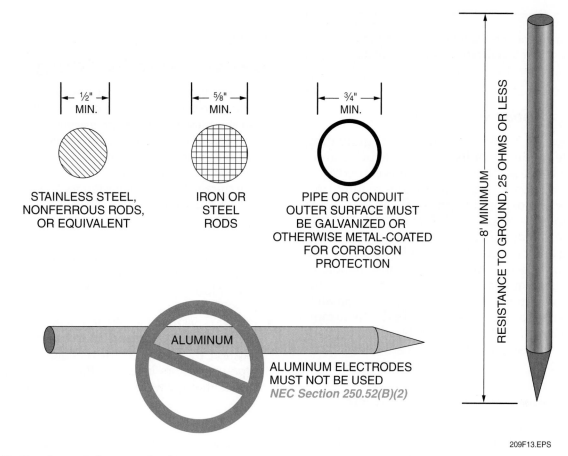

Figure 13 Requirements for ground rods.

Figure 14 Grounding electrode connection for grounded system.

The majority of applications use 75°C copper conductors. This information is necessary when using *NEC Table 310.15(B)(16)*. For a 100A service application, the size of the service conductor is No. 3 AWG copper.

Now that the size of the service conductors is known, the size of the grounding electrode conductor can be determined from *NEC Table 250.66*. Using the No. 3 AWG from the 100A service example, it can be determined from the table that a No. 8 AWG copper grounding electrode conductor is required.

Now, use the following steps to determine the size of the grounding electrode conductor for a 200A, 208V, three-phase system:

*Step 1* Determine the size of the service conductor from *NEC Table 310.15(B)(16)*. It is 3/0 AWG copper.

*Step 2* Go down the left column of *NEC Table 250.66* and find 3/0.

*Step 3* Go across to find the size of the grounding electrode conductor. It is a No. 4 AWG copper conductor.

### 6.2.2 Installation and Protection

*NEC Section 250.64(B)* requires the grounding electrode conductor or its enclosure to be securely fastened to the surface on which it is carried.

No. 4 AWG or larger conductors require protection where exposed to physical damage. If free from exposure to physical damage, No. 6 AWG conductors may be run along the surface of the building and be securely fastened. Otherwise, the conductor must be protected by installation in rigid or intermediate metal conduit, rigid non-metallic conduit, EMT, or cable armor. Smaller conductors shall be protected in conduit or armor. See *NEC Section 250.64(E)* for bonding requirements when using a ferrous enclosure or raceway to enclose a grounding electrode conductor.

### 6.3.0 Grounding Electrode Conductors

The grounding electrode conductor, which connects the grounded conductor and the panelboard neutral bus to ground, must be copper, aluminum, or copper-clad aluminum. Furthermore, the material selected must be resistant to any corrosive condition existing at the installation, or it must be suitably protected against corrosion. The conductor may be either solid or stranded, and covered or bare, but it must be in one continuous length without a splice or joint, except for the following conditions:

- Splices in busbars are permitted.
- Where a service consists of more than one single enclosure, it is permissible to connect taps to the grounding electrode conductor.

**Think About It**

## Sizing Electrode Conductors

*NEC Table 310.15(B)(16)* shows that the service conductor size (copper wire at 75°C) for a 150A service is 1/0 AWG. Which of the three AWG copper electrode conductor sizes listed below would be used with this service?

No. 4
No. 6
No. 8

---

- Bonding jumper(s) from grounding electrode(s) and GECs shall be permitted to be connected to an aluminum or copper busbar not less than ¼" × 2". See *NEC Section 250.64(F)(3)*.
- Per *NEC Section 250.64(C)(1)*, wire-type grounding electrode conductors may be spliced at any location by means of irreversible compression-type connectors listed for the purpose or using the exothermic welding process (*Figure 15*).

The size of grounding electrode conductors depends on service-entrance size; that is, the size of the largest service-entrance conductor or equivalent for parallel conductors. *NEC Table 250.66* gives the proper sizes of grounding electrode conductors for various sizes of electric services. *Figure 16* shows the various protection guidelines.

### 6.4.0 Other Electrodes

When an electrode that meets the requirements of *NEC Section 250.50* is not present, other electrodes may be used. These may be the rod and pipe electrodes discussed earlier, other listed electrodes, plate electrodes, or other local metal underground systems such as piping and underground tanks. *NEC Sections 250.52(A)(5) through 250.52(A)(8)* provide information and requirements for these electrodes. *Figure 17* shows other electrodes.

The specific requirements for other electrodes are:

---

❶ IRREVERSIBLE COMPRESSION-TYPE CONNECTORS

❷ EXOTHERMIC WELDING

UNUSED CHAMBERS MUST BE PLUGGED

SPARK CHAMBER

EXOTHERMIC WELDING BLOCKS

GROUNDING ELECTRODE TO BE WELDED

GROUNDING ELECTRODE TO BE WELDED

UNUSED CHAMBERS MUST BE PLUGGED

WELDED JOINT

BUTT CONNECTION

3-WAY CONNECTION

4-WAY CONNECTION

**WARNING:** NEVER USE AN EXOTHERMIC WELD UNLESS THE MOLD IS HEATED TO REMOVE THE MOISTURE, OTHERWISE MOLD CAN EXPLODE. FOLLOW COMPANY'S SAFETY RULES AND MANUFACTURER'S PROCEDURES.

209F15.EPS

*Figure 15*   Methods of splicing grounding conductors.

**PROTECTION OF GROUNDING ELECTRODE CONDUCTORS**

Grounding electrode conductor or enclosure must be securely fastened to the surface.

No. 4 or larger – If exposed to severe physical damage.

No. 6 – Run along surface, securely fastened or protected.

Smaller than No. 6 – Must be protected from damage (must be enclosed in rigid metal conduit).

Bare aluminum – Not allowed in contact with masonry, where subject to corrosive conditions, or within 18 inches of the earth.

209F16.EPS

*Figure 16*   Protection of grounding electrode conductors.

- *Local systems* – Local metallic underground systems such as piping, tanks, metal well casings, etc.

> **NOTE**
> Underground metal gas piping systems are not permitted to be used as grounding electrodes. However, this does not eliminate the requirement that metal gas piping systems be bonded.

- *Pipe electrodes* – Pipe or conduit electrodes shall be not less than 8' in length nor smaller than ¾"

trade size, and if of iron or steel, shall be galvanized or metal-coated for corrosion protection.

- *Rod and pipe electrodes* – Electrodes of steel or iron shall be at least ⅝" in diameter. Rods of nonferrous metal or stainless steel that are less than ⅝" in diameter shall be listed and must be at least ½" in diameter.
- *Plate electrodes* – Electrodes shall have at least two square feet of surface in contact with exterior soil. If of iron or steel, the plate shall be at least ¼" thick. If of nonferrous metal, it shall be at least 0.06" thick.

LOCAL METALLIC
UNDERGROUND SYSTEMS
SUCH AS PIPING, METAL
WELL CASINGS, AND TANKS

PIPE OR CONDUIT
ELECTRODES NOT LESS
THAN 8' LONG

PLATE ELECTRODES
MINIMUM ¼" THICK IF
IRON OR STEEL; 0.06"
IF NONFERROUS METAL

209F17.EPS

*Figure 17*   Other electrodes.

### 6.4.1 *Installation of Rod, Pipe, and Plate Electrodes*

Rod and pipe electrodes must be installed so that at least 8' of rod is in contact with the soil. Rods must be driven vertically unless rock obstruction is encountered. If an obstruction is encountered, the rod may be driven at not more than a 45° angle to clear the rock. The other possibility is for the rod to be buried in a trench that is at least 2.5' deep.

The upper end of the rod must be at or below surface level. The rod may be above ground, but at least 8' of rod must be beneath the surface.

## 7.0.0 EQUIPMENT GROUNDING

*Figure 18* summarizes the equipment grounding rules for most types of equipment required to be grounded. These general *NEC*® regulations apply to all installations except for specific equipment (special applications) as indicated in the code. The

*NEC*® also lists specific equipment that is to be grounded regardless of voltage.

In all occupancies, major appliances and many hand-held appliances and tools are required to be grounded. The appliances include refrigerators, freezers, air conditioners, clothes dryers, washing machines, dishwashing machines, sump pumps, and electrical aquarium equipment. Other tools likely to be used outdoors and in wet or damp locations must be grounded or have a system of double insulation.

### 7.1.0 Equipment Grounding Conductor

The equipment grounding conductor (EGC) is the conductor used to interconnect and bond the noncurrent-carrying metal parts of equipment, raceways, and other enclosures to the system grounded conductor and/or the grounding electrode conductor at the service equipment or at the source of a separately derived system.

### 7.1.1 *Grounded and Ungrounded Systems*

The EGC or path must extend from the farthest point on the circuit to the service equipment where it is connected to the grounded conductor. *Figure 19* shows an EGC and its path to ground when a ground fault occurs. It is the path for ground faults within the system back to the source that completes the circuit, allowing enough current to flow to ensure that overcurrent devices will trip.

### 7.1.2 *Sizing an EGC*

EGCs carry fault current from the load to the grounded terminal bar of the service equipment. The size of the EGC is determined from the size of the overcurrent device that is protecting that particular system. *NEC Table 250.122* is used to determine the size of the EGC.

For example, if the ampacity is 400A, the overcurrent device is set at 400A. Use *NEC Table 250.122* and go down the left column until

A GROUNDED CIRCUIT CONDUCTOR IS PERMITTED TO GROUND NONCURRENT-CARRYING METAL PARTS OF EQUIPMENT, RACEWAYS, AND OTHER ENCLOSURES ON THE SUPPLY SIDE OF THE SERVICE DISCONNECTING MEANS
*NEC Section 250.142(A)(1)*

METAL ENCLOSURES FOR SERVICE CONDUCTORS AND EQUIPMENT MUST BE GROUNDED TO THE GROUNDED SYSTEM CONDUCTOR
*NEC Section 250.80*

FOR AC WIRING SYSTEMS, THE NEUTRAL CONDUCTOR MUST BE GROUNDED ON SINGLE-PHASE, THREE-WIRE SYSTEMS

EFFECTIVE BONDING JUMPER MUST BE PROVIDED AROUND GROUNDING ELECTRODES (WATER PIPE) AT ANY LOCATION THAT IS LIKELY TO BE DISCONNECTED
*NEC Section 250.68(B)*

AN UNSPLICED MAIN BONDING JUMPER MUST BE USED TO CONNECT THE EQUIPMENT GROUNDING CONDUCTOR TO THE GROUNDED CONDUCTOR*
*NEC Section 250.24(B)*

A GROUNDING ELECTRODE CONDUCTOR MUST BE USED TO CONNECT THE GROUNDED SERVICE CONDUCTOR TO THE GROUNDING ELECTRODE
*NEC Section 250.24(D)*

GROUND CLAMP, TYPICAL

UNION, TYPICAL

WATER METER

*NEC Section 250.68(B)*

GROUNDING ELECTRODE (WATER PIPE IN THIS CASE)

*CHECK WITH LOCAL JURISDICTION FOR REQUIREMENTS

209F18.EPS

*Figure 18*   Equipment grounding summary.

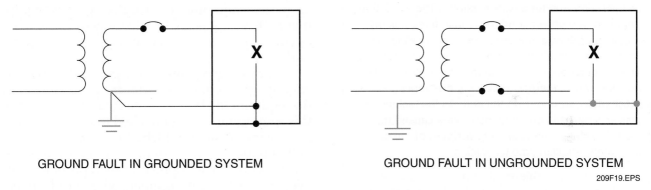

GROUND FAULT IN GROUNDED SYSTEM

GROUND FAULT IN UNGROUNDED SYSTEM

209F19.EPS

*Figure 19*   Ground faults.

the desired ampacity or the next highest ampacity is found. For the 400A system, the size of the EGC is No. 3 copper wire or No. 1 aluminum or copper-clad wire.

> **NOTE**
>
> The size of the EGC is determined by the overcurrent protective device.

Determine the size of the EGC for a flexible metal conduit connection to a motor supplied with a 30A circuit.

*Step 1*   Using *NEC Sections 250.118(5) and 348.60*, verify an EGC is needed.

*Step 2*   Since this is an installation that requires flexibility, an EGC is required.

*Step 3*   The 30A circuit will have a 30A overcurrent protective device feeding the circuit.

*Step 4*   Find 30A in *NEC Table 250.122*.

*Step 5*   A No. 10 copper or No. 8 aluminum or copper-clad EGC is required for 30A.

*Figure 20* shows other requirements for using flexible metal conduit as an equipment ground.

CONNECTORS MUST BE LISTED FOR FLEX AND GROUNDING

**NOTE:**
WHERE USED TO CONNECT EQUIPMENT WHERE FLEXIBILITY IS REQUIRED, AN EQUIPMENT GROUNDING CONDUCTOR MUST BE INSTALLED.

SUITABLE FOR GROUNDING; MAXIMUM 6' IN GROUND RETURN PATH AND MAXIMUM 20 AMPERE OVERCURRENT PROTECTION

209F20.EPS

*Figure 20*   Installation requirements for flexible metal conduit.

### Case History

# Mechanical Grounds

A technician was electrocuted while servicing a refrigeration unit for a walk-in cooler. The insulation on one of the power conductors inside the flexible metal conduit was damaged, resulting in electrical arcing to a conduit connector on the unit starter box. The conduit connection from the unit to the starter box was loose, thereby effectively disconnecting the mechanical ground from the unit. As the technician was servicing the unit, the thermostat caused the starter to close, energizing the surfaces of the unit and fatally shocking the technician.

**The Bottom Line:** This accident might have been prevented if one or more of the following procedures had been implemented:

• Proper preventive maintenance checks of all wiring and connections
• Proper lockout/tagout procedures
• Equipment grounding conductor provided along with the power feed conductors

*Figure 21* details the requirements for using the conduit as the grounding conductor. After determining the requirements and the ampacity of this application, a No. 8 copper EGC would be required.

## 7.2.0 Grounding Enclosures

As previously mentioned, equipment grounding covers the metallic noncurrent-carrying parts of an electrical system. Such parts include metallic conduit, outlet boxes, enclosures, and frames on motors, and other electrically operated equipment. These items are bonded together to ensure operation of overcurrent devices in case of ground faults.

A bare or green-insulated grounding wire is usually attached to the metal frame or cabinet of the equipment. When connected to the circuit, this grounding wire is attached to the equipment grounding system which, in turn, was originally bonded to the system neutral busbar at the service-entrance equipment. When properly connected, if a live ungrounded wire should make contact with the frame or cabinet of a motor, appliance, or other metallic object in the system, a ground fault will occur and open the overcurrent protective device that is protecting the circuit. Consequently, equipment grounding is one of the best methods of protecting life and property should a ground fault occur.

### 7.2.1 Grounding Outlet Boxes and Devices

All receptacles used in residential, commercial, or industrial applications must be of the grounding type, which means that one of the receptacle openings connects the equipment ground to the appliance, tool, or other apparatus that is connected to the receptacles.

FITTINGS LISTED FOR GROUNDING REQUIRED

PERMITTED IN EARTH WHERE LISTED AND MARKED

ACCEPTABLE FOR GROUNDING IF LISTED AND:
• MAX. 20 AMPERE OC FOR ⅜" TO ½"
• MAX. 60 AMPERE OC FOR ¾" TO 1¼"
• FITTINGS LISTED FOR GROUNDING
• WHERE USED TO CONNECT EQUIPMENT WHERE
  FLEXIBILITY IS REQUIRED, AN EQUIPMENT
  GROUNDING CONDUCTOR MUST BE USED

209F21.EPS

*Figure 21*   Installation requirements for liquidtight flexible metal conduit.

Ensuring continuity of the equipment grounding system to each receptacle can be handled in different ways, depending upon the wiring method used. For example, in most residential wiring systems, Type NM cable will be used. If metallic device boxes are used, the bare or green-insulated grounding wire must be attached to the box. This is accomplished by using either a grounding clip, as shown in *Figure 22*, or a screw designed for this purpose that is secured to the outlet box. *NEC Section 250.8* lists acceptable methods for connecting grounding conductors to enclosures.

*Figure 23* shows methods of providing a grounding conductor to duplex receptacles using several wiring methods. For example, if nonmetallic boxes are used with Type NM cable, no connection to the box itself is required. However, the equipment grounding conductor must be connected to the grounding terminal on the receptacle. If two or more cables enter the nonmetallic box, each of the equipment grounding wires must be connected with an approved connector, and one wire must then be attached to the grounding terminal on the receptacle. The terminal itself may not contain more than one grounding wire. Therefore, if more than one grounding wire enters the box, they must be spliced independently from the device, and then only one conductor may be attached to the device.

When metallic boxes are used, these equipment grounding wires must be attached to the

EQUIPMENT
GROUNDING
CONDUCTOR

METALLIC
OUTLET
BOX

GROUNDING CLIP

209F22.EPS

*Figure 22*   Grounding clip.

box, along with the wiring device (receptacle). As mentioned previously, this may be accomplished with either a grounding clip or a grounding screw (*NEC Section 250.8*). See also *NEC Section 250.146*.

Any of the following are recognized by *NEC Section 250.118* as being adequate for use as an equipment grounding conductor:

• A copper or other corrosion-resistant conductor (solid or stranded; insulated, covered, or bare; and in the form of a wire or a busbar of any shape)
• Rigid metal conduit

**NONMETALLIC BOX CABLE WITH INTEGRAL GROUNDING CONDUCTOR**

CONNECTOR, TYPICAL

GROUNDING TERMINAL ON WIRING DEVICE, TYPICAL

**METALLIC BOX CABLE WITH INTEGRAL GROUNDING CONDUCTOR**

GROUNDING SCREW

GROUNDING CLIP

**RACEWAY SYSTEM**

GROUNDING SCREW

GROUNDING CLIP

209F23.EPS

*Figure 23*  Grounding receptacles with different wiring methods.

- Intermediate metal conduit
- Listed liquidtight flexible metal conduit
- Electrical metallic tubing
- Listed flexible metal conduit and tubing as permitted by *NEC Sections 250.118(5), 250.118(6), and 250.118(7)*
- Armor of Type AC cable
- The copper sheath of mineral-insulated, metal-sheathed cable
- The metallic sheath or the combined metallic sheath and grounding conductors of Type MC cable
- Cable trays as permitted in *NEC Section 392.60(A)*
- Cable bus framework as permitted in *NEC Section 370.3*
- Other electrically continuous metal raceways listed for grounding

### 7.3.0 Main Bonding Jumper

When a listed switchboard or panelboard is installed, the main bonding jumper that is provided with the equipment is rated for the size of conductors used for the service. The proper size of the main bonding jumper is verified by the

**Think About It**

## Sheet Metal Screws

Why does the *NEC®* list machine screws as the only type of screw acceptable for connecting grounding conductors to enclosures? Why can't sheet metal screws be used instead?

4 – 250 KCMIL ALUMINUM CONDUCTORS
4 × 250 = 1,000 KCMIL
*NEC Table 250.66*
2/0 COPPER OR 4/0 ALUMINUM

209F24.EPS

*Figure 24*   Main bonding jumper for parallel runs.

listing agency and can be used without calculating the size.

Since the main bonding jumper must carry the full ground fault current of the system back to the grounded service conductor, its size must relate to the rating of the service conductors that supply the service. *NEC Table 250.66* is used to determine the size of the main bonding jumper. *Figure 24* shows how to calculate the size of the main bonding jumper when parallel runs are used.

Where the multiple conductors are larger than the maximum given in *NEC Table 250.66*, the jumper cannot be less than 12.5% of the area at the largest phase conductor [*NEC Section 250.28(D)*].

An example of three 500 kcmil copper conductors gives a total area of 1,500 kcmil. Because this is larger than that of *NEC Table 250.66*, take 12.5% of the total area. This is 187.5 kcmil. Since this is not a standard size conductor, refer to *NEC Chapter 9, Table 8*.

The table shows that the next larger conductor size in circular mils is 211,600. This gives a conductor of 4/0 AWG.

## 8.0.0 BONDING SERVICE EQUIPMENT

Electrical continuity is the key to successful clearing of ground fault currents. This continuity between service equipment and enclosures must be maintained by bonding. The items that are required to be bonded together are listed in *NEC Section 250.92(A)*. They include:

- The service raceways, cable trays, cable bus framework, or service cable armor or sheath
- All service equipment enclosures containing service conductors, including meter fittings, boxes, etc., interposed in the service raceway or armor
- Any metallic raceway or armor that encloses the grounding electrode conductor

*NEC Section 250.92(B)* defines the methods permitted to be used to ensure continuity at service equipment. These methods are:

- Bonding equipment to the grounded service conductor in a manner provided by *NEC Section 250.8*
- Connections utilizing threaded couplings or threaded hubs on enclosures if made up wrenchtight
- Threadless couplings and connectors if made up wrenchtight
- Other approved devices, such as bonding-type locknuts

The exception to this requirement is found in *NEC Section 250.84(A)*. This states that an underground service cable that is metallically connected to the grounded system conductor on the supply side shall not be required to be connected to the grounded system conductor at the building or structure.

## 8.1.0 Bonding Multiple Service Disconnecting Means

Multiple disconnecting means can take several forms. *NEC Article 230* discusses the various forms.

Because this bonding jumper is located on the supply side of the service disconnect, the basic rule for sizing the supply side bonding jumper for bonding these various configurations is found in *NEC Section 250.102(C)*. This section requires the use of *NEC Table 250.66* for selecting the size. However, if the size of the service conductor is larger than 1,100 kcmil copper or 1,750 kcmil aluminum, the bonding jumper must have a cross-sectional area not less than 12.5% of the largest phase conductor. Where the service conductors are paralleled in two or more raceways, the equipment bonding jumper shall be run in parallel. The size of each bonding jumper shall be based on the largest phase conductor in each raceway or cable.

*Table 1* shows various sizes of bonding jumpers based on the size of the service-entrance conductor.

## 8.2.0 Bonding of Enclosures and Equipment

The bonding requirements for an installation depend on the system voltage, number of raceways, and type of equipment in use.

### 8.2.1 Bonding of Services Over 250 Volts

*NEC Section 250.97* defines the methods permitted to be used on circuits over 250V. The methods permitted for these circuits are the same as for services, with the exception of using the service grounded conductors beyond the service.

*Figure 25* shows examples of these bonding methods.

### 8.2.2 Bonding Multiple Raceway Systems

*NEC Section 250.102(D)* allows the use of a single conductor to bond two or more raceways or cables where the bonding jumper is sized in accordance with *NEC Table 250.122* for the largest overcurrent device supplying circuits therein. *Figure 26* shows bonding of multiple raceways.

*Figure 25*  Bonding for circuits over 250 volts.

*Figure 26*  Bonding multiple raceways.

**Table 1**  Bonding of Multiple Service Disconnecting Means

| Service-Entrance Conductor | Bonding Jumper (Copper) |
| --- | --- |
| 500 kcmil in service mast | 1/0 |
| 1,000 kcmil in wireway | 2/0 |
| 300 kcmil | No. 2 |
| 3/0 | No. 4 |
| No. 2 | No. 8 |

## Multiple Load Centers

For installations in which the main service panel is located at a distance from areas that have many circuits and/or a heavy load concentration, subpanels may be installed near these loads. Doing so allows the branch circuit wiring runs to be shorter, resulting in lower line voltage losses than would occur if the branch circuits had been run all the way back to the main panel.

It is acceptable to install one equipment bonding jumper individually from each raceway to the equipment grounding terminals of the equipment. This is shown in *Figure 27*. Each bonding conductor is sized per *NEC Table 250.122*.

### 8.2.3 Installation of Equipment Bonding Jumper

The equipment bonding jumper is permitted to be installed either inside or outside of a raceway or enclosure. When the jumper is installed outside, the length is limited to not more than 6'. The bonding jumper must also follow the raceway routing. This is vital to keep the impedance of the equipment bonding jumper as low as possible.

*Figure 28* shows a bonding jumper installed inside a nonmetallic enclosure.

**NOTE**

The equipment bonding jumper must be sized for the overcurrent device.

*Figure 28*   Bonding jumper.

| OVERCURRENT PROTECTION | EQUIPMENT BONDING JUMPER |
|---|---|
| 400 AMPERES | NO. 3 COPPER - NO. 1 ALUMINUM |
| 300 AMPERES | NO. 4 COPPER - NO. 2 ALUMINUM |
| 225 AMPERES | NO. 4 COPPER - NO. 2 ALUMINUM |
| 125 AMPERES | NO. 6 COPPER - NO. 4 ALUMINUM |

**(A)**

209F27A.EPS

**(B)**

209F27B.EPS

*Figure 27*   Individual bonding jumpers.

## 9.0.0 EFFECTIVE GROUNDING PATH

As previously stated, to have effective grounding for both a grounded and an ungrounded system, the grounding system shall have an **effective ground fault path** to provide a permanent and continuous path, sized to safely conduct any fault current that is likely to occur, and it shall have impedance that is low enough to limit the voltage to ground and to facilitate the operation of the circuit protective devices in the circuit.

Low impedance means that the grounding path and the circuit conductors must always be within the same raceway, or the raceway may provide the grounding path.

### 9.1.0 Clearing Ground Faults

A ground fault is a different circuit that becomes dependent on the impedance of the return path back to the source. The amount of current flow depends on the impedance. The higher the current, the harder it becomes to maintain voltage in the system. It is at this voltage that the short circuit capacity plays its role by attempting to maintain voltage. As the system tries to maintain voltage, its current will increase, and when the current reaches the overcurrent setting of the device, the system will be protected by the opening of the overcurrent device.

## 10.0.0 GROUNDED CONDUCTOR

The neutral in a grounded system serves two main purposes:

- It permits utilization of power at line-to-neutral voltage. This will allow the current-carrying conductor to carry any unbalanced current.
- It provides a low-impedance return path for the flow of fault current to the source to facilitate the operation of the overcurrent devices in the circuit.

*Figure 29* shows the path to ground for this neutral. If the neutral is not needed for voltage requirements, it still must be run to the service and connected to the grounded electrode conductor and equipment grounding conductor at the service.

## 11.0.0 SEPARATELY DERIVED SYSTEMS

*Figure 30* shows a transformer-type separately derived system. This system has no direct electrical connection with the premises' source of electrical power. *NEC Section 250.30* discusses whether or not a separately derived system is required to

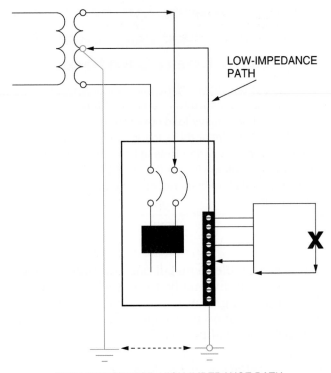

LOW-IMPEDANCE PATH

HIGH-RESISTANCE, HIGH-IMPEDANCE PATH

209F29.EPS

*Figure 29* Grounded service conductor run to service.

be grounded. Each voltage level that is required to be grounded must meet the requirements of *NEC Section 250.30(A)(1)* and have a system bonding jumper that is used to connect the equipment grounding conductors of the separately derived system to the grounded conductor. The grounding electrode conductor is connected to the grounded conductor at the same point where the system bonding jumper is connected. The equipment grounding conductors are found in the feeder panel with the branch circuits supplied by this system.

### On Site

## Separately Derived Systems

The requirements of *NEC Section 250.30* pertaining to the grounding of separately derived AC systems are most commonly applied to 480Y/277V transformers that transform a 480V supply to a 208Y/120V system for lighting and appliance loads. The requirements provide for a low-impedance path to ground, so that line-to-ground faults on circuits supplied by the transformer will result in sufficient current flow to operate the overcurrent devices.

Figure 30 Transformer-supplied separately derived system.

In *Figure 30*, the supply to the transformer will be at one voltage level while the secondary is at either a higher or a lower voltage level.

## 11.1.0 Grounding of Separately Derived Systems

For separately derived systems that are required to be grounded, an equipment grounding conductor must be supplied with the primary circuit. This will provide a low-impedance fault-current path from the transformer case to the main service. Along with the equipment grounding conductor, *NEC Section 250.30(A)(1)* requires a bonding jumper be installed for the derived system required to be grounded. This bonding jumper may be placed in either the transformer or the panel, but not in both. Installing it in both locations would create a parallel path with the grounded conductor, resulting in an unwanted current flow on the grounding conductor. Next, a grounding electrode conductor must be installed and connected at the same point. This is illustrated in *Figure 31*.

The grounding electrode for separately derived systems must be:

• As near as practicable to the system
• The nearest available grounded structural metal member or the nearest available grounded water pipe
• Other electrodes where the above are not available (*NEC Sections 250.50 and 250.52*)

*EQUIPMENT GROUNDING CONDUCTORS NOT SHOWN                209F31.EPS

Figure 31 Separately derived system grounding and bonding locations.

## 11.2.0 Generator-Type Separately Derived System

A simple generator-type separately derived system is shown in *Figure 32*.

The way to determine that a generator is a separately derived system is to examine the transfer switch. If the neutral and all phase conductors are switched, then the system is separately derived. If the neutral is not switched, but solidly connected, then the system is not a separately derived system.

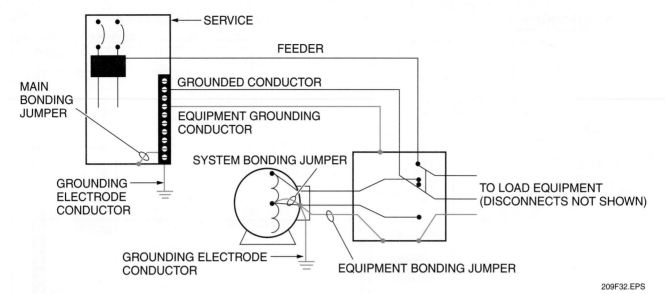

Figure 32  Generator-type separately derived system.

The system bonding jumper must be installed either at the generator or any point in between. A grounding electrode conductor must be connected at the same point as the system bonding jumper.

When the generator is not a separately derived system, the system bonding jumper must be removed from the generator and the neutral must not be grounded. Grounding is accomplished by connecting the neutral of the premises wiring to a grounding electrode.

## 12.0.0 GROUNDING AT MORE THAN ONE BUILDING

*NEC Section 250.32* provides the method for grounding electrical systems at additional buildings on the premises (*Figure 33*).

The grounded circuit conductor and the equipment grounding conductor must both be extended to the second building. The grounded conductor is terminated on an insulated bus isolated from the metal cabinet, while the equipment grounding conductor from the main building is terminated on an equipment grounding terminal bus, directly connected to the cabinet. The equipment grounding conductor would be sized using *NEC Table 250.122* based on the overcurrent device supplying the second structure. All grounding electrodes present at the second building would connect to the grounding bus. The GEC in the second structure is sized using *NEC Table 250.66* based on the size of the feeder conductors. *Table 2* shows the number and type of conductors that must be taken from the first structure to where service is located in the second structure.

When no grounding electrodes are at the additional structures, a grounding electrode must be

**Table 2**  Number of Conductors Between Structures

| System | Ungrounded | Grounded | Equipment Ground |
|--------|-----------|----------|------------------|
| 120V | 1 | 1 | 1 |
| 120/240V | 2 | 1 | 1 |
| 208/120V | 3 | 1 | 1 |
| 480/277V | 3 | 1 | 1 |

installed. These electrodes include underground metal pipes, the grounded metal frame of the structure, or concrete-encased electrodes. Rod, pipe, or plate electrodes may also be permitted, in accordance with *NEC Section 250.52*.

By exception and in existing installations only, the *NEC®* allows a method whereby no equipment ground is extended to the second building per *NEC Section 250.32(B)(1), Exception 1*. In this case, the grounded conductor is extended to and terminated on a terminal bus that is bonded to the equipment, similar to the connection made at the grounded conductor at the service equipment. The grounded system will be connected to all grounding electrodes at the second building. The grounded conductor is sized using *NEC Sections 220.61 and 250.122*. This method may only be used when an equipment grounding conductor is not run with the supply to the second building, there are no continuous metallic paths bonded to the grounding system in both buildings, and ground fault protection has not been installed on the common electrical service in the first building. These additional provisions prevent the possibility of parallel current paths between grounded and grounding conductors, as well as nuisance tripping of a ground fault protective device.

**Figure 33**  Grounding neutral at second building.

209F33.EPS

## 13.0.0 SYSTEMS OVER 1kV

*NEC Article 250, Part X* discusses the grounding requirements for systems over 1kV. It also lists additional requirements for derived neutral systems, solidly grounded neutral systems, and impedance-grounded neutral systems.

## 14.0.0 TESTING FOR EFFECTIVE GROUNDS

An earth ground resistance tester (*Figure 34*) may be used to make soil **resistivity** measurements or to measure the resistance of the installed grounding electrode system to the earth.

> **NOTE**
>
> An ordinary ohmmeter is not used to measure the resistance of a grounding electrode to the earth because it does not provide for adequate levels of current and voltage.

> **WARNING!**
>
> Ground testers can be hazardous to both equipment and personnel if improperly used. Always check with your supervisor before using a ground tester.

One use of the ground tester is for testing electrical systems after they are installed and before normal voltage is applied. This test is made after all the conductors, fuses or circuit breakers,

209F34.EPS

**Figure 34**  Earth ground resistance tester.

panelboards, outlets, etc., are in place and connected. The current used for testing is produced by a small generator within the ground tester that generates DC power, either by turning a crank handle (also a part of the ground tester), or by using a small electric DC motor within the ground tester.

The test is made by connecting the terminals to the two points between which the test is to be made and then rapidly turning the handle on the ground tester. The resistance in ohms can then be read from the meter dial. Satisfactory insulation resistance values will vary under different conditions, and the charts supplied with the ground tester should be consulted for the proper value for a particular installation.

## 15.0.0 MEASURING EARTH RESISTANCE

An earth ground is commonly used as an electrical conductor for system returns. Although the resistivity of the earth is high compared to a metal conductor, its overall resistance can be quite low because of the large cross-sectional area of the electrical path.

Connections to the earth are made with grounding electrodes, ground grids, and ground mats. The resistance of these devices varies proportionately with the earth's resistivity, which in turn depends on the composition, compactness, temperature, and moisture content of the soil.

A good grounding system limits system-to-ground resistance to an acceptably low value. This protects personnel from a dangerously high voltage during a fault in the equipment. Furthermore, equipment damage can be limited by using this ground current to operate protective devices.

Ground testers measure the ground resistance of a grounding electrode or ground grid system. Some of the major purposes of ground testing are to verify the adequacy of a new grounding system, detect changes in an existing system, and determine the presence of hazardous step voltage and touch voltage.

In addition to personnel safety considerations, ground testing also provides information for equipment insulation ratings. Equipment can be damaged by an overvoltage that exceeds the rating of the insulation system. *Figure 35* depicts a poorly grounded system where the ground resistance ($R_1$) is 10Ω.

Assuming a power source resistance ($R_2$) of 40Ω, a short circuit between the 5,000V power line and the steel tower would produce 100A of short circuit current (I).

$$I = \frac{E}{R_1 + R_2} = \frac{5,000}{40 + 10} = 100A \text{ of short circuit current}$$

A person touching the tower would be subjected to the voltage (E') developed across the ground resistance:

$$E' = IR = 100 \times 10 = 1,000 \text{ volts}$$
between power and ground

Statistics vary widely concerning what may be considered a dangerous voltage. This depends largely on body resistance and other conditions. However, to limit the touch voltage for this situation to 100V, the ground resistance for the tower would have to be less than 1Ω.

SHORT CIRCUIT POINT

$(R_1)$ 40Ω

5,000V

POWER SOURCE

1,000V

10Ω $(R_2)$

EARTH GROUND

209F35.EPS

*Figure 35* Poorly grounded system.

NCCER — *Contren® Learning Series* 26209-11

# Step Voltage

In September of 1999, a single stroke of lightning is believed to have killed a herd of 56 elk on Mount Evans in Colorado. This happened because when lightning strikes the earth, it disperses in concentric rings of equal potential (equipotential zones). Because the earth has resistance, the voltage drops as it travels away from the strike point. Step voltage is the difference in potential that exists when you step between these equipotential zones. In this case, the hind legs of the elk were at a different potential than their front legs, resulting in current flow.

**Bottom Line:** During a lightning strike or any other condition that creates zones of different potential, your best bet is to remain in one spot, crouched low with your feet together and your head bent forward. If you must move, hop with your feet together, moving much like a bird on a wire.

209SA03.EPS

## 15.1.0 How the Ground Tester Works

Several methods are used for measuring the resistance to earth of a grounding electrode; one of the most common is the fall-of-potential method. See *Figure 36*.

In this method, **auxiliary electrodes** 1 and 2 are placed at sufficient distances from grounding electrode X. A current (I) is passed through the earth between the grounding electrode X and auxiliary current electrode 2 and is measured by the ammeter. The voltage drop (E) between the grounding electrode X and the auxiliary potential electrode 1 is indicated on the voltmeter. Resistance (R) can therefore be calculated as follows:

$$R = \frac{E}{I}$$

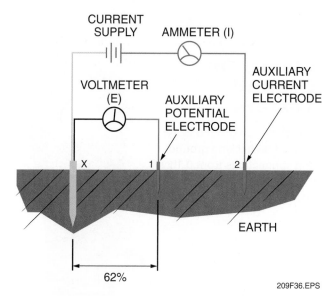

*Figure 36*  Fall-of-potential method of testing.

Certain problems may arise in measuring with the simple system shown in *Figure 36*:

- Natural currents in the soil caused by electrolytic action can cause the voltmeter to read either high or low, depending on polarity.
- Induced currents in the soil, instrument, or electrical leads can cause vibration of the meter pointer, interfering with readability.

# Touch Voltage

If a system is ungrounded and the metal parts become energized, a potential difference will exist between the system and ground. This is known as touch voltage. For example, a 10-year-old boy was swimming at a community pool and exited the pool to buy a snack from a vending machine. The grounding prong had been removed from the plug and the cord was pinched under the machine's metal leg, exposing the conductor and causing the metal frame of the unit to become energized. As soon as he touched the metal, he received a fatal shock.

**Bottom Line:** Make sure the entire grounding system is intact for cord and plug connected equipment. In case of damage to the cord, the equipment grounding conductor could be damaged and deficient. When servicing any motor, check to ensure that the cord is not pinched. Never remove the grounding prong from a three-prong plug. Because of this and similar incidents, *NEC Section 422.51* now requires GFCI protection for cord and plug connected vending machines.

## Clamp-On Ground Testers

This clamp-on ground tester allows for checking ground grid connections from 0.10Ω to 1,200Ω without the need to disconnect the conductors.

209SA04.EPS

• Resistance in the auxiliary electrode and electrical leads can introduce error into the voltmeter reading.

Most ground testers use a null balance metering system. Unlike the separate voltmeter and ammeter method, this instrument provides a readout directly in ohms, thus eliminating calculation. Although the integrated systems of the ground testers are sophisticated, they still perform the basic functions for fall-of-potential testing.

### 15.2.0 Current Supply

As in the simple circuit, the ground tester also has a current supply circuit (see *Figure 37*). This may be traced from grounding electrode X through terminal X, potentiometer R1, the secondary of power transformer T1, terminal 2, and auxiliary current electrode 2. This produces a current in the earth between electrodes X and 2.

When switch S1 is closed, battery B energizes the coil of vibrator V. Vibrator reed V1 begins oscillating, thereby producing an alternating

209F37.EPS

*Figure 37*   Three-point testing using a ground tester.

# Step or Touch Potential

current in the primary and secondary windings of T1. The negative battery terminal is connected alternately across first one and then the other half of the primary winding.

## 15.3.0 Voltmeter Circuit

The voltmeter circuit can be traced from grounding electrode X through terminal X, the T2 secondary, switch S2, resistors R2 and R3 (paralleled by the meter and V2 contacts), capacitor C, terminal 1, and auxiliary potential electrode 1. The current in the earth between grounding electrode X and auxiliary current electrode 2 creates a voltage drop due to the earth's resistance. With auxiliary potential electrode 1 placed at any distance between grounding electrode X and auxiliary current electrode 2, the voltage drop causes a current in the voltmeter circuit through balanced resistors R2 and R3. The voltage drop across these resistors causes galvanometer M to deflect from zero center scale.

Vibrator reed V2 operates at the same frequency as V1, thereby functioning as a mechanical rectifier for galvanometer M. The vibrator is tuned to operate at 97.5 hertz (Hz), a frequency unrelated to commercial power line frequencies and their harmonics. Thus, currents induced in the earth by power lines are rejected by most ground testers and have virtually no effect on their accuracy. Stray direct current in the earth is blocked out of the voltmeter circuit by capacitor C.

The current in the primary of T2 can be adjusted with potentiometer R1. Primary current in T2 induces a voltage in the secondary of T2, which is opposite in polarity to the voltage drop caused by current in the voltmeter circuit.

With R1 adjusted so the primary and opposing secondary voltages of T2 are equal, current in the voltmeter circuit is zero, and the galvanometer reads zero. The resistance of grounding electrode X can then be read on the calibrated dial of the potentiometer.

With no current in the voltmeter circuit, the lead resistance of the auxiliary potential electrode 1 has no bearing on accuracy. (With no current,

there is no voltage drop in the leads.) Resistance to earth of the current electrode results only in a reduction of current, and consequently, a loss of sensitivity. Therefore, the auxiliary electrodes need only be inserted into the earth 6" to 8" to make sufficient contact. In some locations, where the soil is very dry, it may be necessary to pour water around the current electrode to lower the resistance to a practical value.

## 15.4.0 Nature of Earth Electrode Resistance

Current in a grounding system is primarily determined by the voltage and impedance of the electrical equipment. However, the resistance of the grounding system is very important in determining the voltage rise between the ground electrode and the earth as well as the voltage gradients that will occur in the vicinity when current is present.

Three components constitute the resistance of a grounding system:

- Resistance of the conductor connecting the ground electrode
- Contact resistance between the ground electrode and the soil
- Resistance of the body of earth immediately surrounding the electrode

Resistance of the connecting conductor can be dealt with separately since this is a function of the conductor cross-sectional area and length. Contact resistance is usually negligibly small if the electrode is free from paint or grease. Therefore, the main resistance is that of the body of earth immediately surrounding the electrode. Current from a grounding electrode flows in all directions via the surrounding earth. It is as though the current flows through a series of concentric spherical shells, all of equal thickness.

The shell immediately surrounding the electrode has the smallest cross-sectional area, and therefore its resistance is highest. As the distance from the electrode is increased, each shell becomes correspondingly larger; thus, the resistance becomes smaller. Finally, a distance from the electrode is reached where additional shells do not add significantly to the total resistance. From a theoretical viewpoint, total resistance is included only when the distance is infinite. For practical purposes, only the volume that contributes the major part of the resistance need be considered. This is known as the effective resistance area and depends on electrode diameter and driven depth.

## 16.0.0 THREE-POINT TEST FOR SINGLE ELECTRODE/TRIAD

To measure the resistance of a grounding electrode with most ground testers, an auxiliary current electrode and an auxiliary potential electrode are required (see *Figures 38* and *39*). The current electrode is placed a suitable distance from the grounding electrode under test, and the potential electrode is then placed at 62% of the current electrode distance.

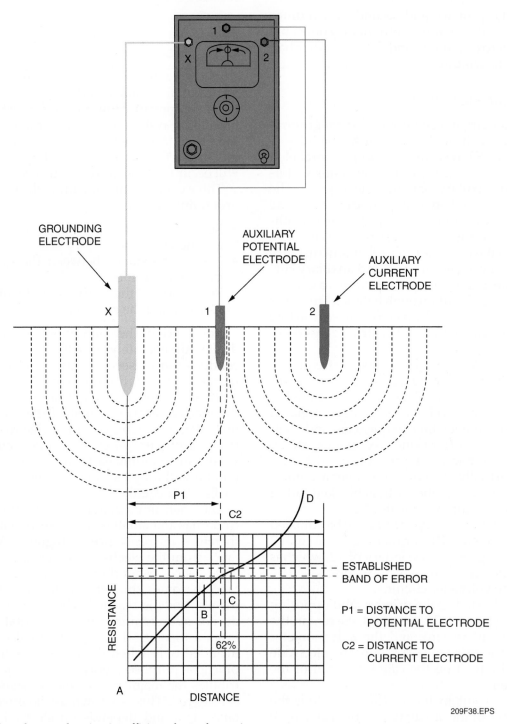

P1 = DISTANCE TO POTENTIAL ELECTRODE

C2 = DISTANCE TO CURRENT ELECTRODE

209F38.EPS

*Figure 38* Plotted curve showing insufficient electrode spacing.

## Think About It

# Size and Depth of Ground Electrode

What effect do the diameter and driven depth of a ground electrode have on its resistance?

*Figure 39* Plotted curve showing adequate electrode spacing.

209F39.EPS

If the current electrode is too near the grounding electrode, their effective resistance areas will overlap, as shown in *Figure 38*. If a series of measurements are made with the potential electrode driven at various distances in a straight line between the current electrode and grounding electrode, the readings will yield a curve as illustrated in *Figure 38*.

In *Figure 39*, a curve is plotted with the current electrode at a sufficient distance from the grounding electrode. Note that the curve is relatively flat between points B and C, which are usually considered to be at ±10' with respect to the 62% point. Usually, a tolerance is established for the maximum allowable deviation for the second and third readings with relation to the initial reading at 62%. This tolerance is a certain ±% of the initial reading, such as ±1%, ±2%, etc.

No definite distance from the current electrode can be forecast since the optimum distance is based on the homogeneity of the earth, depth of the grounding electrode, diameter, etc. However, for a starting point for a single driven grounding electrode, the effective radius of the equivalent hemisphere can be computed. The curve of *Figure 40* can then be used for initial placement of the auxiliary electrodes. Using the practical method of moving the auxiliary potential electrode 10' to either side of the 62% point, a curve can be plotted

to determine if the current electrode spacing is adequate, as depicted by the curve flattening out between points B and C in *Figure 39*.

For example, assume that a 1" (2.54 cm) diameter grounding electrode is buried 10' (3.0 m) deep. The equivalent hemisphere radius is 1.7' (51.8 cm). Using *Figure 40*, the current electrode would be established at approximately 90' (27 m), and the potential electrode at approximately 55' (17 m) for the initial reading. The results obtained when the potential electrode is moved (to 45' [14 m] and then to 65' [20 m]) will determine if the current electrode, and consequently the potential electrode, must be spaced at a greater distance.

It is usually advisable to plot a complete curve for each season of the year. See *Figure 41*. These curves should be retained for comparison purposes. Measurements at established intervals in the future need only be made at the 62% point

---

### Think About It

# Poor Soil Conductivity

What can be done to lower the ground electrode (rod) resistance when it is driven into soil with poor conductivity?

---

209F40.EPS

*Figure 40*   Auxiliary electrode distance/radii chart.

---

and, if desired, 10' on each side, providing there is no erratic deviation from the original curve. Serious deviation, other than seasonal, could mean corrosion has eaten away some of the electrode.

> **NOTE**
>
> Safety grounds should be applied in such a way that a zone of equal potential is formed in the work area. This equipotential zone is formed when fault current is bypassed around the work area by metallic conductors. In this situation, the worker is bypassed by the low-resistance metallic conductors of the safety ground.

209F41.EPS

*Figure 41*  Typical grounding resistance curve to be recorded and retained.

## SUMMARY

There is no subject in the electrical industry more important than grounding. It is the chief means of protecting life and property from electrical hazards. It also ensures proper operation of the system and helps other protective devices to function properly.

Before a system is grounded, it must first be bonded together. Bonding is the interconnection of all noncurrent-carrying metal components in a system. A bonded system is then connected to a grounding electrode.

The term *grounded* means connected to earth (ground) by a conductor or to some conducting body that extends the ground connection, such as a ground rod. The earth as a whole is properly classed as a conductor. For convenience, its electric potential is assumed to be zero. When a metal object is grounded, it too is thereby forced to take the same zero potential as the earth. Therefore, the main purpose of grounding is to ensure that the grounded object cannot take on a potential differing sufficiently from earth potential to be hazardous.

Refer to *Figure 1* when answering Questions 1 through 4.

**Figure 1**

1. *Item B* in the figure represents the _____.
   a. grounded conductor
   b. ungrounded conductors
   c. grounding electrode conductor
   d. main bonding jumper

2. *Item E* in the figure represents the _____.
   a. grounded conductor
   b. ungrounded conductors
   c. grounding electrode conductor
   d. main bonding jumper

3. *Item C* in the figure represents the _____.
   a. grounded conductor
   b. ungrounded conductors
   c. grounding electrode conductor
   d. main bonding jumper

4. *Item F* in the figure represents the _____.
   a. grounded conductor
   b. ungrounded conductors
   c. grounding electrode conductor
   d. main bonding jumper

5. While grounding prevents excessive voltage due to lightning or line surges, it does *not* stabilize voltage during normal operation.
   a. True
   b. False

6. Which conductor(s) are grounded on a single-phase, three-wire, 120/240V system?
   a. Hot
   b. Current-carrying
   c. Neutral
   d. Black

7. The neutral conductor is always larger than the ungrounded conductors.
   a. True
   b. False

8. Residential electrical services are grounded by connecting the grounded conductor to a grounding electrode at the _____.
   a. pole-mounted transformer
   b. service drop connection
   c. neutral bus in the main panelboard
   d. meter enclosure

9. Which of the following is an *NEC*® violation when used as a grounding electrode?

    a. Copper cold-water pipe
    b. Galvanized pipe
    c. An underground gas line
    d. A grounding ring

10. The maximum distance allowed by the *NEC*® to connect a grounding conductor to a water pipe after the pipe enters the building is _____.

    a. 5'
    b. 10'
    c. 15'
    d. 20'

11. Which of the following is used to connect a ground rod with a grounding conductor?

    a. A grounding clip
    b. A butt-and-slide connector
    c. A wire nut
    d. A grounding clamp

12. The conducting connection, whether intentional or accidental, between any of the conductors of an electrical system is known as a _____.

    a. ground fault
    b. bonding
    c. grounded conductor
    d. short circuit

13. All of the following are suitable materials for ground rods *except* _____.

    a. ½" solid stainless steel rod
    b. ¾" galvanized pipe
    c. ⅝" copperweld solid rod
    d. ½" solid aluminum rod

14. Which of the following copper AWG sizes should be used for the grounding conductor for a 200A service using 3/0 copper conductors?

    a. No. 2
    b. No. 4
    c. No. 6
    d. No. 8

15. When a metal water pipe is used as a grounding electrode, what must be provided around the water meter?

    a. Bonding jumpers
    b. A floor drain
    c. A grounding locknut
    d. A grounding bushing

1. A short circuit is a conducting connection, whether intentional or accidental, between any of the _____ of an electrical system.

2. True or False? The neutral or grounded conductor is a system or circuit conductor that is intentionally grounded.

3. A conductor used to connect equipment or the grounded circuit of a wiring system to a grounding electrode is called a(n) _____ conductor.

4. The two types of grounding systems are system grounding and _____ _____.

5. If the loads were only 240V on a single-phase 240V system with a center-tapped transformer, the neutral conductor would carry _____ amperes.

6. Systems less than 50V are required to be grounded if the supply voltage to the transformer exceeds _____ to ground.

7. Circuits that cannot be grounded include certain circuits for cranes, electrolytic cells, secondary circuits on lighting systems, and isolated power systems in _____.

8. Systems are solidly grounded to limit the _____ during normal operation and to prevent excessive voltages due to lightning and line surges.

9. Grounding methods include an underground water pipe in direct contact with the earth for no less than _____.

10. The size of the grounding electrode conductor is based on the size of the _____ _____.

11. The size of the EGC is determined by the _____.

12. All receptacles used in residential applications must be of the _____ type.

13. Electrical _____ is the key to successful clearing of ground faults.

14. The size of the bonding jumper on the supply side is based on the size of the _____ in each raceway.

15. The two main purposes of the neutral in a grounded system are to permit utilization of power at line-to-neutral voltage and _____

_____

_____

# Trade Terms Introduced in This Module

**Auxiliary electrodes:** Metallic electrodes pushed or driven into the earth to provide electrical contact for the purpose of performing measurements on grounding electrodes or ground grid systems.

**Bonding:** Connected to establish electrical continuity and conductivity.

**Effective ground fault path:** An intentionally constructed, permanent, low-impedance electrically conductive path designed and intended to carry current under ground-fault conditions from the point of a ground fault on a wiring system to the electrical supply source and that facilitates the operation of the overcurrent protective device or ground-fault detectors on a high-impedance grounded system.

**Equipment bonding jumper:** The connection between two or more portions of the equipment grounding conductor.

**Equipment grounding conductor:** The conductive path installed to connect the normally noncurrent-carrying metal parts of equipment together and to the system grounded conductor, or to the grounding electrode conductor, or both.

**Ground:** A conducting connection, whether intentional or accidental, between an electrical circuit or equipment and the earth.

**Ground current:** Current in the earth or grounding connection.

**Ground grids:** System of grounding electrodes interconnected by bare cables buried in the earth to provide lower resistance than a single grounding electrode.

**Ground mats:** System of bare conductors, on or below the surface of the earth, connected to a ground or ground grid to provide protection from dangerous touch voltage.

**Ground resistance:** The ohmic resistance between a grounding electrode and a remote or reference grounding electrode that are spaced such that their mutual resistance is essentially zero.

**Ground rod:** A metal rod or pipe used as a grounding electrode.

**Grounded:** Connected to ground or to a conductive body that extends the ground connection.

**Grounded conductor:** A system or circuit conductor that is intentionally grounded.

**Grounding clip:** A spring clip used to secure a bonding conductor to an outlet box.

**Grounding conductor:** A conductor used to connect equipment or the grounded circuit of a wiring system to a grounding electrode or electrodes.

**Grounding connections:** Connections used to establish a ground; they consist of a grounding conductor, a grounding electrode, and the earth surrounding the electrode.

**Grounding electrode:** A conducting object through which a direct connection to earth is established.

**Grounding electrode conductor:** A conductor used to connect the system grounded conductor or the equipment to a grounding electrode or to a point on the grounding electrode system.

**Main bonding jumper:** The connection between the grounded circuit conductor and the equipment grounding conductor at the service.

**Neutral:** The conductor connected to the neutral point of a system that is intended to carry current under normal conditions.

**Resistivity:** Resistance between opposite faces of a unit cube. Expressed in ohm-centimeters or ohms per cubic centimeter.

**Separately derived system:** A premises wiring system whose power is derived from a source of electric energy or equipment other than a service. Such systems have no direct electrical connection, including a solidly connected ground circuit conductor, to supply conductors originating in another system.

**Short circuit:** An often unintended low-resistance path through which current flows around, rather than through, a component or circuit.

**Step voltage:** The potential difference between two points on the earth's surface separated by a distance of one pace, or about three feet.

**System grounding:** Intentional connection of one of the circuit conductors of an electrical system to ground potential.

**Touch voltage:** The potential difference between a grounded metallic structure and a point on the earth's surface equal to the normal maximum horizontal reach—approximately three feet.

**Ungrounded conductors:** Conductors in an electrical system that are not intentionally grounded.

---

# Additional Resources

This module presents thorough resources for task training. The following resource material is suggested for further study.

*National Electrical Code® Handbook*, Latest Edition. Quincy, MA: National Fire Protection Association.

# Figure Credits

# Circuit Breakers and Fuses

**Inn at the Ballpark, Houston, TX**

HOAR Construction won an ABC Excellence in Construction award in the Renovation division for its work in completely gutting a 150,000-square-foot condemned down-town office space and rebuilding it into a 201-room, baseball-theme boutique hotel.

## 26210-11

Trainees with successful module completions may be eligible for credentialing through NCCER's National Registry. To learn more, go to **www.nccer.org** or contact us at **1.888.622.3720.** Our website has information on the latest product releases and training, as well as online versions of our *Cornerstone* newsletter and Pearson's Contren® product catalog.

Your feedback is welcome. You may email your comments to **curriculum@nccer.org,** send general comments and inquiries to **info@nccer.org,** or use the User Update form at the back of this module.

 V.1 7/11

## Objectives

When you have completed this module, you will be able to do the following:

1. Explain the necessity of overcurrent protection devices in electrical circuits.
2. Define the terms associated with fuses and circuit breakers.
3. Describe the operation of a circuit breaker.
4. Apply the *National Electrical Code®* (*NEC®*) requirements for overcurrent devices.
5. Describe the operation of single-element and time delay fuses.

## Performance Tasks

Under the supervision of the instructor, you should be able to do the following:

1. Identify the following on one or more circuit breaker(s) and fuse(s):
   - Number of poles
   - Load rating
   - Voltage rating
   - Amperage interrupting rating

## Trade Terms

Cartridge fuse
Circuit breaker
Dual-element fuse
Edison-base
Fault current

Frame size
Fuse
Fuse link
Molded-case circuit breaker
Nonrenewable fuse

Overcurrent protection
Plug fuse
Pole
Short circuit

## Required Trainee Materials

1. Pencil and paper
2. Appropriate personal protective equipment
3. Copy of the latest edition of the *National Electrical Code®*

Note: *NFPA 70®*, *National Electrical Code®*, and *NEC®* are registered trademarks of the National Fire Protection Association, Inc., Quincy, MA 02269. All *National Electrical Code®* and *NEC®* references in this module refer to the 2011 edition of the *National Electrical Code®*.

# Circuit Breaker Current Interrupting Rating

*NEC Section 240.83(C)* requires that circuit breakers having a current interrupting rating other than 5,000A have the rating clearly shown on the circuit breaker. By definition, circuit breakers that have not been marked with an interrupting rating have a rating of 5,000A.

interrupting rating of electrical equipment is now divided into two parts:

- Current at fault (short circuit) levels
- Current at operating levels

Most people are familiar with the normal current-carrying ampere rating of fuses and circuit breakers. If an overcurrent protective device is designed to open a circuit when the circuit load exceeds 20A for a given time period, as the current approaches 20A, the overcurrent protective device begins to overheat. If the current barely exceeds 20A, the circuit breaker will open normally or a fuse link will melt after a given period of time with little, if any, arcing. If 40A of current are instantaneously applied to the circuit, the overcurrent protective device will open faster, but again with very little arcing. However, if a fault current occurs that runs the amperage up to 5,000A, an explosion effect would occur within the protective device. One simple indication of this explosion effect is demonstrated by the blackened windows of blown plug fuses.

If this fault current exceeds the interrupting rating of a fuse or circuit breaker, the protective device can be damaged or destroyed; such current can also cause severe damage to equipment and injure personnel. Therefore, selecting overcurrent protective devices with the proper interrupting capacity is extremely important in all electrical systems.

Consider the following analogy using a dammed stream as an example of interrupting rating (*Figure 4*). The reservoir capacity represents the available fault current in an electrical circuit; the flood gates (located downstream from the dam) represent the overcurrent protective device in the circuit rated at 10,000 gallons per minute (10,000A); and the stream of water coming through the discharge pipes in the dam represents the normal load current. *Figure 4* shows a normal flow of 100 gallons per minute (100A).

Note the bridge downstream from the flood gates. This bridge represents downstream circuit components or equipment connected to the circuit.

*Figure 5* shows this same diagram with a fault in the dam, creating a water short circuit that allows 50,000 gallons per minute to flow (50,000 fault circuit amperes). Such a situation destroys the flood gates because of inadequate interrupting rating. The overcurrent protective device in the circuit will also be destroyed. With the flood gates damaged, this surge of water continues downstream, wrecking the bridge. Similarly, the downstream components may not be able to withstand the let-through current in an electrical circuit.

*Figure 6* shows the same situation, but with adequate interrupting capacity. Note that the flood gates have adequately contained the surge of water and restricted the let-through current to an amount that can be withstood by the bridge or the components downstream.

There are several factors that must be considered when calculating the required interrupting capacity of an overcurrent protection device. *NEC Section 110.10* states that the overcurrent protective devices, the total impedance, the component short circuit current ratings, and other characteristics of the circuit to be protected shall be so selected and coordinated to permit the circuit protective devices used to clear a fault to do so without extensive damage to the electrical components of the circuit. This fault shall be assumed to be either between two or more of the circuit conductors, or between any circuit conductor and the equipment grounding conductor(s) permitted in *NEC Section 250.118*.

The component short circuit rating is a current rating given to conductors, switches, circuit breakers, and other electrical components, which, if exceeded by fault currents, will result in extensive damage to the component. Short circuit damage can be the result of heat generated or the electromechanical force of a high-intensity magnetic field. The rating is expressed in terms of time intervals and/or current values.

# Current Interrupting Rating

When replacing a circuit breaker, is it necessary to take into account the current interrupting rating of the replacement breaker? If so, why?

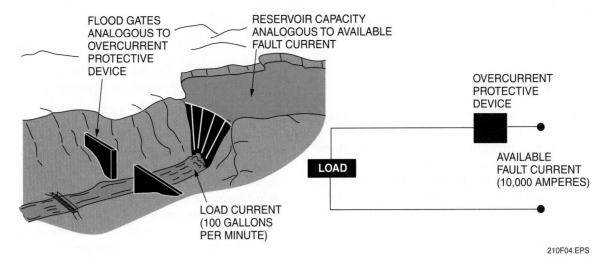

Figure 4 Normal current operation.

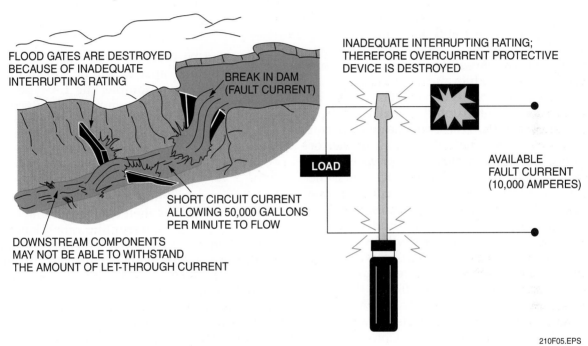

Figure 5 Short circuit operation with inadequate interrupting rating.

Figure 6 Short circuit operation with adequate interrupting rating.

The intent of the *NEC®* is that the design of a system must be such that short circuit currents cannot exceed the short circuit current ratings of the components selected as part of the system. Given specific system components and the level of available short circuit currents that could occur, overcurrent protection devices (mainly fuses and/or circuit breakers) must be used that will limit the let-through current to levels within the withstand ratings of the system components.

In most large commercial and industrial installations, it is necessary to calculate the available short circuit current at various points in a system to determine if the equipment meets the requirements of *NEC Sections 110.9 and 110.10*. There are a number of methods used to determine the short circuit requirements in an electrical system. Some give approximate values; others require extensive computations and are quite exacting. A simple, yet accurate method is the point-by-point method, which will be discussed in more detail later in this module.

An overcurrent protection device must be selected with an interrupting capacity that is equal to or greater than the available short circuit current at the point where the circuit breaker or fuse is applied in the system. The breaker interrupting capacity is based on tests to which the breaker is subjected. There are two such tests; one is set up by Underwriters Laboratories, Inc. (UL) and the other by the National Electrical Manufacturers Association (NEMA). The NEMA tests are self-certification tests, while the UL tests are certified by unbiased witnesses. UL tests have been limited to a maximum of 10,000A in the past, so the emphasis was placed on the NEMA tests with higher ratings. The UL tests now include the NEMA tests plus other ratings. Consequently, emphasis is now placed on the UL tests. It is important to remember that NEMA does not list devices. Two organizations that both test and list devices are UL and the Canadian Standards Association (CSA).

UL requires that molded-case circuit breakers open within a certain period of time during overcurrent conditions. For example, UL requires that a 240V molded-case circuit breaker trip at 300% of its rated continuous current within 50 seconds. If the breaker were rated at 600V, it would be required to trip within 70 seconds at 300% rated current. Currents between 7 and 15 times the rated current (depending on voltage rating and frame size) are handled as overcurrents by the thermal magnetic trip element. Currents above this level activate the magnetic trip element.

The interrupting capacity of a circuit breaker is based on its rated voltage. Where the circuit

breaker can be used on more than one voltage, the interrupting capacity will be shown for each voltage level. For example, the LA-type circuit breaker has 42,000A symmetrical interrupting capacity at 240V, 30,000A symmetrical at 480V, and 22,000A symmetrical at 600V.

### 2.2.1 Standard Interrupting Capacity

Standard interrupting capacity circuit breakers can be identified by their black operating handles and black printed interrupting rating labels. The interrupting rating of a circuit breaker is as important in application as the voltage and current ratings and should be considered each time a breaker is applied. In residential applications, the available fault current is seldom higher than 10,000A or even near this value. The QO and A1 breakers (15A–150A) with the interrupting rating of 10,000A are used in these applications. Higher interrupting ratings are available when required.

### 2.2.2 High Interrupting Capacity

If a circuit breaker requires higher interrupting capacity than the standard ratings, the FH (H for high interrupting capacity) is available (*Figure 7*). The FH-type circuit breaker has an interrupting rating of 65,000A symmetrical at 240VAC. The continuous current ratings are duplicated in these breakers (15A–100A), but the interrupting capacity has been increased to satisfy the need for greater interrupting capacity. This type of breaker is applied in installations where higher fault currents are available (such as large industrial plants). These breakers are built with a case material that will withstand the higher shocks from heat and interrupting forces. Circuit breakers with high

210F07.EPS

*Figure 7*   High interrupting capacity circuit breaker.

interrupting capacity are available in the FH, KH, LH, MH, NH, and PH types.

Three ampere ratings (15A, 20A, and 30A) are also available in the QH high interrupting capacity breakers. These breakers satisfy conditions of lighting circuits supplied from these high available fault current circuit breakers.

### 2.2.3 Current-Limiting Circuit Breakers

The need for current limitation is the result of the increasingly higher available fault currents associated with the growth and interconnection of modern power systems. To meet this need, current-limiting circuit breakers have been developed. This type of breaker operates extremely fast to provide downstream protection for other types of fuses with as little as 10,000 AIC on systems with 100,000A available fault current.

There are circuit breakers available for almost every need in any electrical system. For example, in circuits where the breaker must be highly sensitive to amperage changes (in an environment where the temperature changes are great), an ambient-compensating circuit breaker may be used. Its design permits it to compensate for temperature variation.

A current-limiting circuit breaker is basically a conventional, common-trip, thermomagnetic circuit breaker with an independent limiter section in series with each pole. For purposes of explanation, each limiter can be represented by a set of contacts shunted by a transformable resistor.

At currents below a threshold of about 1,000A, the current-limiting circuit breaker performs in a manner similar to conventional thermomagnetic circuit breakers. In the event of an overload or minor fault current, only the breaker section

contacts open to interrupt the circuit. The limiter contacts remain closed.

For 100A frame current-limiting circuit breakers, the threshold of current limitation occurs in the range of 1,000A to 2,000A. Above this point, the limiter contacts open first—starting the limiting action within ¼ of a millisecond of fault initiation. A substantial arc voltage is generated across the limiter contacts. Rapid generation of this arc voltage is essential to drastically limit the peak let-through current in the faulted circuit.

The rapid rising equivalent resistance of the arc across the limiter contacts forces the current to transfer to the alternate path provided by the limiter resistor. This special resistor has a high positive temperature coefficient of resistance. As current is transferred to it, the temperature of the resistor increases rapidly with a consequent rapid rise in resistance. The resistor dissipates and limits let-through energy and increases the short circuit power factor of the circuit nearly to unity, resulting in an easier, more reliable, transient-free interruption.

During the extremely short period of operation of the limiter section, the breaker section tripping mechanism is activated and its contacts begin to open. After the arc between the limiter contacts has been extinguished and all the current is flowing through the limiter resistor, the breaker contacts interrupt the already drastically limited fault current. Because the fault current and system voltage are nearly in phase at this point, complete interruption of the circuit takes place near the first voltage zero of the AC wave.

When a current-limiting circuit breaker trips to clear a faulted circuit, it can be reset in the same manner as a conventional molded-case circuit breaker. This is true whether tripping was due to low-level overload currents or major faults involving the breaker's current-limiting mode of operation. The limiter contacts are automatically closed after circuit interruption. Breaker contacts are reclosed by moving the handle to the extreme Off position and then to On. There are no fusible elements in this type of circuit breaker.

Regular maintenance of circuit breakers and their enclosures is necessary to obtain the best service and performance.

Since every circuit breaker failure represents a potential hazard to other equipment in the system, it is difficult to calculate the risks involved in prolonging maintenance inspections. One of the chief causes of circuit breaker failure is high heat caused by loose connections at the load side of the breaker. Another cause of circuit breaker failure is defective loads; that is, electrical equipment that cycles too frequently, which in turn causes the circuit breaker to overheat. In the majority of circuit breaker failures, heat is the most likely cause.

## 2.3.0 Labeling Considerations

*NEC Sections 110.22(A) and 240.86(A)* require special marking for UL-recognized, series-rated systems (see *Figure 8*). For example, *NEC Section 110.22(B)* states that where equipment enclosures for circuit breakers or fuses applied in compliance with the series combination ratings selected under engineering supervision in accordance with *NEC Section 240.86(A)* shall be legibly marked in the field as directed by the engineer to indicate the equipment has been applied with a series combination rating. The marking shall be readily visible and state the following:

> CAUTION — SERIES COMBINATION SYSTEM RATED _____ AMPERES. IDENTIFIED REPLACEMENT COMPONENTS REQUIRED.

Complete coverage of engineered electrical distribution systems giving complete details of these *NEC®* requirements are presented in *Electrical Level Three* and will not be covered in depth in this module.

## 2.4.0 *NEC®* Regulations

The following are critical *NEC®* sections with which you should become familiar. These are not the only *NEC®* requirements of importance when dealing with overcurrent protection devices, but they are considered by many to be the most important.

- *NEC Section 110.3(B), Installation and Use* – Listed or labeled equipment shall be used and installed in accordance with any instructions included in the listing or labeling.
- *NEC Section 110.9, Interrupting Rating* – Equipment intended to interrupt current at fault levels shall have an interrupting rating not less than the nominal circuit voltage and the current that is available at the line terminals of the equipment. Equipment intended to interrupt current at other than fault levels shall

Circuit breakers rated at 100 amperes or less and 600 volts or less shall have the ampere rating molded, stamped, etched, or similarly marked into their handles or escutcheon areas.
*NEC Section 240.83(B)*

Circuit breakers must be marked with their ampere rating in a manner that will be durable and visible after installation. The breaker rating must meet or exceed the panel rating or the entire panel must be derated.
*NEC Section 240.83(A)*

INTERRUPTING RATING

200,000 AMPS
FUSETRON
TIME DELAY
CURRENT LIMIT
CLASS R FUSE
FRN R100

OFF
20

10,000
A/C

CURRENT INTERRUPTING RATING 10,000

NORMAL CURRENT RATING

**FUSE**

Every circuit breaker having an interrupting rating other than 5,000 amperes shall have its interrupting rating shown on the circuit breaker.
*NEC Section 240.83(C)*

**CIRCUIT BREAKER**

210F08.EPS

*Figure 8*  Most overcurrent protective devices are labeled with two current ratings.

have an interrupting rating at nominal circuit voltage not less than the current that must be interrupted.

- *NEC Section 110.10, Circuit Impedance, Short Circuit Current Ratings, and Other Characteristics* – The overcurrent protection devices, total impedance, component short circuit current ratings, and other characteristics of the circuit to be protected shall be so selected and coordinated as to permit the circuit protective devices that are used to clear a fault to do so without extensive damage to the electrical components of the circuit. The fault shall be assumed to be either between two or more of the circuit conductors, or between any circuit conductor and the equipment grounding conductor(s) permitted in *NEC Section 250.118*.

- *NEC Section 240.1, Scope (Informational Note)* – Overcurrent protection for conductors and equipment is provided to open the circuit if the current reaches a value that will cause an excessively dangerous temperature in conductors or conductor insulation. See also *NEC Sections 110.9 and 110.10* for requirements for interrupting ratings and protection against fault currents.

- *NEC Section 240.2, Definition of Current-Limiting Overcurrent Protective Devices* – A current-limiting overcurrent protection device is a device which, when interrupting currents in its current-limiting range, reduces the current flowing in the faulted circuit to a magnitude that is substantially less than that obtainable in the same circuit if the device were replaced with a solid conductor having comparable impedance.

- *NEC Section 240.83(C), Interrupting Rating* – Every circuit breaker having an interrupting rating other than 5,000A shall have its interrupting rating shown on the circuit breaker.

- *NEC Section 240.83(E), Voltage Marking* – Circuit breakers shall be marked with a voltage rating that is not less than the nominal system voltage that is indicative of their capability to interrupt fault currents between phases or from phase to ground.

- *NEC Section 240.85, Applications* – A circuit breaker with a straight voltage rating, such as 240V or 480V, may be applied in a circuit in which the nominal voltage between any two conductors does not exceed the circuit breaker's voltage rating. A two-pole circuit breaker cannot be used for protecting a three-phase corner-grounded delta circuit unless it is marked single-phase/three-phase to indicate such suitability. A circuit breaker with a slash rating such as 120V/240V or 480Y/277V shall be permitted to be applied on a solidly grounded circuit where the nominal voltage of any conductor to ground does not exceed the lower of the two values of the circuit breaker's voltage rating, and the nominal voltage between any two conductors does not exceed the higher value of the circuit breaker's voltage rating.

- *NEC Section 250.4(A)(1), Electrical System Grounding* – Electrical systems that are grounded shall be connected to earth in a manner that will limit the voltages imposed by lightning, line surges, or unintentional contact with higher voltage lines, and that will stabilize the voltage to earth during normal operation. System equipment and circuit conductors are solidly grounded to the system grounded conductor to facilitate overcurrent device operation in case of fault currents.

## Think About It

## Installing Circuit Breakers

In addition to using the correct size circuit breaker (trip current and interrupt current), what are some other factors that should be taken into consideration when selecting and installing a circuit breaker in a panel?

210SA02.EPS

---

## 3.0.0 GROUND FAULT CURRENT CIRCUIT PROTECTION

Fault current protection is extremely important to anyone who works with and uses electrically operated equipment. A fault current exists when an unintended path is established between an ungrounded conductor and ground. This situation can occur not only from worn or defective electrical equipment, but also from accidental misuse of equipment that is in good working order.

Will a conventional overcurrent device (fuse or circuit breaker) detect a fault current and open the circuit before irreparable harm is done? Before answering this question, a study of the effects of current on the human body is in order.

The hand-to-hand body resistance of an adult lies between 1,000Ω and 4,000Ω, depending on moisture, muscular structure, and voltage. The average value is 2,100Ω at 240VAC and 2,000Ω at 120VAC.

Using Ohm's law, the current resulting from the above-listed average hand-to-hand resistance values is 120 milliamperes (mA) at 240VAC and 60mA at 120VAC. The effects of 60Hz alternating current on a normal healthy adult are as follows (note that the current is in milliamperes):

- *More than 5mA* – Generally painful shock.
- *More than 15mA* – Sufficient to cause freezing to the circuit for 50% of the population.
- *More than 30mA* – Breathing difficulty (possible suffocation).

- *50mA to 100mA* – Possible ventricular fibrillation or very rapid uncoordinated contractions of the ventricles of the heart resulting in loss of synchronization between the heartbeat and pulse beat. Once ventricular fibrillation occurs, it usually continues and death will happen within a few minutes unless special defibrillation equipment is used.
- *100mA to 200mA* – Certain ventricular fibrillation.
- *Over 200mA* – Severe burns and muscle contractions. The heart is more likely to stop than fibrillate.

A conventional overcurrent device will not open a circuit without causing harm. Here is why.

The current that would flow from a defective electric drill through the metal housing and through the human body to ground would be 60mA, calculated using the above value for 120VAC. The 60mA current is less than one-half of 1% of the rating of 15A circuit breaker or fuse, and yet it approaches the current level, which may produce ventricular fibrillation. Obviously, the standard circuit breaker or fuse will not open the circuit under such low levels of current flow.

Ground fault circuit interrupters (GFCIs) are Class A protective devices built in accordance with *UL Standard No. 943* for ground fault circuit interrupters. UL defines a Class A device as one that will trip when a fault current to ground is 6mA or more. Also, Class A devices must not trip below 4mA.

Class A GFCIs provide a self-contained means of testing the fault current circuitry, as required by UL. To test, simply push the test button, and the device will respond with a trip indication. UL requires that the current generated by the test circuit shall not exceed 9mA. Also, UL requires the device to be functional at 85% of the rated voltage.

Knowing these facts, OSHA and other codes require the use of fault current circuit interrupters to be used on certain circuits, namely, all construction sites where temporary service is in use for electric power tools. Other required uses include:

- Circuits for electrically operated pool covers
- Power or lighting circuits for swimming pools, fountains, and similar locations
- Receptacles in both commercial and residential garages
- Receptacles in residential bathrooms
- Receptacles installed outdoors in public spaces
- Receptacles installed in residential crawl spaces or unfinished basements

# GFCI Circuit Protection

Electricity to operate portable power tools at a construction site was supplied by a temporary service. A panel box on the pole had two duplex receptacles in waterproof boxes. One was a regular receptacle and the other a GFCI receptacle. The pole had not been inspected by the city and was not in compliance with code requirements. (It was not grounded.)

A worker used extension cords plugged into the regular receptacle to supply a power tool. The site was wet, humidity was high, and the worker was sweating. Reportedly, he was getting shocks whenever he used the tool. Unknown to the worker, the tool had a ground fault. As he was climbing down to the ground using a piece of metal floor truss as a makeshift ladder, he received a shock, causing him to fall from the floor truss into a puddle of water while still holding onto the tool. He was electrocuted.

**The Bottom Line:** This accident could have been prevented if the temporary electrical service at the construction site had complied with all local regulations and OSHA standards, including proper grounding. Also, OSHA requires that all 120V, single-phase 15A and 20A receptacles at construction sites that are not part of the permanent wiring be GFCI protected.

Another factor that contributed to this accident was the continued use of a power tool that should have been taken out of service immediately at the occurrence of the first shock. Also, the worker should have used an approved ladder instead of the floor truss to climb down to the ground.

- All kitchen countertop receptacles or any receptacles mounted within 6' of a wet bar sink
- Receptacles installed in bathhouses
- Receptacles installed in bathrooms of commercial or industrial buildings
- Receptacles installed on the roof of any building
- Dockside receptacles
- Branch circuits derived from autotransformers

There are two types of GFCI receptacles. The feed-through type will provide GFCI protection to the remainder of the outlets in a circuit, similar to a GFCI circuit breaker. The second, a nonfeed-through type, offers protection only at the point of installation. Both types are available in various styles.

When installed properly, GFCIs continuously monitor the current in the two conductor wires of the circuit—both the grounded and ungrounded conductors. The current in these two conductors should always be equal. If the GFCI senses a difference between them of more than 6mA, it assumes that the difference is due to fault current and automatically trips the circuit. Power is interrupted within $\frac{1}{40}$ of a second or less, which is fast enough to prevent injury to anyone in normal health. Note that the ungrounded and grounded conductors are constantly being monitored. If at any time the sensor detects a difference in current flow between these conductors, this difference will be amplified. See *Figure 9*. The amplified signal will then cause a circuit breaker to open to de-energize the circuit.

**(A) DIAGRAM OF A GFCI CIRCUIT MONITORING A TWO-WIRE CIRCUIT**

210F09A.EPS

**(B) GFCI CIRCUIT BREAKER**

210F09B.EPS

*Figure 9*   GFCI protection.

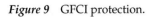

# 4.0.0 FUSES

A fuse is the simplest device for opening an electric circuit when excessive current flows due to an overload or such fault conditions as grounds and short circuits. A fusible link or links encapsulated in a tube and connected to contact terminals comprise the fundamental elements of the basic fuse. The electrical resistance of the link is so low that it simply acts as a conductor, and every fuse is intended to be connected in series with each phase conductor so that current flowing through the conductor or to any load must also pass through the fuse. The continuous current rating of the fuse in amps establishes the maximum amount of current the fuse will carry without opening. When the circuit current flow exceeds this value, an internal element (link) in the fuse melts due to the heat of the current flow and opens the circuit.

Fuses are manufactured in a wide variety of types and sizes with different current ratings, different abilities to interrupt fault currents, various speeds of operation (either quick-opening or time delay types), different internal and external constructions, and voltage ratings for both low-voltage (600V and below) and medium-voltage (over 600V) circuits.

- *Voltage rating* – Most low-voltage power distribution fuses have 250V or 600V ratings (other ratings are 125V and 300V). The voltage rating of a fuse must be at least equal to the circuit voltage. It can be higher but never lower. For instance, a 600V fuse can be used in a 208V circuit. The voltage rating of a fuse is a function of or depends upon its ability to open a circuit under an overcurrent condition. Specifically, the voltage rating determines the ability of the fuse to suppress the internal arcing that occurs after a fuse link melts and an arc is produced. If a fuse is used with a voltage rating lower than the circuit voltage, arc suppression will be impaired and, under some fault current conditions, the fuse may not safely clear the overcurrent.
- *Ampere rating* – Every fuse has a specific ampere rating. In selecting the ampacity of a fuse, consideration must be given to the type of load and code requirements. The ampere rating of a fuse should normally not exceed the current-carrying capacity of the circuit. For instance, if a conductor is rated to carry 20A, a 20A fuse is the largest that should be used in the conductor circuit. However, there are some specific circumstances under which the ampere rating is permitted to be greater than the current-carrying capacity of the circuit. A typical

example is the motor circuit; dual-element fuses are generally permitted to be sized up to 175% and non-time delay fuses up to 300% of the motor full-load amperes. Generally, the ampere rating of a fuse and switch combination should be selected at 125% of the load current. There are exceptions, such as when the fuse-switch combination is approved for continuous operation at 100% of its rating. A protective device must be able to withstand the destructive energy of a short circuit. If a fault current exceeds a level beyond the capability of the protective device, the device may actually rupture and cause severe damage. Thus, it is important when applying a fuse or circuit breaker to use one that can sustain the largest potential short circuit currents. The rating that defines the capacity of a protective device to maintain its integrity when reacting to fault currents is termed its interrupting rating. The interrupting rating of most branch circuit, molded-case circuit breakers typically used in residential service entrance boxes is 10,000A. The rating is usually expressed as 10,000 AIC. Larger, more expensive circuit breakers may have ratings of 14,000 AIC or higher. In contrast, most modern, current-limiting fuses have an interrupting capacity of 200,000A and are commonly used to protect the lower-rated circuit breakers. *NEC Section 110.9* requires equipment intended to interrupt current at fault levels to have an interrupting rating not less than the current that must be interrupted.

- *Time delay ratings* – The time delay rating of a fuse is established by standard UL tests. All fuses have an inverse time current characteristic. That is, the fuse will open quickly on high currents and after a period of time delay on low overcurrents. Specific types of fuses are made to have specific time delays. The basic UL requirement for Class RK-1, RK-15, and J fuses that are marked time delay is that the fuse must carry a current equal to five times its continuous rating for a period of not less than ten seconds. UL has not developed time delay tests for all fuse classes. Fuses are available for use where a time delay is needed along with current limitation on high-level short circuits. In all cases, the manufacturer's literature should be consulted to determine the degree of time delay in relation to the operating characteristics of the circuit being protected.

### 4.1.0 Plug Fuses

Plug fuses have a screw-shell base and are commonly used in dwellings for circuits that supply lighting, heating, and appliances. Plug fuses are supplied with standard screw bases (Edison-base) or Type S bases. An Edison-base fuse consists of a strip of fusible (capable of being melted) metal in a small porcelain or glass case, with the fuse strip or link visible through a window in the top of the fuse. The screw base corresponds to the base of a standard medium-base incandescent lamp. Edison-base fuses are permitted only as replacements in existing installations; all new work must use Type S fuses in accordance with *NEC Section 240.51(B)*, which states that plug fuses of the Edison-base type may only be used for replacements in existing installations where there is no evidence of overfusing or tampering.

The chief disadvantage of the Edison-base plug fuse is that it is made in several ratings (from 0 to 30A), all with the same size base—permitting unsafe replacement of one rating by a higher rating. Type S fuses were developed to reduce the possibility of over-fusing a circuit (inserting a fuse with a rating greater than that required by the circuit). There are 15 classifications of Type S fuses, from 0 to 30A. Each Type S fuse has a base of a different size and a matching adapter. Once an adapter is screwed into a standard Edison-base fuseholder, it locks into place and is not readily removed without destroying the fuseholder. As a result, only a Type S fuse with a size the same as that of the adapter may be inserted. Two types of plug fuses are shown in *Figure 10*; a Type S adapter is also shown.

Plug fuses are also made in time delay types that permit a longer period of overload flow before operation, such as on motor inrush current and other higher-than-normal currents. They are available in ratings up to 30A, both in Edison-base and Type S.

EDISON-BASE FUSE
(MAXIMUM RATING 30 AMPERES)

TYPE S
FUSE ADAPTER

TYPE S FUSE
(MAXIMUM RATING 30 AMPERES)

210F10.EPS

*Figure 10*   Standard Edison-base fuse on top; Type S with adapter shown on bottom.

Their principal use is in motor circuits, where the starting inrush current to the motor is much higher than the running or continuous current. The time delay fuse will not open on the inrush of high starting current. If, however, the high current persists, the fuse will open the circuit just as if a short circuit or heavy overload current had developed. All Type S fuses are time delay fuses.

Plug fuses are normally permitted to be used in circuits of no more than 125V between phases, but they may also be used where the voltage between any ungrounded conductor and ground is not more than 150V. The screwshell of the fuseholder for plug fuses must be connected to the load side circuit conductor; the base contact is connected to the line side or conductor supply. A disconnecting means (switch) is not required on the supply side of a plug fuse.

The plug fuse is a nonrenewable fuse, or one-time fuse; that is, once it has opened the circuit because of a fault or overload, it cannot be used again or renewed. For safe and effective restoration of circuit operation, it must be replaced by a new fuse of the same rating and characteristics.

## 4.2.0 Cartridge Fuses

In most industrial and commercial applications, **cartridge fuses** are used because they have a wider range of types, sizes, and ratings than plug fuses. Older cartridge fuses were provided with a means to renew the fuse by unscrewing the end caps and replacing the links. These fuses presented safety hazards because it was easy to tamper with them to override the safety feature. Their production has been discontinued. Two types of cartridge fuses are shown in *Figure 11*.

*Single-element cartridge fuses* – The basic component of this fuse is the link. Depending upon the ampere rating of the fuse, the single-element fuse may have one or more links. They are electrically connected to the end blades (or ferrules) and enclosed in a tube or cartridge surrounded by an arc-quenching filler material.

Under normal operation, when the fuse is operating at or near its ampere rating, it simply functions as a conductor. However, if an overload current occurs and persists for more than a short interval of time, the temperature of the link eventually reaches a level that causes a restricted segment of the link to melt; as a result, a gap is formed and an electric arc established. As the arc causes the link metal to burn back, the gap becomes progressively larger. The electrical resistance of the arc eventually reaches such a high level that the arc cannot be sustained and is extinguished; the fuse will have then completely cut off all current flow in the circuit. Suppression or quenching of the arc is accelerated by the filler material.

Single-element fuses have a very high speed of response to overcurrents. They provide excellent short circuit component protection. However, temporary harmless overloads or surge currents may cause nuisance openings unless these fuses are oversized. They are best used, therefore, in

*Figure 11*   Cartridge fuses.

circuits that are not subject to heavy transient surge currents and the temporary overload of circuits with inductive loads such as motors, transformers, and solenoids. Because single-element fuses have a high speed of response to short circuit currents, they are particularly suited for the protection of circuit breakers with low interrupting ratings.

*Dual-element cartridge fuses* – Unlike single-element fuses, the dual-element fuse can be applied in circuits subject to temporary motor overload and surge currents to provide both high performance short circuit and overload protection. Oversizing in order to prevent nuisance openings is not necessary. The dual-element fuse contains two distinct types of elements. Electrically, the two elements are series connected. The fuse links, similar to those used in the single-element fuse, perform the short circuit protection function; the overload element provides protection against low-level overcurrents or overloads and will hold an overload that is five times greater than the ampere rating of the fuse for a minimum time of ten seconds.

The overload section consists of a copper heat absorber and a spring-operated trigger assembly. The heat-absorber strip is permanently connected to the first short circuit link and to the second short circuit link on the opposite end of the fuse by the S-shaped connector of the trigger assembly.

The connector electronically joins the one short circuit link to the heat absorber in the overload section of the fuse. These elements are joined by a calibrated fusing alloy. An overload current causes heating of the short circuit link connected to the trigger assembly. Transfer of heat from the short circuit link to the heat absorber in the midsection of the fuse begins to raise the temperature of the heat absorber. If the overload is sustained, the temperature of the heat absorber eventually reaches a level that permits the trigger spring to fracture the calibrated fusing alloy and pull the connector free. The short circuit link is electrically disconnected from the heat absorber, the conducting path through the fuse is opened, and the overload current is interrupted. A critical aspect of the fusing alloy is that it retains its original characteristic after repeated temporary overloads without degradation.

Dual-element fuses may also be used in circuits other than motor branch circuits and feeders, such as lighting circuits and those feeding mixed lighting and power loads. The low-resistance construction of the fuses offers cooler operation of the equipment, which permits higher loading of fuses in switches and panel enclosures without heat damage and without nuisance openings from accumulated ambient heat.

### 4.3.0 Fuse Markings

It is a requirement of the *NEC*® that any cartridge fuses used for branch circuit or feeder protection must be plainly marked, either by printing on the fuse barrel or by a label attached to the barrel that shows the following:

- Ampere rating
- Voltage rating
- Interrupting rating (if other than 10,000A)
- Current limiting, where applicable
- The name or trademark of the manufacturer

### 4.4.0 UL Fuse Classes

Fuses are tested and listed by Underwriters Laboratories, Inc., in accordance with established standards of construction and performance. There are many varieties of miscellaneous fuses used for special purposes or for supplementary protection of individual types of electrical equipment. However, most common fuses are used for the protection of branch circuits and feeders on systems operating at 600V or below. Cartridge fuses in this category include UL Classes H, K, G, J, R, T, CC, and L.

## 5.0.0 OVERCURRENTS

An overcurrent is either an overload current or a short circuit current. The overload current is an excessive current relative to normal operating current, but one that is confined to the normal conductive paths provided by the conductors and other components and loads of the electrical system.

A short circuit (*Figure 12*) is probably the most common cause of electrical problems. It is an undesired current path that allows the electrical current to bypass the load on the circuit. Sometimes the short is between two wires due to faulty insulation, or it can occur between a wire and a grounded object, such as the metal frame of a motor.

### 5.1.0 Overloads

Overloads are most often between one and six times the normal current level. Usually, they are caused by harmless temporary surge currents that occur when motors are started up or transformers are energized. Since they are of brief duration, any temperature rise is trivial and has no harmful effect on the circuit components. Therefore, it is important that protective devices (fuses and circuit breakers) do not react to them.

Continuous overloads can result from motor defects (such as worn motor bearings), overloaded equipment, or too many loads on one circuit. Such sustained overloads are destructive and must be cut off by protective devices before they damage the electrical distribution system or affect the system loads. However, since they are of relatively low magnitude compared to short circuit currents, removal of the overload current within a few seconds will generally prevent equipment damage. A sustained overload

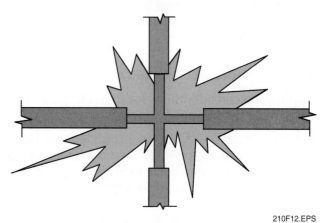

210F12.EPS

*Figure 12*   Short circuit.

## Fuse Classes

Cartridge fuses are classified by Underwriters Laboratories, Inc., according to their maximum voltage and amperage ranges. Some typical values assigned for common fuse classifications are:

Class G    480V, 1 to 60A, interrupting rating of 100,000 rms symmetrical; listed as current limiting

Class H    250/600V, 1 to 600A, interrupting rating of 10,000 rms symmetrical

Class J    600V, 1 to 600A, interrupting ratings ranging from 200,000 to 300,000 rms symmetrical; listed as current limiting

Class K    250/600V, 1/10 to 600A, interrupting ratings ranging from 50,000 to 200,000 rms symmetrical

Class L    600V, 601 to 6,000A, interrupting ratings ranging from 200,000 to 300,000 rms symmetrical; listed as current limiting

Class R    250/600V, 1/10 to 600A, interrupting rating of 200,000 rms symmetrical; listed as current limiting

Class T    300/600V, 1 to 1200A, interrupting rating of 200,000 rms symmetrical; listed as current limiting

Class CC    600V, 1/10 to 30A, interrupting rating of 200,000 rms symmetrical; listed as current limiting

---

current results in overheating of conductors and other components and will cause deterioration of insulation, which may eventually result in severe damage and short circuits if not interrupted.

## 6.0.0 GUIDE FOR SIZING FUSES

General guidelines for sizing fuses are given here for most circuits that will be encountered on conventional systems. Some specific applications may warrant other fuse sizing; in these cases, the load characteristics and appropriate *NEC®* sections should be considered. The selections shown here are not, in all cases, the maximum or minimum ampere ratings permitted by the *NEC®*. Demand factors as permitted by the *NEC®* are not included here.

### 6.1.0 Dual-Element Time Delay Fuses

Application considerations for dual-element time delay fuses include the following:

- *Main service* – Each ungrounded service-entrance conductor shall have a fuse in series with a rating not higher than the ampacity of the conductor, except as permitted in *NEC Section 230.90(A), Exceptions.* As specified in *NEC Section 230.91,* the service overcurrent device must be part of the service disconnecting means or located immediately adjacent to it.

- *Feeder circuit with no motor load* – The fuse size must be at least 125% of the continuous load plus 100% of the non-continuous load.
- *Feeder circuit with all motor loads* – Select the largest branch circuit overcurrent device and add 100% of the full-load current of all other motors connected on the same feeder. The *NEC®* will not permit you to size up; you must size down.
- *Feeder circuit with mixed loads* – Select the largest motor branch circuit fuse and add 100% of the full-load current of all other motors plus 125% of the continuous, non-motor load plus 100% of the non-continuous, non-motor load.
- *Branch circuit with no motor load* – The fuse size must be at least 125% of the continuous load plus 100% of the non-continuous load.

### 6.2.0 Non-Time Delay Fuses

Application considerations for non-time delay fuses include the following:

- *Main service* – Service-entrance conductors shall have a short circuit protective device in each ungrounded conductor, either on the load side of or as an integral part of the service-entrance switch. The protective device shall be capable of detecting and interrupting all values of current in excess of its trip setting or melting point, which can occur at its location.

---

## Current-Limiting Fuses

Current-limiting fuses have a much faster response to short circuit currents and are typically used to provide protection to mains, feeders, circuit breakers, and other components that lack an adequate interrupting rating.

A fuse rated in continuous amperes not to exceed three times the ampacity of the conductor, or a circuit breaker with a trip setting of not more than six times the ampacity of the conductors, shall be considered as providing the required short circuit protection.

- *Feeder circuit with no motor load* – The fuse size must be at least 125% of the continuous load plus 100% of the non-continuous load.

- *Feeder circuit with all motor loads* – Select the largest branch circuit overcurrent device and add 100% of the full-load current of all other motors connected on the same feeder. The *NEC*® will not permit you to size up; you must size down.

- *Feeder circuit with mixed loads* – Select the largest motor branch circuit fuse and add 100% of the full-load current of all other motors plus 125% of the continuous, non-motor load plus 100% of the non-continuous, non-motor load.

- *Branch circuit with no motor load* – The fuse size must be at least 125% of the continuous load plus 100% of the non-continuous load.

When sizing fuses for a given application, a schematic drawing of the system will help tremendously. Such drawings do not have to be detailed; a single-line schematic will suffice, such as the one shown in *Figure 13*.

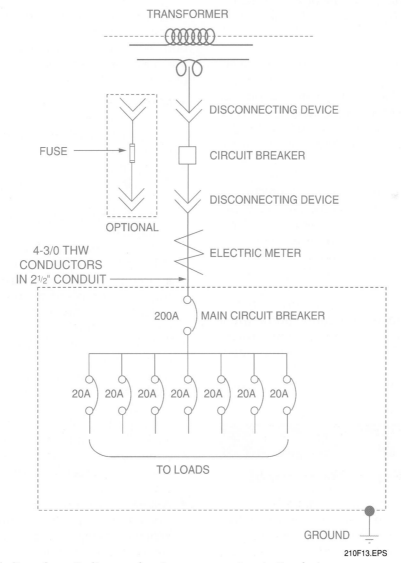

210F13.EPS

*Figure 13*  Typical single-line schematic diagram showing overcurrent protective devices.

For example, determine the feeder ampacity, motor overload protection, branch circuit short circuit and fault current protection, and feeder protection for one 25hp squirrel-cage induction motor (full starting voltage, nameplate current 31.6A, service factor 1.15, Code letter F), and two 30hp wound rotor induction motors (nameplate primary current 36.4A, nameplate secondary current 65A, 40°C rise), on a 460V, three-phase, 60Hz electrical supply.

- *Feeder ampacity* – The full-load current value used to determine the ampacity of conductors for the 25hp motor is 34A [*NEC Section 430.6(A) and Table 430.250*]. A full-load current of 34A × 1.25 = 42.5A (*NEC Section 430.22*). The full-load current value used to determine the ampacity of primary conductors for each 30hp motor is 40A [*NEC Section 430.6(A) and Table 430.250*]. A full-load primary current of 40A × 1.25 = 50A (*NEC Section 430.22*). A full-load secondary current of 65A × 1.25 = 81.25A [*NEC Section 430.23(A)*]. The feeder ampacity will be (40A × 1.25) + 40A + 34A = 124A (*NEC Section 430.24*).

- *Overload* – Where protected by a separate overload device, the 25hp motor with a nameplate current of 31.6A must have overload protection of not over 39.5A [*NEC Sections 430.6(A) and 430.32(A)(1)*]. Where protected by a separate overload device, the 30hp motor with a nameplate current of 36.4A must have overload protection of not over 45.5A [*NEC Section 430.6(A)*]. If the overload protection is not sufficient to start the motor or carry the load, it may be increased according to *NEC Section 430.32(C)*. For a motor marked thermally protected, overload protection is provided by the thermal protector [*NEC Sections 430.7(A)(13) and 430.32(A)(2)*].

- *Branch circuit short circuit and fault current* – The branch circuit of the 25hp motor must have short circuit and fault current protection of not over 300% for a non-time delay fuse (*NEC Table 430.52*) or 3.00 × 34A = 102A. The next smallest standard size fuse

is 100A. The fuse size may be increased to 110A [*NEC Section 430.52(C)(1), Exception No. 1 and NEC Section 240.6*]. Where the maximum value of branch circuit short circuit and fault current protection is not sufficient to start the motor, the value for a non-time delay fuse may be increased to 400% [*NEC Section 430.52(C)(1), Exception No. 2(a)*]. If a time delay fuse is to be used, see *NEC Section 430.52(C)(1), Exception No. 2(b)*.

- *Feeder circuit* – The maximum rating of the feeder short circuit and fault current protection device is based on the sum of the largest branch circuit protective device (110A fuse) plus the sum of the full-load currents of the other motors, or 110A + 40A + 40A = 190A. The nearest standard fuse that does not exceed this value is 175A [*NEC Section 430.62(A)*].

## 7.0.0 SAFETY

Overcurrent protection offers one of the greatest means of protection against electrical faults to both equipment and personnel. However, there are certain precautionary measures that must be observed at all times:

- Make certain the switch or circuit breaker is open before making any inspections or changes to either the overcurrent device or the circuit it is protecting.
- Make certain that all components are compatible with each other and with the system on which they are used; that is, manufacturer, type, normal current ratings, interrupting capacity, voltage, etc.
- Use the proper tools for the job.
- When an overcurrent protection device opens a circuit, determine the reason before changing fuses or resetting the circuit breaker. A severe fault may not damage the circuit or equipment the first time, but repeated hits with high inrush current may do so the second or third time.
- Make sure that adjustable circuit breakers are calibrated correctly (refer to the manufacturer's instructions for calibration details). An improperly calibrated circuit breaker is considered by many to be the most dangerous of circuit breaker faults.
- Use the test breaker to trip devices one phase at a time.
- Never substitute a fuse or circuit breaker that is rated higher than the circuit conductors.
- When working on circuits protected by overcurrent protective devices, lock out the device so that it cannot be inadvertently closed. If

**Think About It**

# Cartridge Fuse Removal

After the On-Off handle of a general-purpose disconnect switch is set to Off and the door has been opened, is there any danger involved in the removal or installation of the cartridge fuses?

locking out the device or switch is not possible, tagging is the next best means to help prevent someone from closing the device while the circuit is being worked on.

## 8.0.0 COORDINATION

Coordination is the name given to the time-current relationship among a number of overcurrent devices connected in series, such as fuses in a main feeder, subfeeder, and branch circuits. Safety is the prime consideration in the operation of fuses; however, coordination of the characteristics of fuses is also an important factor in large and complex electrical systems. Every fuse must be properly rated for continuous current and overloads and for the maximum short circuit current the electrical system could feed into a fault

on the load side of the fuse, but this is not enough. It might still be possible for a fault on a feeder to open the main service fuse before the feeder fuse opens. Alternately, a branch circuit fault might open the feeder fuse before the branch circuit device opens. Such applications are said to be uncoordinated or nonselective—the fuse closest to the fault is not faster than the one farthest from the fault. When a fault on a feeder opens the main service fuse instead of the feeder fuse, the entire electrical system is taken out of service instead of just the one faulted feeder. Effective coordination minimizes the extent of electrical outage when a fault occurs. It therefore minimizes loss of production, interruption of critical continuous processes, and loss of vital facilities.

Selective coordination is the selection of overcurrent devices with time/current characteristics that ensure the clearing of a fault or short circuit by the device nearest the fault on the line side of the fault. A fault on a branch circuit is cleared by the branch circuit device. The subfeeder, feeder, and main service overcurrent devices will not operate. Alternately, a fault on a feeder is opened by the feeder fuse without opening any other fuse on the supply side of the feeder. With selective coordination, only the faulted part of the system is taken out of service, which represents the condition of minimum outage.

With proper selective coordination, every device is rated for the maximum fault current it might be called upon to open. Coordination is achieved by studying the curve of current versus the time required for the operation of each device. The selection is then made so that the device nearest any load is faster than all devices closer to the supply, and each device going back to the service entrance is faster than all devices closer to the supply. The main service fuses must have the longest opening time for any branch or feeder fault.

## SUMMARY

Personal safety is the most important issue when completing electrical installations, and the protection of buildings and equipment comes second. Much of this protection is provided by overcurrent protection. This is why the *NEC®*, UL, NEMA, and other organizations put so much effort into regulating the type and methods of overcurrent protection.

To fully understand the purpose of overcurrent protection devices, you must first understand the meaning and importance of common electrical terms relating to overcurrent protection. Furthermore, you must understand the key *NEC®* requirements regarding overcurrent protection. When making electrical installations, always check the working drawings to make certain that all parts of the electrical system—that is, short circuit currents, fault currents, interrupting capacities, grounding electrode conductors, electrical grounding conductors, and the like—comply with the latest edition of the *NEC®*.

Also verify that all conductors (circuit, feeder, service, grounding electrode conductors, equipment grounding conductors, and bonding conductors) have adequate capacity to safely conduct any fault current that is likely to be imposed on them.

Determine let-through current values (peak and rms) when current-limiting overcurrent devices are used. Furthermore, apply current-limiting data to protect downstream electrical components that have interrupting ratings and/or withstand ratings less than the available fault current at any given point in the system.

1. When a common molded-case circuit breaker trips, in what position will the operating handle be?

   a. Halfway between the On and Off positions
   b. In the On position
   c. In the Off position
   d. In the Reset position

2. What is the highest voltage rating for circuit breakers used on DC systems that UL recognizes?

   a. 240V
   b. 277V
   c. 480V
   d. 600V

3. Where is the ampere rating normally marked on circuit breakers?

   a. On the side of the molded case
   b. On the handle
   c. Beneath the terminal lug
   d. On the bottom of the molded case

4. Which of the following is the most important consideration when working with overcurrent protection devices?

   a. The brand name
   b. The type of project
   c. The correct interrupting capacity
   d. The age of the device

5. What is normally the highest short circuit current in residential applications?

   a. 10,000A
   b. 30,000A
   c. 50,000A
   d. 100,000A

6. Every circuit breaker must have its interrupting rating shown on the circuit breaker, unless it is rated for _____.

   a. 1,000A
   b. 5,000A
   c. 10,000A
   d. 65,000A

7. Which of the following best describes where the *NEC*® allows the use of Edison-base fuses?

   a. On new installations only
   b. On circuits rated at over 40A
   c. On circuits rated under 40A
   d. For replacements in existing installations

8. What must be used in conjunction with Type S plug fuses when they are installed in a fuse holder?

   a. A Type S adapter
   b. A copper penny
   c. A circuit breaker connected in series
   d. A setscrew holding device

9. On renewable cartridge fuses, which part of the fuse is normally renewed?

   a. The barrel
   b. The link
   c. The ferrule
   d. The ferrule ring

10. Which of the following best describes a dual-element fuse?

    a. A fuse having an overcurrent limit and a time delay before activation
    b. A fuse with two attachment points
    c. A fuse having two ferrules at each end
    d. A fuse with two barrels

1. A standard circuit breaker contains _____.

   a. a mechanical linkage
   b. a thermal element
   c. a magnetic element
   d. All of the above.

2. All of the following are standard breaker ratings *except* _____.

   a. 10A
   b. 45A
   c. 175A
   d. 1,600A

3. The amperage interrupting capacity (AIC) is the _____

   _____.

4. A current-limiting circuit breaker provides a downstream protection for _____

   _____.

5. A ground fault circuit interrupter will trip when a fault current _____.

   a. is 1A or more
   b. is 2mA
   c. is 6mA or more
   d. reaches the rated ampacity of the circuit

6. A dual-element fuse can be applied in circuits to provide both short circuit and _____ protection.

7. An overcurrent condition is _____.

   a. an overload current
   b. a short circuit current
   c. a harmless condition that exists on larger circuits
   d. Both a and b.

8. Each ungrounded service entrance conductor shall have a _____ in series with a rating not higher than the ampacity of the conductor.

9. Coordination is the name given to the _____.

   a. sizing of all overcurrent devices from the service to branch circuit
   b. sizing of panelboards to hold the proper circuit breakers
   c. sizing of S-type fuses
   d. time-current relationship of overcurrent devices connected in series with each other

10. True or False? Edison-base fuses are outdated and should not be used in new installations.

# Trade Terms Introduced in This Module

**Cartridge fuse:** A fuse enclosed in an insulating tube to confine the arc when the fuse blows.

**Circuit breaker:** A device that is designed to open a circuit automatically at a certain overcurrent. It can also be used to manually open and close the circuit.

**Dual-element fuse:** A fuse having two fuse characteristics; the usual combination is an overcurrent limit and a time delay before activation.

**Edison-base:** The standard screw base used for ordinary lamps and Edison-base plug fuses.

**Fault current:** The current that exists when an unintended path is established between an ungrounded conductor and ground.

**Frame size:** A method used to classify circuit breakers according to given current ranges.

**Fuse:** A protective device that opens a circuit when the fusible element is severed by heating due to a fault current or overcurrent passing through it.

**Fuse link:** The fusible part of a cartridge fuse.

**Molded-case circuit breaker:** A circuit breaker enclosed in an insulating housing.

**Nonrenewable fuse:** A fuse that must be replaced after it interrupts a circuit.

**Overcurrent protection:** De-energizing a circuit whenever the current exceeds a predetermined value; the usual devices are fuses, circuit breakers, or magnetic relays.

**Plug fuse:** A type of fuse that is held in position by a screw thread contact instead of spring clips, as is the case with a cartridge fuse.

**Pole:** That portion of a device associated exclusively with one electrically separated conducting path of the main circuit or device.

**Short circuit:** The current that exists when an unintended path is created between any two components in a circuit.

# Additional Resources

This module presents thorough resources for task training. The following resource material is suggested for further study.

*National Electrical Code® Handbook*, Latest Edition. Quincy, MA: National Fire Protection Association.

# Figure Credits

Associated Builders and Contractors, Inc., Module opener

John Traister, Figures 1, 2, and 10

Topaz Publications, Inc., SA01, SA02, Figures 9B and 11

Tim Dean, Figure 3

Courtesy of Cooper Bussmann, division of Cooper Industries plc, Figures 4–6

Tim Ely, Figure 7

# CONTREN® LEARNING SERIES — USER UPDATE

NCCER makes every effort to keep its textbooks up-to-date and free of technical errors. We appreciate your help in this process. If you find an error, a typographical mistake, or an inaccuracy in NCCER's Contren® materials, please fill out this form (or a photocopy), or complete the online form at www.nccer.org/olf. Be sure to include the exact module number, page number, a detailed description, and your recommended correction. Your input will be brought to the attention of the Authoring Team. Thank you for your assistance.

*Instructors* – If you have an idea for improving this textbook, or have found that additional materials were necessary to teach this module effectively, please let us know so that we may present your suggestions to the Authoring Team.

**NCCER Product Development and Revision**
3600 NW 43rd Street, Building G, Gainesville, FL 32606

**Fax:** 352-334-0932
**Email:** curriculum@nccer.org
**Online:** www.nccer.org/olf

☐ Trainee Guide   ☐ AIG   ☐ Exam   ☐ PowerPoints   Other _____

Craft / Level: _____   Copyright Date: _____

Module Number / Title: _____

Section Number(s): _____

Description: _____

_____

_____

_____

Recommended Correction: _____

_____

_____

_____

Your Name: _____

Address: _____

_____

Email: _____   Phone: _____

# Control Systems and Fundamental Concepts

**Cathedral of the Blessed Sacrament**

Harbison, Mahony, Higgins Builders, Inc., won an Aon Build America Award for its restoration of an 1889 cathedral, located one block from California's state capitol. The Harbison, Mahoney, Higgins team used new technologies and innovative techniques to tackle this project, which required extensive seismic upgrading to prevent its collapse in the event of an earthquake.

**26211-11**

Trainees with successful module completions may be eligible for credentialing through NCCER's National Registry. To learn more, go to **www.nccer.org** or contact us at **1.888.622.3720.** Our website has information on the latest product releases and training, as well as online versions of our *Cornerstone* newsletter and Pearson's Contren® product catalog.

Your feedback is welcome. You may email your comments to **curriculum@nccer.org,** send general comments and inquiries to **info@nccer.org,** or use the User Update form at the back of this module.

 V.1 7/11

# Control Systems and Fundamental Concepts

## Objectives

When you have completed this module, you will be able to do the following:

1. Describe the operating principles of contactors and relays.
2. Select contactors and relays for use in specific electrical systems.
3. Explain how mechanical contactors operate.
4. Explain how solid-state contactors operate.
5. Install contactors and relays according to the *NEC®* requirements.
6. Select and install contactors and relays for lighting control.
7. Read wiring diagrams involving contactors and relays.
8. Describe how overload relays operate.
9. Connect a simple control circuit.
10. Test control circuits.

## Performance Tasks

Under the supervision of the instructor, you should be able to do the following:

1. Mount and connect a 120V lighting contactor with a three-wire pushbutton control.

## Trade Terms

Ambient temperature
Break
Inductive load
Instantaneous
Latch coil

Make
Mechanically held relay
Normally closed (N.C.) contacts

Normally open (N.O.) contacts
Shading coil
Solid-state relay
Unlatch coil

## Required Trainee Materials

1. Pencil and paper
2. Appropriate personal protective equipment
3. Copy of the latest edition of the *National Electrical Code®*

Note: *NFPA 70®*, *National Electrical Code®*, and *NEC®* are registered trademarks of the National Fire Protection Association, Inc., Quincy, MA 02269. All *National Electrical Code®* and *NEC®* references in this module refer to the 2011 edition of the *National Electrical Code®*.

# Contents

*Topics to be presented in this module include:*

# Figures and Tables

## 1.0.0 INTRODUCTION

The general classification of contactor covers a type of electromagnetic apparatus that is designed to handle relatively high currents. The conventional contactor is identical in appearance, construction, and current-carrying ability to the equivalent National Electrical Manufacturers Association (NEMA) size magnetic motor starter. The magnet assembly and coil, contacts, holding circuit interlock, and other structural features are the same.

The significant difference is that the contactor does not provide overload protection. Contactors, therefore, are used to *make* or *break* the circuit to high-current, non-motor loads, or are used in motor circuits if overload protection is separately provided. A typical application of the latter is in a reversing motor starter.

A relay is an electromagnetic device whose contacts are used in the control circuits of magnetic starters, contactors, solenoids, timers, and other relays. Relays are generally used to amplify the contact capability or multiply the switching functions of a pilot device.

A tremendous variety of relays are used for lighting, motor, and HVAC control circuits. The magnetic relay is still common, although solid-state control devices are rapidly finding their way into all types of applications.

Besides controlling electrical circuits, some types of relays are also used for circuit and equipment protection. For example, thermal overload relays are intended to protect motors, controllers, and branch circuit conductors against excessive heating due to prolonged motor overcurrents up to and including locked rotor currents. These relays sense motor current by converting this current to heat in a resistance element. The heat

generated is used to open a set of contacts in series with a starter coil, causing the motor to be disconnected from the line.

There are also a great many electrical protective relays installed in modern power systems, including:

- Overcurrent (time delay and *instantaneous*)
- Over-/under-voltage
- Directional overcurrent
- Percentage differential

The operating characteristics of these protective relays are all similar and are covered in this module. However, the sensing, adjustment, and actuating functions are quite different and are beyond the scope of this module. For specific instructions, always consult the manufacturer's literature.

## 2.0.0 MAGNETIC CONTACTORS

The typical magnetic contactor (*Figure 1*) contains a coil of wire wound around an iron core or a laminated iron core (as shown) and an armature. The coil acts as an electromagnet, and when the proper electric current is applied, it attracts or picks up the armature. Since the movable contacts are attached via an iron bar to the armature, when the armature moves upwards, the movable contacts make contact with a set of stationary contacts to close and complete the circuit.

When power is being applied to the coil, the device is said to be energized. The motion of the armature moves the contact points together or apart, according to their arrangement. When the power is disconnected from the coil, a spring returns the armature to its original position. When no power is being applied, the device is said to be in its de-energized or normal condition. Contacts that are together when the relay is de-energized

211F01.EPS

*Figure 1*   A magnetic contactor.

are said to be normally closed (N.C.) contacts. Contacts that are apart when the relay is de-energized are said to be normally open (N.O.) contacts.

Notice that a shading coil has been added to the iron core in *Figure 1*. Shading coils are used with AC relays to prevent contact chatter and hum. They operate in the same way as in the shaded-pole motor. In general, the shading coil is used to provide a continuous magnetic flow to the armature when the voltage of the AC waveform is zero.

The relay operates on the solenoid principle; that is, it converts electrical energy to linear motion, as shown in *Figure 2*. In this example, a coil of wire is wound around an iron core. When current flows through the coil, a magnetic field develops in the iron core. The magnetic field of the iron core attracts the movable arm, known as the armature. In its present position, the movable contact makes connection with a stationary contact. This contact is normally closed. When the armature is attracted to the iron core, the movable contact breaks the connection with the normally closed stationary contact and makes the connection with another stationary contact. This contact is normally open. In a wiring diagram, the movable contact would

be the common, and the stationary contacts would be labeled N.O. and N.C.

## 2.1.0 Lighting Contactors

Filament-type lamps (e.g., tungsten, infrared, and quartz) have inrush currents of approximately 15 to 17 times their normal operating currents. To prevent welding of their contacts on the high initial current, standard motor control contactors must be derated if used to control this type of load.

*Table 1* lists electrical ratings for AC magnetic contactors and starters. As shown, a NEMA Size 1 contactor has a continuous current rating of 27A, but if it is used to switch certain lighting loads, it must be derated to 15A. The standard contactor, however, need not be derated for resistance heating or fluorescent lamp loads that do not impose as high an inrush current.

Lighting contactors differ from standard contactors in that the contacts are rated for high inrush currents. A holding circuit interlock is not normally provided because this type of contactor is frequently controlled by a two-wire pilot device, such as a time clock or photoelectric relay.

Unlike standard contactors, lighting contactors are not horsepower-rated or categorized by NEMA size, but are designated by ampere ratings (20A, 30A, 60A, 100A, 200A, and 300A). It should be noted that lighting contactors are specialized

211F02.EPS

*Figure 2*  Contactor operating on the solenoid principle.

in their application and should not be used on motor loads.

A basic circuit using a lighting contactor and serving as a control for a high-amperage lamp is shown in *Figure 3*. Its purpose is to control the lamp from two different locations. The relay coil is indicated by the standard circular symbol and the relay contacts are indicated in the diagram by two parallel lines. The contacts are drawn separated from the coil to show their function. In reality, they are next to the coil in the relay itself.

Closing the N.O. pushbutton energizes the coil of relay R1, and both sets of normally open R1 contacts close. The R1-2 contacts provide a closed circuit to light the lamp. The R1-1 contacts act as holding or seal-in contacts for the relay coil, so that the coil remains energized when the N.O. pushbutton is released.

The lamp will remain lighted until either of the N.C. pushbuttons is momentarily pressed. Pressing either of these buttons opens the circuit to the coil. The coil will drop out or become de-energized and will open the contacts of the relay. The circuit through the lamp is now open and the lamp is turned off. The lamp remains off until one of the N.O. pushbuttons is pressed again. One set of N.O. and N.C. pushbuttons (on and off) is placed at one location where remote control of the lamp is desired, and another set is placed in another convenient location. The lamp can be turned on and off from either of these locations.

Another application of a magnetic relay is its use in an emergency lighting circuit. The lamps in such a circuit should be lighted instantly after the failure of the main power supply. A simple emergency lighting system controlling one lamp is shown in *Figure 4*. The main power is

*Figure 3*  Pushbutton control circuit with a magnetic relay.

*Figure 4*  Emergency lighting circuit using a relay for operation.

**Table 1** Electrical Ratings for AC Magnetic Contactors and Starters *(Reprinted with the Permission of the National Electrical Manufacturers Association)*

| NEMA Size | Volts | Maximum Horsepower Rating—Nonplugging and Nonjogging Duty | | Maximum Horsepower Rating—Plugging and Jogging Duty | | Continuous Current Rating, Amperes—600 Volt Max. | Service-Limit Current Rating, Amperes | Tungsten and Infrared Lamp Load, Amperes—250 Volts Max. | Resistance Heating Loads, kW—other than Infrared Lamp Loads | | KVA Rating for Switching Transformer Primaries at 50 to 60 Cycles | | 3-Phase Rating for Switching Capacitors |
|---|---|---|---|---|---|---|---|---|---|---|---|---|---|
| | | Single Phase | Poly-Phase | Single Phase | Poly-Phase | | | | Single Phase | Poly-Phase | Single Phase | Poly-Phase | Kvar |
| 00 | 115 | 1/3 | — | — | — | 9 | 11 | 5 | — | — | — | — | — |
| | 200 | — | 1½ | — | — | 9 | 11 | 5 | — | — | — | — | — |
| | 230 | 1 | 1½ | — | — | 9 | 11 | 5 | — | — | — | — | — |
| | 380 | — | 1½ | — | — | 9 | 11 | — | — | — | — | — | — |
| | 460 | — | 2 | — | — | 9 | 11 | — | — | — | — | — | — |
| | 575 | — | 2 | — | — | 9 | 11 | — | — | — | — | — | — |
| 0 | 115 | 1 | — | 1/2 | — | 18 | 21 | 10 | — | — | 0.9 | 1.2 | — |
| | 200 | — | 3 | — | 1½ | 18 | 21 | 10 | — | — | — | 1.4 | — |
| | 230 | 2 | 3 | 1 | 1½ | 18 | 21 | 10 | — | — | 1.4 | 1.7 | — |
| | 380 | — | 5 | — | 1½ | 18 | 21 | — | — | — | — | 2 | — |
| | 460 | — | 5 | — | 2 | 18 | 21 | — | — | — | 1.9 | 2.5 | — |
| | 575 | — | 5 | 1 | 2 | 18 | 21 | — | — | — | 1.9 | 2.5 | — |
| 1 | 115 | 2 | — | — | — | 27 | 32 | 15 | 3 | 5 | 1.4 | 1.7 | — |
| | 200 | — | 7½ | — | 3 | 27 | 32 | 15 | — | 9.1 | — | 3.5 | — |
| | 230 | 3 | 7½ | 2 | 3 | 27 | 32 | 15 | 6 | 10 | 1.9 | 4.1 | — |
| | 380 | — | 10 | — | 5 | 27 | 32 | — | — | 16.5 | — | 4.3 | — |
| | 460 | — | 10 | — | 5 | 27 | 32 | — | 12 | 20 | 3 | 5.3 | — |
| | 575 | — | 10 | — | 5 | 27 | 32 | — | 15 | 25 | 3 | 5.3 | — |
| 2 | 115 | 3 | — | 1½ | — | 36 | 42 | 24 | 5 | — | 1 | — | — |
| | 230 | 5 | — | 3 | — | 36 | 42 | 24 | 10 | — | 4.6 | — | — |
| | 115 | 3 | 10 | 2 | 7½ | 45 | 52 | 30 | — | 8.5 | — | 4.1 | 11.3 |
| | 200 | — | 15 | 5 | 10 | 45 | 52 | 30 | — | 15.4 | — | 6.6 | 13 |
| | 230 | 7½ | 25 | — | 15 | 45 | 52 | 30 | — | 17 | — | 7.6 | 21 |
| | 380 | — | 25 | — | 15 | 45 | 52 | — | — | 28 | — | 9.9 | 26 |
| | 460 | — | 25 | — | 15 | 45 | 52 | — | 20 | 34 | 5.7 | 12 | 33 |
| | 575 | — | 25 | — | 15 | 45 | 52 | — | 25 | 43 | 5.7 | 12 | — |

| Size | Volts | 7½ | | | | | | | | | | | |
|---|---|---|---|---|---|---|---|---|---|---|---|---|---|
| 3 | 115 | – | – | – | – | 90 | 104 | 60 | 10 | 17 | 4.6 | 7.6 | – |
| | 200 | 15 | 25 | – | 15 | 90 | 104 | 60 | – | 31 | – | 13 | 23.4 |
| | 230 | – | 30 | – | 20 | 90 | 104 | 60 | 20 | 34 | 8.6 | 15 | 27 |
| | 380 | – | 50 | – | 30 | 90 | 104 | – | – | 56 | – | 19 | 43.7 |
| | 460 | – | 50 | – | 30 | 90 | 104 | – | 40 | 68 | 14 | 23 | 53 |
| | 575 | – | 50 | – | 30 | 90 | 104 | – | 50 | 86 | 14 | 23 | 67 |
| 4 | 200 | – | 40 | – | 25 | 135 | 156 | 120 | – | 45 | – | 20 | 34 |
| | 230 | – | 50 | – | 30 | 135 | 156 | 120 | 30 | 52 | 11 | 23 | 40 |
| | 380 | – | 75 | – | 50 | 135 | 156 | – | – | 86.7 | – | 38 | 66 |
| | 460 | – | 100 | – | 60 | 135 | 156 | – | 60 | 105 | 22 | 46 | 80 |
| | 575 | – | 100 | – | 60 | 135 | 156 | – | 75 | 130 | 22 | 46 | 100 |
| 5 | 200 | – | 75 | – | 60 | 270 | 311 | 240 | – | 91 | – | 40 | 69 |
| | 230 | – | 100 | – | 75 | 270 | 311 | 240 | 60 | 105 | 28 | 46 | 80 |
| | 380 | – | 150 | – | 125 | 270 | 311 | – | – | 173 | – | 75 | 132 |
| | 460 | – | 200 | – | 150 | 270 | 311 | – | 120 | 210 | 40 | 91 | 160 |
| | 575 | – | 200 | – | 150 | 270 | 311 | – | 150 | 260 | 40 | 91 | 200 |
| 6 | 200 | – | 150 | – | 125 | 540 | 621 | 480 | – | 182 | – | 79 | 139 |
| | 230 | – | 200 | – | 150 | 540 | 621 | 480 | 120 | 210 | 57 | 91 | 160 |
| | 380 | – | 300 | – | 250 | 540 | 621 | – | – | 342 | – | 148 | 264 |
| | 460 | – | 400 | – | 300 | 540 | 621 | – | 240 | 415 | 86 | 180 | 320 |
| | 575 | – | 400 | – | 300 | 540 | 621 | – | 300 | 515 | 86 | 180 | 400 |
| 7 | 230 | – | 300 | – | – | 810 | 932 | 720 | 180 | 315 | – | – | 240 |
| | 460 | – | 600 | – | – | 810 | 932 | – | 360 | 625 | – | – | 480 |
| | 575 | – | 600 | – | – | 810 | 932 | – | 450 | 775 | – | – | 600 |
| 8 | 230 | – | 450 | – | – | 1215 | 1400 | 1080 | – | – | – | – | 360 |
| | 460 | – | 900 | – | – | 1215 | 1400 | – | – | – | – | – | 720 |
| | 575 | – | 900 | – | – | 1215 | 1400 | – | – | – | – | – | 900 |

## Troubleshooting Contactors

If the contacts of a contactor are defective, it can usually be spotted during a visual inspection. Another way to check is to place a voltmeter across the closed contacts with voltage applied. If the voltage drop is more than 5% of the rated value of the contactor, the contacts are defective.

## Emergency Lighting Circuits

A typical emergency lighting circuit uses an automatic transfer switch. A potential transformer in the switch senses the absence of normal voltage and switches on the emergency generator. When it senses power from the generator, it closes the contactor to the main circuit.

supplied through the power supply terminals to the relay coil. The relay has a set of normally open contacts and a set of normally closed contacts. As long as the main power is available, it energizes the relay coil, which in turn keeps the N.O. contacts closed. The power is therefore supplied to the transformer, whose primary is connected in series with the N.O. contacts. The secondary of the transformer keeps the storage battery charged through the rectifier. Since the coil is energized, the N.C. contacts are open and the emergency lamp is not lighted.

In the event of main power failure or low voltage in the line, the relay will drop out, the N.O. contacts will open, and the N.C. contacts will close. The lamp will be connected to the battery and give light as long as the relay coil remains de-energized. Such relay and circuit arrangements are frequently used in the emergency battery packs found in many commercial, industrial, and public buildings of all types.

The following example demonstrates how a relay and one single-pole, 15A toggle switch can be used to control 60A of lighting.

*Figure 5* shows a diagram of a lighting contactor controlled by a single-pole switch. The contactor coil operates six sets of contacts, each of which is connected to a lighting circuit with lamps totaling 1,200W (10A). When the single-pole switch is turned to the On position (contacts closed), the coil is energized and moves the armature, which in turn closes all of the contacts simultaneously. When these contacts are closed, this completes the circuit to all of the lamps. Thus, all of the lamps will come on at the same time.

Six single-pole switches could have been used in place of the contactor, but simultaneous switching of the lamps would not be possible.

### 2.2.0 Reversing Motor Starter

Reversing the direction of motor shaft rotation is often required. Three-phase squirrel cage motors can be reversed by reconnecting any two of the three line connections to the motor. By interwiring two contactors, an electromagnetic method of making the reconnection can be obtained.

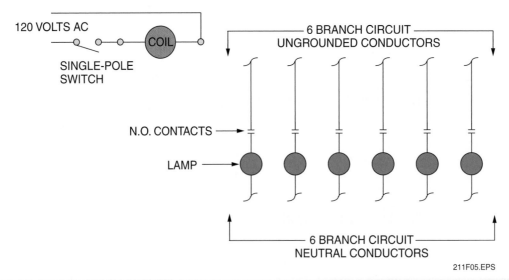

*Figure 5* Several lighting circuits controlled by one contactor.

As seen in the power circuit in *Figure 6*, the contacts (F) of the forward contactor, when closed, connect lines L1, L2, and L3 to the motor terminals T1, T2, and T3, respectively. As long as the forward contacts are closed, mechanical and electrical interlocks prevent the reverse contactor from being energized.

When the forward contactor is de-energized, the second contactor can be picked up, closing its contacts (R), which reconnect the lines to the motor. Note that by running through the reverse contacts, L1 is connected to motor terminal T3, and L3 is connected to motor terminal T1. The motor will now run in the opposite direction.

Whether operating through either the forward or reverse contactor, the power connections are run through an overload relay assembly, which provides motor overload protection. A magnetic reversing starter, therefore, consists of a starter and a set of contactors, suitably interwired, with electrical and mechanical interlocking to prevent the coil of both units from being energized at the same time.

Manual reversing starters (employing two manual starters) are also sometimes used. As in the magnetic version, the forward and reverse switching mechanisms are mechanically interlocked, but since coils are not used in manually operated equipment, electrical interlocks are not furnished.

## 2.3.0 Mechanically Held Contacts

In a conventional contactor, current flow through the coil creates a magnetic pull to seal in the armature and maintain the contacts in a switched position; that is, the N.O. contacts will be held closed while the N.C. contacts will be held open. Because

the contactor action is dependent on the current flow through the coil, the contactor is described as electrically held. As soon as the coil is de-energized, the contacts will revert to their initial position.

Versions of **mechanically held relays** and contactors are also used in some applications. The action is accomplished through the use of two coils and a latching mechanism. Energizing one coil (called the **latch coil**) through a momentary signal causes the contacts to switch, and a mechanical latch holds the contacts in this position even though the initiating signal is removed. The coil is then de-energized. To restore the contacts to their initial position, a second coil (called the **unlatch coil**) is momentarily energized.

Mechanically held relays and contactors are used where the hum of an electrically held device would be objectionable, such as in auditoriums, hospitals, churches, etc.

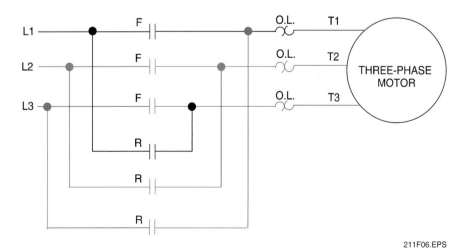

*Figure 6*    Three-pole reversing starting diagram.

## 3.0.0 RELAYS

A control relay is an electromagnetic device that is similar in operating characteristics to a contactor. The contactor, however, is generally employed to switch power circuits or relatively high-current loads.

Relays, with few exceptions, are used in control circuits, and consequently, their lower ratings (15A maximum at 600V) reflect the reduced current levels at which they operate.

Contactors generally have from one to five poles. Although both normally open and normally closed contacts are frequently encountered, the great majority of applications use normally open contacts. There is little, if any, conversion of contact operation on the job.

As compared to contactors, it is not uncommon to find relays used in applications requiring 10 or 12 poles per device, with various combinations of normally open and normally closed contacts. In addition, some relays have convertible contacts, permitting changes to be made in the field from N.O. to N.C. operation, or vice versa, without requiring modification kits or additional components.

### 3.1.0 Control Relays

A control relay is an electromagnetic device whose contacts are used in the control circuits of magnetic starters, solenoids, timers, and other relays. Relays are generally used to amplify the contact capability or multiply the switching functions of a pilot device.

The wiring diagrams in *Figures 7* and *8* demonstrate how a relay amplifies contact capacity. *Figure 7* represents a current amplification. The relay and starter coil voltages are the same, but the ampere rating of the temperature switch is too low to handle the current drawn by the starter coil (M). A relay is interposed between the temperature switch and the starter coil. The current drawn by the relay coil (CR) is within the rating of the temperature switch, and the relay contact (CR) has a rating that is adequate for the current drawn by the starter coil.

*Figure 8* represents a voltage amplification. A condition may exist in which the voltage rating of the temperature switch is too low to permit its direct use in a starter control circuit operating at a higher voltage. In this application, the coil of the interposing relay and the pilot device are both wired to a low-voltage source of power that is compatible with the rating of the pilot device. The relay contact, with its higher voltage rating, is then used to control the operation of the starter.

*Figure 9* represents another use of relays, which is to multiply the switching functions of a pilot device with a single or limited number of contacts. In this circuit, a single-pole pushbutton contact can, through the use of an interposing six-pole relay, control the operation of a number of different loads such as a pilot light, starter, contactor, solenoid, and timing relay.

Relays are commonly used in complex controllers to provide the logic required to set up and initiate the proper sequencing and control of a number of interrelated operations.

*Figure 7* Relay used to amplify contact capacity.

(A)

(B)

211F08.EPS

***Figure 8*** Relay used for voltage amplification.

211F09.EPS

***Figure 9*** Relay used to multiply the switching functions of a pilot device.

Relays differ in voltage ratings (up to 600V), number of contacts, contact convertibility, physical size, and attachments for accessory functions such as mechanical latching and timing.

In selecting a relay for a particular application, one of the first steps should be a determination of the control voltage at which the relay will operate. Once the voltage is known, the relays that have the necessary contact rating can be further reviewed and a selection made on the basis of the number of contacts and other characteristics required.

*Figures 10* and *11* show an 8-pin plug-in relay with its base and wiring diagram.

### 3.2.0 Timers and Timing Relays

A pneumatic timer or timing relay is similar to a control relay, except that it contains contacts that are designed to operate at a specific time interval after the coil is energized or de-energized. A delay on energization is also referred to as an on delay. A delay on de-energization is called an off delay.

A timed function is useful in applications such as the lubricating system of a large industrial machine, in which a small oil pump must deliver lubricant to the bearings of the main motor for a set period of time before the main motor starts.

In pneumatic timers, the timing is accomplished by the transfer of air through a restricted orifice. The amount of restriction is controlled by an adjustable needle valve, permitting changes to be made in the timing period.

PLUG-IN, 8-PIN RELAY          BASE FOR PLUG-IN, 8-PIN RELAY

2.00" MAX.

1.41" MAX

1.41" MAX

(A)

211F10A.EPS

211F10B.EPS

(B)

*Figure 10*  Plug-in relays.

DOUBLE-POLE, DOUBLE-THROW

SCREW AND CAPTIVE WASHERS

WIRING DIAGRAM

211F11.EPS

*Figure 11*   Wiring diagram for a plug-in relay.

Timing relays are often motor-driven by a clock-type mechanism, but solid-state timing circuits will also be encountered.

## 4.0.0 SOLID-STATE RELAYS

Like magnetic relays, solid-state relays are also used in switching applications (*Figure 12*). They are available in different styles and ratings, including time-delay relays. One common use for a solid-state relay is the I/O track of a programmable controller, which is a controller that can be changed by reprogramming rather than requiring an extensive rewiring of the control system. This type of relay can be used to switch a 120V signal to a 5VDC signal.

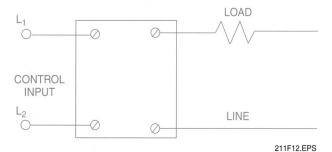

211F12.EPS

*Figure 12*   Basic solid-state relay.

Some of the advantages of solid-state relays include the following:

- No moving parts
- Resistant to shock and vibration

- Sealed against dirt and moisture
- Control input voltage isolated from relay-controlled circuit

*Figure 13* shows typical solid-state relay control circuits. These include:

- *Power transistors for DC loads* – In this DC circuit, a power transistor is used to connect the load to the line, as shown in *Figure 13(A)*. This is an opto-isolated relay with a light-emitting diode (LED) connected to the input or control voltage. This is a common circuit arrangement in solid-state relays, and relays that use this coupling method are referred to as optoisolated because the load side of the relay is optically isolated from the control side of the relay. Since a light beam (rather than electricity) is used to control the relay, no voltage spikes or electrical noise produced on the load side of the relay can be transmitted to the control side of the relay. When the LED is activated by the input voltage, a photo-detector connected to the base of the transistor turns the transistor on and connects the load to the line via this optical coupling.
- *Triac-controlled AC relays* – In this AC circuit, a triac is connected to the load in the place of a power transistor. Refer to *Figure 13(B)*. This circuit also uses an LED as the control device. When the photo-detector detects the LED, it triggers the gate of the triac and connects the load to the line.
- *Reed-controlled AC relays* – Another method of controlling solid-state relays in AC circuits is to use a small reed relay to control the output, as shown in *Figure 13(C)*. In this circuit, small sets of reed contacts are connected to the gate of the triac and the control circuit is connected to the coil of the reed relay. When the control voltage sends a current through the coil, the magnetic field produced around the coil of the relay closes the reed contacts and activates the triac. Unlike opto-isolated relays, this type of solid-state relay uses a magnetic field to isolate the control circuit from the load circuit.

Solid-state relays normally have control voltages from 3V to 32V. Triac-controlled relays commonly have load voltage ratings of 120VAC to 240VAC, with current ratings between 5A and 25A.

(A) POWER TRANSISTOR USED TO CONTROL A DC LOAD

(B) TRIAC USED TO CONTROL AN AC LOAD

(C) REED RELAY CONTROLS THE OUTPUT

211F13.EPS

*Figure 13* Solid-state relay circuits.

**On Site**

## Plug-In Relays

Plug-in relays make troubleshooting and replacement of relays much easier. Instead of having to connect and disconnect wiring, a new relay can simply be snapped into place.

211SA01.EPS

Many solid-state relays incorporate a zero switching feature, which means that if the relay is programmed to turn off when the AC voltage is in the middle of a cycle, it will continue to conduct until the AC voltage returns to zero and turns off. Zero switching applications include some **inductive loads** such as transformers. In normal operation, the core material of a transformer can be left saturated on one end of the flux swing if power is removed from the primary winding when the AC voltage is at its positive or negative peak. This can cause inrush currents of up to 600% of the normal operating current when power is restored to the primary. If the circuit is equipped with a zero switching feature, the power will return to zero before the relay is de-energized.

## 5.0.0 OVERLOAD RELAYS

To protect a motor and its related circuits from accidental or prolonged overloads, either the starter or the motor should be equipped with automatic devices that will open the circuit in the event of an overload. This protection may be provided by fuses, circuit breakers, or overload relays.

Overcurrent protection must be provided in the line of every motor circuit, but additional protection should be provided in the form of magnetic overload relays. These are used in both manual and automatic starters.

*Figure 14* shows a simplified bimetallic overload relay. The load current passes through a heater element in the overload relay. The heater element is replaceable and contains a resistance. When the current exceeds a certain value, the heat radiated from the element causes a bimetallic strip to spring away, interrupting the control circuit or other mechanical action. The heater element is sized to the full load current of the motor. A small overload may have a long delay, while a larger overload will have a shorter delay.

Most overload relays used on modern motor starters are thermally operated and usually consist of two strips of metal that are welded together. Each has a different degree of thermal expansion. If this bimetallic strip is heated, it will deflect sufficiently to trip two normally closed contacts, which in turn will open the circuit connected to the holding coil of the magnetic contactor, causing the main contacts to open. An advantage of this protective device is that it provides a time delay, which prevents the circuit from being opened by momentary high starting currents and short overloads. At the same time, it protects the motor from prolonged overloading.

Conditions of motor overload may be caused by an overload on driven machinery, by a low line voltage, or by an open line in a poly-phase system, which results in single-phase operation. Under any conditions of overload, a motor draws excessive current that causes overheating. Since the motor winding insulation will deteriorate when

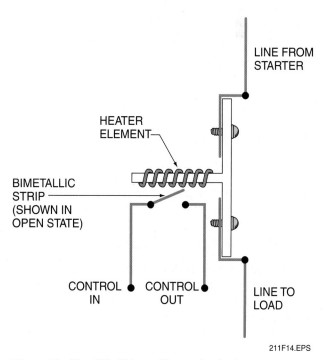

211F14.EPS

*Figure 14* Simplified bimetallic overload relay.

subjected to overheating, there are established limits on motor operating temperatures. To protect a motor from overheating, overload relays are employed on a motor control to limit the amount of current drawn. This is known as overload protection or running protection.

The ideal overload protection for a motor is an element with current-sensing properties that are very similar to the heating curve of the motor it protects and will act to open the motor circuit when the full-load current is exceeded. The operation of the protective device should be such that the motor is allowed to carry harmless overloads but is quickly removed from the line when an overload has persisted for too long.

Fuses are not the best choice to provide overload protection. Their basic function is to protect against short circuits (overcurrent protection). Motors draw a high inrush current when starting, and conventional single-element fuses have no way of distinguishing between this temporary and harmless inrush current and a damaging overload. Such fuses, if chosen on the basis of motor full-load current, would blow every time the motor started. On the other hand, if a fuse were chosen so that it was large enough to pass the starting or inrush current, it would not protect the motor against small, sustained overloads that might occur later.

Dual-element or time-delay fuses can provide motor overload protection but have the disadvantage of being nonrenewable and must be replaced once tripped.

The overload relay is the heart of motor protection. It has inverse trip time characteristics, permitting it to hold in during the accelerating period (when inrush current is drawn) yet providing protection on small overloads above the full-load current when the motor is running. Unlike dual-element fuses, overload relays are renewable and can withstand repeated trip and reset cycles without needing replacement. They cannot, however, take the place of overcurrent protective equipment.

An overload relay consists of a current-sensing unit connected in the line to the motor, plus a mechanism that is actuated by the sensing unit and serves to directly or indirectly break the circuit. In a manual starter, an overload trips a mechanical latch, causing the starter contacts to open and disconnect the motor from the line. In magnetic starters, an overload opens a set of contacts within the overload relay itself. These contacts are wired in series with the starter coil in the control circuit of the magnetic starter. Breaking the coil circuit causes the starter contacts to open, disconnecting the motor from the line.

Overload relays can be classified as being either thermal or magnetic. Magnetic overload relays react only to current excesses and are not affected by temperature. As the name implies, thermal overload relays rely on the rising temperatures caused by the overload current to trip the overload mechanism. Thermal overload relays can be further subdivided into two types: melting alloy (eutectic) and bimetallic.

The melting alloy assembly of a heater element and solder pot are shown in *Figure 15*. The motor load current passes through the heater element. When it exceeds the designated capacity of the heater, it melts an alloy solder pot. The ratchet

HEAT WINDING
(HEAT-PRODUCING ELEMENT)

**FRONT**

SOLDER POT
(HEAT-SENSITIVE ELEMENT)

**BACK**

211F15.EPS

*Figure 15*  Melting alloy thermal unit.

wheel will then be allowed to turn in the molten pool, and a tripping action of the starter control circuit results, stopping the motor. A cooling-off period is required to allow the solder pot to harden before the overload relay assembly may be reset and the motor service restored.

Melting alloy thermal units are interchangeable and of a one-piece construction, which ensures a constant relationship between the heater element and solder pot and allows for factory calibration, making them virtually tamperproof in the field. These important features are not possible with any other type of overload relay construction. Many types of interchangeable thermal units are available to provide precise overload protection from any full-load current.

Bimetallic overload relays have two major advantages: first, the automatic reset feature is desirable when devices are mounted in locations that are not easily accessible for manual operation; and, second, these relays can easily be adjusted to trip within a range of 85% to 115% of the nominal trip rating of the heater unit. This feature is useful when the recommended heater size might result in unnecessary tripping, while the next larger size would not give adequate protection.

The **ambient temperature** affects overload relays operating on the principle of heat. Ambient-compensated bimetallic overload relays were designed for one particular situation, that is, when the motor is at a constant temperature and the controller is located separately and has a varying temperature. In this case, if a standard thermal overload relay were used, it would not trip consistently at the same level of motor current if the controller temperature changed. A thermal overload relay is always affected by the surrounding temperature. To compensate for temperature variations, an ambient-compensated overload relay is applied. Its trip point is not affected by temperature and it performs consistently at the same value of current.

Melting alloy and bimetallic overload relays are designed to approximate the heat actually generated in the motor. As the motor temperature increases, so does the temperature of the thermal unit. Motor and relay heating curves show this relationship. From such graphs (obtainable from the manufacturers of overload relays), you can see that, no matter how high the current drawn, the overload relay will provide protection, yet the relay will not trip out unnecessarily.

When selecting thermal overload relays, the following must be considered:

- Motor full-load current
- Type of motor
- Difference in ambient temperature between the motor and controller

Motors of the same horsepower and speed do not all have the same full-load current, and the motor nameplate must always be referred to and checked to obtain the full-load amperes for a particular motor. Thermal unit selection tables are published on the basis of continuous-duty motors operating under normal conditions with a 1.15 service factor. These tables are shown in manufacturers' catalogs and also appear on the inside of the door or cover of the motor controller. These values are selected to properly protect the motor and allow it to develop its full horsepower, allowing for the service factor if the ambient temperature is the same at the motor as at the controller. If the temperatures are not the same, or if the motor service factor is less than 1.15, a special procedure is required to select the proper thermal unit.

Standard overload relay contacts are closed under normal conditions and open when the relay trips. An alarm signal is sometimes required to indicate when a motor has stopped due to an overload trip. Also, with some machines, particularly those associated with continuous processing, it may be required to signal an overload condition, rather than have the motor and process stop automatically. This is done by fitting the overload relay with a set of contacts that close when the relay trips, thus completing the alarm circuit. These contacts are appropriately called alarm contacts.

A magnetic overload relay has a movable magnetic core inside a coil that carries the motor current. The flux set up inside the coil pulls the core upward. When the core rises far enough, it trips a set of contacts on the top of the relay. The movement of the core is slowed by a piston working in an oil-filled dashpot mounted below the coil. This produces an inverse-time characteristic. The effective tripping current is adjusted by moving the core on a threaded rod. The tripping time is varied by uncovering oil bypass holes in the piston. Because of the time and current adjustments, the magnetic overload relay is sometimes used to protect motors having long accelerating times or unusual duty cycles.

## 6.0.0 Protective Enclosures

The correct selection and installation of an enclosure for a particular application can contribute considerably to extended service life and trouble-free operation. To shield live parts from accidental contact, some form of enclosure is always necessary. This function is usually filled by a general-purpose sheet steel cabinet. Frequently, however, dust, moisture, or explosive gases make it necessary to employ a special enclosure to protect the motor controller from corrosion or the surrounding equipment from explosion. In selecting and installing control apparatus, it is always necessary to carefully consider the conditions under which the apparatus must operate; there are many applications where a general-purpose enclosure does not afford protection.

Underwriters Laboratories, Inc., has defined the requirements for protective enclosures according to hazardous conditions, and the National Electrical Manufacturers Association (NEMA) has standardized enclosures from these requirements. These enclosures were discussed in detail in an earlier module.

## 7.0.0 LOW-VOLTAGE REMOTE CONTROL SWITCHING

In applications where lighting must be controlled from several points, where there is a complexity of lighting or power circuits, or where flexibility is desirable in certain systems, low-voltage remote control relay systems have been applied. Basically, these systems use special low-voltage components operated from a transformer to switch relays, which in turn control the standard line voltage circuits. Because the control wiring does not carry the line load directly, small, lightweight cable can be used. It can be installed wherever and however convenient—placed behind moldings, stapled to woodwork, buried in shallow plaster channels, or installed in holes bored in wall studs. A basic circuit of a remote control switching system is shown in *Figure 16*. In both circuits, the relay permits positive control for on and off. It can be located near the load or installed in centrally located distribution panel boxes, depending on the application.

Because no line voltage flows through the control circuits and low voltage is used for all switch and relay wiring, it is possible to place the controls at a great distance from the source or load, thus offering many advantages.

### 7.1.0 Advantages of Remote Control Switching

Since branch circuits go directly to the loads in this type of system, no line voltage switch legs are required. This saves costly runs of larger conductor through all switches and saves installation time and costs if multipoint switches are used.

*Figure 16*  Basic circuit of a remote control switching system.

211F16.EPS

Under certain conditions, the *NEC®* allows low-voltage conductors to be run in open spaces above ceilings and through wall partitions without further protection. An outlet box is not required at the switch locations. When the electricians rough-in the wiring, they merely secure a plaster ring of the correct depth at each switch location and usually wrap the low-voltage cable around a nail driven behind the plate. The switch and its cover are then installed after the wall is finished.

Low-voltage switching is especially useful in rewiring existing buildings, because the small cables are as easy to run as telephone wires. They are easy to hide behind baseboards or even behind quarter-round molding. The cable can be run exposed without being very noticeable because of its small size. The small, flexible wires are easily fished in partitions.

*Figure 17* shows several types of remote control circuits. In the circuits shown, any number of on-off switches can be connected to provide control from many remote points. In a typical installation, outside lights could be controlled from any area within the building without the need to run full-size cable for three- or four-way switch legs through the building. By running a low-voltage line from each relay to a central point, such as an exit door, all relays can be operated from one spot using a selector switch; this allows quick,

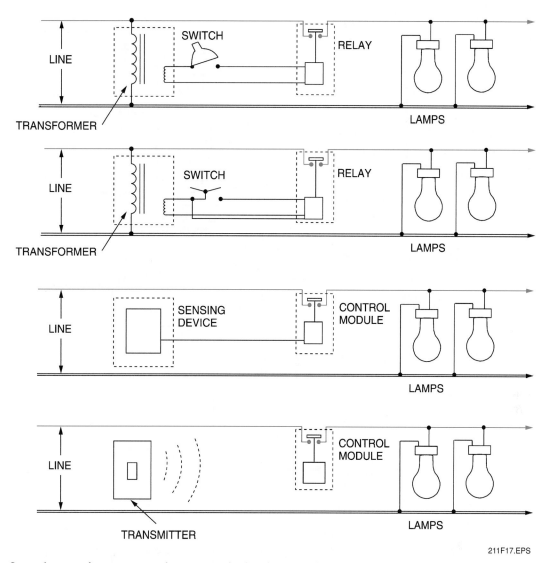

*Figure 17*   Several types of remote control circuits and related components.

convenient control of all exterior and interior lighting and power circuits.

A motor-driven rotary switch is available that allows control of 25 (or more) separate circuits with the push of one button. Dimming can also be accomplished by using the motorized control unit together with a modular incandescent dimming system.

### 7.2.0 System Components

The relay is the heart of a remote control switching system. It employs a split low-voltage coil to move the line voltage contact armature to the On (latched) position. As illustrated in *Figure 18*, the On coil moves the armature to the right when a 24VAC control signal is impressed across its leads. This is similar to a magnet attracting the handle of a standard single-pole switch to the On position when energized. The armature (handle)

*Figure 18* Relay operation.

211F18.EPS

latches in the On position and will remain there until the Off coil is energized, drawing the armature into the Off position. *Figure 19* shows how a low-voltage relay is installed in an outlet box.

### 7.2.1 Power Supplies

Transformers designed for use on remote control switching systems supply 24VAC power. This is rectified to 24VDC to operate the relays and pilot

**Figure 19** Relay installed in an outlet box.

lights. The AC input voltage to the power supply can vary by manufacturer, with the more common voltages being 120V, 208V, 240V, and 277V. The type of voltage used depends on the system. *Figure 20* shows a control transformer along with connection details.

**Figure 20** Control transformer with power-supply wiring diagram.

### 7.2.2 Switches

The standard low-voltage switch uses a rocker or two-button configuration to provide a momentary single-pole, double-throw action. Pushing the On button completes the circuit to the On coil of the relay, shifting the contact armature to the corresponding position. When the button is released, the relay remains in that position.

The pulse operation allows any number of switches to be wired in parallel, as shown in *Figure 21*. A group of relays could also be wired for common switch control by paralleling their control leads. Those relays would operate as a group.

Pilot switches include a lamp wired between the switch common (white) and a pilot terminal. The auxiliary contact in the relay provides power to drive this lamp when the relay is energized.

### 7.2.3 Master Sequencer

The master sequencer allows relays to be controlled as a group while still allowing individual switch control for each. When the master switch is turned On, the sequencer pulses each of its On relay outputs sequentially. A local switch can control an individual relay without affecting any other.

A second input channel allows time clocks, building automation system outputs, photocells, or other maintained contact devices to also control the sequencer. This provides simple automation coupled with local override of individual loads. See *Figure 22*.

### 7.3.0 Basic Operation of Remote Control Switching

*Figure 23* shows a single-switch control of a light or a group of lights in one area. This is the basic circuit of relay control and is similar to single-pole conventional switch circuits. The schematic wiring diagram of this circuit is shown in *Figure 24*.

The addition of a time clock and/or photoelectric cell allows applications in which any number of control circuits are turned on or off at desired intervals (e.g., the dusk-to-dawn control of outdoor lighting circuits).

Multipoint switching, or the switching of a single circuit from two or more switches, is shown in *Figure 25*, along with its related wiring diagram. In conventional wiring, multipoint switching requires costly three-way and four-way switches, plus the extension of switch legs and traveler wires through all switches. One of the greatest advantages of relay switching is the low cost of adding additional switch points.

TO OTHER RELAY GROUPS

ANY NUMBER OF SWITCHES

ARMATURE

211F21.EPS

*Figure 21* Switch operation.

POWER INPUT

(MAXIMUM
OF THREE
IN PARALLEL)

TO RECTIFIER

LOCAL SWITCH

TO TRANSFORMER

"ON" OUTPUTS TO RELAYS

"OFF" OUTPUTS TO RELAYS

A  B  COM    A  B  COM

MASTER
SWITCH(ES)

ON

OFF

TIME CLOCK

211F22.EPS

*Figure 22* Master sequencer connections.

**Figure 23** Single-switch control.

**Figure 24** Schematic wiring diagram of single switches controlling each circuit.

**Figure 25** Diagram of multipoint switching of a single circuit.

*Figure 26* shows many lighting circuits being controlled from one location. In conventional line voltage switches, this is accomplished by ganging the individual switches together. The same procedure can be employed with remote control switches. When more than six switches are required, it is usually desirable to install a master-selector switch that permits the control of individual circuits.

Master control of many individual and isolated circuits is possible. Where more than 12 circuits are to be controlled from a single location and the selection of individual circuits is not required, a motor master control unit automatically sweeps 25 circuits from either On or Off at the touch of a single master switch, which can be placed at one or more locations.

RELAYS

**Figure 26** Switching circuits being controlled from one location.

211F26.EPS

## 7.4.0 Planning a Remote Control Switching System

Low-voltage lighting control systems can be installed in both residential and commercial installations. The systems installed in each type of location may differ. The manufacturer's instructions and recommendations should always be consulted and adhered to, as should the blueprint specifications. The first step in the installation procedure, whether the building is residential or commercial, is to develop a plan. When laying out a remote control switching system for a building, drawing symbols such as the ones shown in *Figure 27* are normally used to identify the various components. The drawing may be simple, requiring only the following:

- The location of each low-voltage control with a dashed line drawn from the switch to the outlet to be controlled
- A schematic wiring diagram showing all components
- The following note:
  *Furnish and install complete remote control wiring system for control of lighting and other equipment as indicated on the drawings, diagrams, and*

*schedules. System shall be complete with transformers, rectifiers, relays, switches, master-selector switches, wallplates, and wiring. All remote control wiring components shall be of the same type and installed in accordance with the recommendations of the manufacturer. Except where otherwise indicated, all remote control wiring shall be installed in accordance with NEC Article 725, Class 2.*

## 7.5.0 Design of Low-Voltage Switching

Many times the low-voltage lighting control system will be designed by architects, especially in new installations. However, in existing buildings, the electrician may be expected to design and install the system. If this situation occurs, work from the set of job blueprints, and use the following general procedure:

*Step 1* Assign each light or receptacle a specific relay. This can be accomplished by using the letters R1 for relay 1, R2 for relay 2, and so on.

*Step 2* Group together all the lights and outlets that are to be switched together.

*Step 3* Determine the number of relays that can be controlled by each control module.

*Step 4* Divide the number of relays by the number computed in Step 3. This is the number of control modules needed to control the relays.

*Step 5* Determine the location of the switch points on the building blueprints.

*Step 6* Indicate what lights and receptacles are being controlled by what relay. This step may be accomplished by indicating the group of lights on the blueprint. However, it is easier if a switch schedule is used.

*Step 7* Review the blueprints or schedule to determine the number and size of the switching stations.

*Figure 28* shows the floor plan of a building. Symbols are used to show the location of all lighting fixtures just as they are used to show the location of conventional switches. Line voltage circuits feeding the lighting outlets are indicated in the same way (by solid lines) that they are in conventional switching. All of the remote control switches, however, are indicated by the symbol $S_L$, instead of S, $S_3$, $S_4$, etc. Lines from these remote control switches to the lighting fixtures that they control are shown by dashed lines instead of the conventional solid line.

| Symbol | Description | Symbol | Description |
|--------|-------------|--------|-------------|
| ◇ (diamond) | THE DIAMOND IS A WIDELY RECOGNIZED SYMBOL THAT IS USED FOR DENOTING LOW VOLTAGE. | | |
| — · — · · | Remote control low-voltage wire | $S_T$ | Low-voltage switch for interchangeable-type bracket |
| T | Low-voltage transformer | $S_X$ | Extension switch receptacle |
| ▶ (rectifier) | Rectifier or rectifier assembly | $MS_8$ | Master-selector pushbutton switch; 8 position |
| $C_A$ | Components cabinet; A refers to first cabinet | $MS_{P8}$ | Master-selector pushbutton switch with pilot lights; 8 position |
| $R_1$ | Relay, standard type, spec. grade; 1 refers to number of relay | $MM_R$ | Motor master; red for ON |
| $R_{P2}$ | Relay with pilot switch, spec. grade; 2 refers to number of relay | $MM_B$ | Motor master; black for OFF |
| S | Low-voltage switch, spec. grade, standard type | $P_L$ | Separate pilot light |
| $S_P$ | Low-voltage switch, spec. grade, with pilot light | | **NOTE:** At controlled receptacles (ceiling or wall) use a standard relay (e.g., RA6, etc. which designates component cabinet A, relay #6). |

211F27.EPS

*Figure 27*   Recommended symbols for low-voltage drawings.

*Figure 28*  Floor plan of a residence using low-voltage switching.

211F28.EPS

The schematic wiring diagram in *Figure 29* shows further details of all components and the related wiring connections. This wiring diagram aids the contractor or electrician in the installation of the system and leaves little doubt as to exactly what is required.

A brief note or specification, such as described earlier, completes the design of the low-voltage remote control system.

On the floor plan in *Figure 28*, notice that the door switches controlling the closet lights use conventional switches since low-voltage switches of this type are uncommon.

## 7.6.0 Installations

Before starting, consult both the *NEC®* and local codes for installation requirements. Once the layout plan has been developed and materials are acquired, the installation of the system can begin. The rough-in instructions presented here are only an example. As with the previous sections of this module, it is important to follow the manufacturer's recommendations and blueprint specifications.

- *Enclosure* – The enclosure should be placed in a location that is easily accessible but hidden from sight. If at all possible, install a unit that is prewired. This will decrease installation time and also limit the possibility of mistakes. Note that even though the panel may be prewired, the supply cables will still have to be installed. This should be done according to the manufacturer's recommendations.

Since heat will be produced by the enclosed control modules, vents are placed around the enclosure to facilitate cooling. The location chosen for the enclosure should not hinder the cooling capabilities for which the panels are designed. The manufacturer may require that the high-voltage portion of the enclosure panel be mounted in the upper right-hand corner. This is due to the *NEC®* requirement that is designed to keep low-voltage conductors separated from high-voltage conductors. The panel itself should be mounted at eye level between two studs.

The power supply should be mounted on the enclosure and the power connections made. The number of power supplies needed per system will depend on the size of the system and the capabilities of the power supply. Some power supplies are capable of supplying up to 10 modules if the average number of LEDs per channel does not exceed 2 or 2.5. Be sure to check the manufacturer's specifications to ensure that the power supply will not be overloaded.

- *Switch points* – The first step when installing the switch points is to mark their location. As with ordinary switches, the switch points should be located as specified by the plans or specifications. Once the location has been determined, mount either a plaster ring or a nonmetallic box. A box is not needed in most frame construction installations.

A plaster ring or Romex® staple should then be positioned partway in the stud, behind the plaster ring or nonmetallic box. The preparation

211F29.EPS

*Figure 29*  Schematic wiring diagram of lighting circuits.

for running the low-voltage wiring is the same as for running 120V wiring. If the edge of the hole is closer than 1¼" from the nearest edge of the stud, a 1/16" steel nail plate or bushing must be used to protect the cable from future damage by nails or screws.

The wires are then pulled from the switch station to the panel enclosure. They should come from the left side if the manufacturer recommends that the enclosure be installed with the high-voltage section in the upper right-hand corner. There should be approximately 10" to 12" of extra wire left at each switch location and 12" to 18" left at the panel enclosure.

Multiconductor cable is frequently used due to its simpler installation. Run one wire (low-voltage lead) from the enclosure to the farthest switching point and anchor the wire to the plaster ring. This wire will supply the low-voltage power to each of the switches. Working back toward the enclosure, loop the wire down and anchor it to each switch location. Take care to leave a sufficient amount of wire for connecting the switch. The low-voltage lead is then connected to the appropriate power supply lead at the enclosure. The process is repeated for the DC common lead. When this wiring method is used, the DC positive and common lead will have the same color throughout the system and future connection mistakes will be minimized.

The next step is to select a multiple-switch circuit. Pull a different colored wire from the enclosure to the farthest switching location in this circuit. Working back toward the panel enclosure as before, loop the wire to the other switch locations in the multiple-switch circuit. Thus, there are two leads going to each switch with one of the leads common to all the switches in the system. However, if the switches contain a pilot light, two additional No. 18 wires must be run to the LED. The wires are connected to the appropriate pilot light transformer located in the panel enclosure.

If possible, use different colored wires for each run. This will limit confusion in the future when connecting the individual circuits to power. Wire schedules should be used to limit hookup mistakes and provide a clearer picture of the system.

Remember that the colors for the control wiring should not be the same as the colors for the DC power and common conductors. Additional marking methods should also be used for future reference. Note that if the wires are marked using adhesive tabs or punch cards, they should be placed in a plastic bag and shoved behind the ring or securely into the plastic box. This will prevent the markings from being covered by paint or plaster when the walls are finished.

The low-voltage wiring should also be kept apart from the high-voltage wiring whenever possible. If the cables must be placed in parallel runs, they should be separated by at least 6", otherwise the noise generated from the high-voltage conductors could seriously interfere with the operation of the low-voltage system.

- *Power and finish trim* – Connecting the circuits to power is now a fairly simple task. Each wire from the switch is connected to a relay with the other side of the relay connected to the common of the DC supply.

When the walls of the building are finished, the switches are connected to the conductors located at each switch point location. Wall plates are then added. A common type of wall plate used with low-voltage lighting control systems requires no screws since it can be snapped into place.

## 8.0.0 RELAY TROUBLESHOOTING

Electromagnetic contactors and relays of various types are used to make and break circuits that control and protect the operation of such devices as electric motors, combustion and process controls, alarms, and annunciators. Successful operation of a relay is dependent on maintaining the proper interaction between its solenoid and its mechanical elements—springs, hinges, contacts, dashpots, and the like.

The operating coil is usually designed to operate the relay from 80% to 110% of rated voltage. At low voltage, the magnet may not be strong enough to pull in the armature against the armature spring and gravity, while at high voltages excessive coil temperatures may develop. Increasing the armature spring force or the magnet gap will require higher pull-in voltage or current values and will result in higher drop-out values.

When sufficient voltage is applied to the operating coil, the magnetic field builds up, and the armature is attracted and begins to close. The air gap is shortened, increasing the magnetic attraction and accelerating the closing action, so that the armature closes with a snap, closing normally open contacts and opening normally closed contacts. Binding at the hinges or excessive armature spring force may cause a contactor or relay that normally has snap action to make and break sluggishly. This condition is often encountered in relays that are rarely operated. Contactors and relays should be operated and observed from time to time to make sure their parts are working freely with proper clearances and spring actions.

Troubleshooting charts are provided in *Table 2*. Knowing these symptoms and their causes should help you to quickly spot and repair contactor and relay problems.

**Table 2**  Contactor and Relay Troubleshooting Chart (1 of 2)

| Symptoms | Probable Cause | Action or Items to Check |
|---|---|---|
| Failure to pull in | Either no voltage or low voltage at coil terminals | Blown fuse, open line switch, break in wiring<br>Line voltage below normal<br>Overload relay open or set too low<br>Tripping toggle (non-automatic breakers) fouled<br>Control level or start button in off position<br>Pull-in circuit open, shorted, or grounded<br>Contacts in protective or controlling circuit open or one of their pigtail connections broken |
| | Operating coil open or grounded | Inspect and test coil |
| | Loose or disconnected coil lead wire | Inspect and correct |
| | Excessive magnet gap, improper alignment | Inspect and correct |
| | Armature obstructed or gumming deposits between armature and pole face | Inspect and correct |
| | Binding caused by deformed or gummy hinge | Replace if bent; degrease if gummed up |
| | Excessive armature spring force | Lessen spring force |
| | Normally closed contacts welded together | Replace contacts |
| Failure of equipment to start with contactor closed | One contact not closing | Replace set of contacts |
| | Contacts burned | Replace |
| | Contact pigtail connection broken | Replace |
| Failure to drop out | Operating coil is energized | Contacts in controlling or protective tripping circuits closed, shorted, or shunted<br>Tripping devices defective, such as overload tripping toggles do not strike release, undervoltage relay plunger stuck or out of adjustment, defective stop button, or defective time-delay escape mechanism (closed air vent)<br>Current supplied over an unintentional path due to grounds, defective insulation, pencil markings, moisture, or lacquer chipped off relay base |
| | Residual magnetism excessive due to armature closed tightly against pole face | Adjust or replace |
| | Armature obstructed or gumming deposits between armature and pole face | Clear and clean<br>Replace or degrease |
| | Binding caused by deformed or gummy hinge | Replace or adjust |
| | Contact pressure spring or armature spring too weak or improperly adjusted | Mount correctly |
| | Improper mounting position (upside down) | Replace contacts |
| | Normally open contacts welded together | |

**Table 2** Contactor and Relay Troubleshooting Chart (2 of 2)

| Symptoms | Probable Cause | Action or Items to Check |
|---|---|---|
| Time-delay relays operate too fast | Air escapes too freely due to holes in bellows or open air vent<br>Non-magnetic shim in air gap too large or armature spring too strong<br>Magnets out of adjustment | Escape mechanism faulty<br>Worn dashpot plunger, dashpot oil too thin<br>Adjust or replace<br>Adjust |
| Contacts pitted or discolored | Contacts overheated from overload<br>Contacts not fitted properly<br>Barriers broken from rough usage or breakers closing with too much force<br>Wiping action of contacts on closing is insufficient<br>Excessive chatter or hum<br>Exposure to weather, dripping water, salt air, or vibration | Reduce load and replace contacts<br>Refit<br>Replace barriers; check for high voltage<br>Adjust<br>Check causes listed below<br>Use suitable NEMA enclosure |
| Excessive chatter or hum | Vibration of surrounding devices communicated to relay<br>Relay receiving contradictory signals<br>Bouncing of controlling or protective contacts<br>Line voltage fluctuating or insufficient<br>Excessive coil circuit resistance<br>Armature spring or contact pressure spring too strong<br>Excessive coil voltage drop on closing<br>Missing shading coil<br>Improper antifreeze pin location<br>Free movement of armature hindered due to deformed parts or dirt | Arrange differently<br>Correct<br>Adjust<br>Consider using buck-and-boost transformers<br>Lessen resistance<br>Lessen<br>Adjust<br>Replace the contactor<br>Locate in correct position<br>Clean and replace appropriate parts |
| Excessive coil temperature | Excessive current or voltage<br>Short circuit in coil | Reduce load or adjust taps on transformer<br>Replace coil |

# SUMMARY

The general classification of contactor covers a type of electromagnetic apparatus that is designed to handle relatively high currents and is identical in appearance, construction, and current-carrying ability to the equivalent NEMA size magnetic starter. The magnet assembly and coil, contacts, holding circuit interlock, and other structural features are the same. The significant difference is that the contactor does not provide overload protection.

A relay is an electromagnetic device whose contacts are used in the control circuits of magnetic starters, contactors, solenoids, timers, and other relays. Relays are also used extensively in remote control switching, especially where an outlet or a group of outlets is to be controlled from several locations.

1. The function of a shading coil in a contactor is to _____.

   a. keep ultraviolet light off the contacts
   b. prevent contact chatter and hum
   c. act as a holding circuit to keep the contactor energized when power is removed
   d. help the contacts return to their normal position when the contactor becomes de-energized

2. What is the major difference between a contactor and a motor starter?

   a. Contactors are always larger than motor starters for the same current-carrying capacity.
   b. All contactors are suitable for full-load lighting loads; starters are not.
   c. A contactor always has a disconnect while a starter does not.
   d. The starter provides overload protection while a contactor does not.

3. Which of the following best describes a control relay?

   a. A device designed to handle relatively high currents
   b. A fuse with two attachment points
   c. An electromagnetic device used in control circuits
   d. A fuse with two barrels

4. Which of the following best describes a relay that is used for motor protection?

   a. Amplifying relay
   b. Thermal overload relay
   c. Solenoid
   d. Contactor

5. The highest voltage rating for magnetic relays is _____.

   a. 240V
   b. 277V
   c. 480V
   d. 600V

6. The approximate inrush current of filament-type lamps is how many times the normal operating current?

   a. 15 to 17
   b. 20 to 30
   c. 30 to 40
   d. 50 to 60

7. What is the normal position of a stop push-button that is used in conjunction with contactors?

   a. Normally open
   b. Normally closed
   c. Normally closed 50% of the time and open 50% of the time
   d. Normally open 70% of the time and closed 30% of the time

8. The normal maximum amperage rating of control relays at 600V is _____.

   a. 15A
   b. 20A
   c. 0A
   d. 65A

9. When selecting overload relays for motors, which of the following must be considered?

   a. The motor manufacturer
   b. The cost of the motor
   c. The length of time the motor has been in use
   d. The motor full-load current

10. Special enclosures are required to protect motor controllers from corrosion.

    a. True
    b. False

# Supplemental Exercises

1. What does the word *normally* refer to in regard to relay contacts?
   a. The placement of the contacts when the circuit is energized
   b. The placement of the contacts when the circuit is de-energized
   c. The way the contact is most of the time
   d. The way the contact is when it has to operate

2. Because filament-type lamps have large inrush currents, standard motor control contactors must be _____ if used to control this type of load.

3. If a solid-state relay is used to control a DC circuit, a(n) _____ is used to connect the load to the line.

4. Solid-state relays can be controlled by _____.
   a. the opto-isolation method
   b. a reed relay
   c. a latch coil
   d. Both a and b.

5. A(n) _____ is used to provide a continuous magnetic flow to the armature when the voltage of the AC waveform is zero.

6. Mechanically held relays and contactors use two coils and a(n) _____ mechanism to provide for quiet operation of the circuit.

7. True or False? Most overload relays used on modern motor starters are thermally operated.

8. If the equipment fails to start with the contactor closed, probable causes would be one contact not closing, burned contacts, or a broken _____.

9. Unlike contactors, which are generally used to switch power circuits on relatively high-current loads, relays are normally used in low-current _____ circuits.

10. On electrical drawings, the _____ is the symbol used for denoting low-voltage devices.
    a. square
    b. circle
    c. triangle
    d. diamond

# Trade Terms Introduced in This Module

**Ambient temperature:** The temperature of the medium (usually air) surrounding a device and into which the heat losses in the device are dissipated.

**Break:** The opening of closed contacts to interrupt an electrical circuit.

**Inductive load:** A load in which the voltage leads the current. Electromagnet coils such as those used in relays and starters are inductive loads.

**Instantaneous:** Contacts that make or break a circuit immediately upon coil energization or de-energization.

**Latch coil:** The coil of a latching relay that, when energized, causes the relay to assume the switched or latched position.

**Make:** The closing of open contacts to establish an electrical circuit.

**Mechanically held relay:** A latching relay in which the switched or latched position is maintained by the interference of a mechanical part that moves to a non-interfering position when the unlatch coil is energized, allowing the relay to return to the normal or unlatched position.

**Normally closed (N.C.) contacts:** A combination of a stationary contact and a movable contact, which are engaged when the coil is de-energized; in the case of a latching relay, the contacts that are engaged after the unlatch coil is energized.

**Normally open (N.O.) contacts:** A combination of a stationary contact and a movable contact, which are disengaged when the coil is de-energized; in the case of a latching relay, the contacts that are disengaged after the unlatch coil is energized.

**Shading coil:** A shorted turn surrounding a portion of the pole of an AC electromagnet, producing by mutual inductance a delay in the change of the magnetic field in that part, thereby tending to prevent chatter and reduce hum.

**Solid-state relay:** A semiconductor switch having no moving parts. These devices control power to the load and are analogous to relay contacts.

**Unlatch coil:** The coil of a latching relay that, when energized, causes the relay to assume the normal or unlatched position.

# Additional Resources

This module presents thorough resources for task training. The following resource material is suggested for further study.

*National Electrical Code® Handbook*, Latest Edition. Quincy, MA: National Fire Protection Association.

# Figure Credits

AGC of America, Module opener

Reprinted with permission of the National Electrical Manufacturers Association, Table 1

Topaz Publications, Inc., Figure 10B, SA01, and SA02

Tim Dean, Figure 15

# CONTREN® LEARNING SERIES — USER UPDATE

NCCER makes every effort to keep its textbooks up-to-date and free of technical errors. We appreciate your help in this process. If you find an error, a typographical mistake, or an inaccuracy in NCCER's Contren® materials, please fill out this form (or a photocopy), or complete the online form at www.nccer.org/olf. Be sure to include the exact module number, page number, a detailed description, and your recommended correction. Your input will be brought to the attention of the Authoring Team. Thank you for your assistance.

*Instructors* – If you have an idea for improving this textbook, or have found that additional materials were necessary to teach this module effectively, please let us know so that we may present your suggestions to the Authoring Team.

**NCCER Product Development and Revision**

3600 NW 43rd Street, Building G, Gainesville, FL 32606

**Fax:** 352-334-0932
**Email:** curriculum@nccer.org
**Online:** www.nccer.org/olf

☐ Trainee Guide  ☐ AIG  ☐ Exam  ☐ PowerPoints  Other _____

Craft / Level: _____ Copyright Date: _____

Module Number / Title: _____

Section Number(s): _____

Description: _____

_____

_____

_____

Recommended Correction: _____

_____

_____

_____

Your Name: _____

Address: _____

_____

Email: _____ Phone: _____

# Glossary

**AL-CU:** An abbreviation for aluminum and copper, commonly marked on terminals, lugs, and other electrical connectors to indicate that the device is suitable for use with either aluminum conductors or copper conductors.

**Ambient temperature:** The temperature of the medium (usually air) surrounding a device and into which the heat losses in the device are dissipated.

**Amperage capacity:** The maximum amount of current that a lug can safely handle at its rated voltage.

**Approximate ram travel:** The distance the ram of a hydraulic bender travels to make a bend. To simplify and speed bending operations, many benders are equipped with a scale that shows ram travel. Using a simple table (supplied with many benders), the degree of bend can easily be converted to inches of ram travel. This measurement, however, can only be approximated because of the variation in springback of the conduit being bent.

**Armature:** The rotating windings of a DC motor.

**Auxiliary electrodes:** Metallic electrodes pushed or driven into the earth to provide electrical contact for the purpose of performing measurements on grounding electrodes or ground grid systems.

**Back-to-back bend:** Any bend formed by two 90° bends with a straight section of conduit between the bends.

**Ballast:** A circuit component in fluorescent and HID lighting fixtures that provides the required voltage surge at startup and then controls the subsequent flow of current through the lamp during operation.

**Barrier strip:** A metal strip constructed to divide a section of cable tray so that certain kinds of cable may be separated from each other.

**Basket grip:** A flexible steel mesh grip that is used on the ends of cable and conductors for attaching the pulling rope. The more force exerted on the pull, the tighter the grip wraps around the cable.

**Bending protractor:** Made for use with benders mounted on a bending table and used to measure degrees; also has a scale for 18, 20, 21, and 22 shots when using it to make a large sweep bend.

**Bending shot:** The number of shots needed to produce a specific bend.

**Bonding:** Connected to establish electrical continuity and conductivity.

**Branch circuit:** The circuit conductors between the final overcurrent device protecting the circuit and the outlet(s).

**Break:** The opening of closed contacts to interrupt an electrical circuit.

**Brush:** A conductor between the stationary and rotating parts of a machine. It is usually made of carbon.

**Cable grip:** A device used to secure ends of cables to a pulling rope during cable pulls.

**Cable pulley:** A device used to facilitate pulling conductor in cable tray where the tray changes direction. Several types are available (single, triple, etc.) to accommodate almost all pulling situations.

**Capacitance:** The storage of electricity in a capacitor; capacitance produces an opposition to voltage change. The unit of measurement for capacitance is the farad (F) or microfarad (μF).

**Capstan:** The turning drum of a cable puller on which the rope is wrapped and pulled. An increase in the number of wraps increases the pulling force of the cable puller.

**Cartridge fuse:** A fuse enclosed in an insulating tube to confine the arc when the fuse blows.

**Circuit breaker:** A device designed to open and close a circuit by nonautomatic means and to open the circuit automatically on a predetermined overcurrent without injury to itself when properly applied within its rating.

**Circuit breaker:** A device that is designed to open a circuit automatically at a certain overcurrent. It can also be used to manually open and close the circuit.

**Clevis:** A device used in cable pulls to facilitate connecting the pulling rope to the cable grip.

**Color rendering index (CRI):** A measurement of the way a light source reproduces color. The higher the index number (0–100), the closer the colors are to how an object appears in full sunlight or incandescent light.

**Commutator:** A device used on electric motors or generators to maintain a unidirectional current.

**Concentric bending:** The process of making 90° bends in parallel runs of conduit. This requires increasing the radius of each conduit from the inside of the bend toward the outside.

**Conductor support:** The act of providing support in vertical conduit runs to support the cables or conductors. The *NEC*® gives several methods in which cables may be supported, including wedges in the tops of conduits, supports to change the direction of cable in pull boxes, etc.

**Conduit:** Piping designed especially for pulling electrical conductors. Types include rigid, IMC, EMT, PVC, aluminum, and other materials.

**Conduit body:** A separate portion of a conduit or tubing system that provides access through a removable cover (or covers) to the interior of the system at a junction of two or more sections of the system or at a terminal point of the system. Boxes such as FS and FD or larger cast or sheet metal boxes are not classified as conduit bodies.

**Conduit piston:** A cylinder of foam rubber that fits inside the conduit and is then propelled by compressed air or vacuumed through the conduit run to pull a line, rope, or measuring tape. Also called a mouse.

**Connection:** That part of a circuit that has negligible impedance and joins components or devices.

**Connector:** A device used to physically and electrically connect two or more conductors.

**Continuous duty:** Operation at a substantially constant load for an indefinitely long time.

**Controller:** A device that serves to govern, in some predetermined manner, the electric power delivered to the apparatus to which it is connected.

**Cross:** A four-way section of cable tray used when the tray assembly must branch off in four different directions.

**Degree indicator:** An instrument designed to indicate the exact degree of bend while it is being made.

**Developed length:** The amount of straight pipe needed to bend a given radius. Also, the actual length of the conduit that will be bent.

**Dip tolerance:** The ability of an HID lamp or lighting fixture circuit to ride through voltage variations without the lamp extinguishing and cooling down.

**Direct rod suspension:** A method used to support cable tray by means of threaded rods and hanger clamps. One end of the threaded rod is secured to an overhead structure, while the other end is connected to hanger clamps that are attached to the cable tray side rails.

**Drain wire:** A wire that is attached to a coaxial connector to allow a path to ground from the outer shield.

**Dropout:** Cable leaving the tray assembly and traveling directly downward; that is, the cable is not routed into a conduit or channel.

**Dropout plate:** A metal plate used at the end of a cable tray section to ensure a greater cable bending radius as the cable leaves the tray assembly.

**Dual-element fuse:** A fuse having two fuse characteristics; the usual combination is an overcurrent limit and a time delay before activation.

**Duty:** Describes the length of operation. There are four designations for circuit duty: continuous, periodic, intermittent, and varying.

**Edison-base:** The standard screw base used for ordinary lamps and Edison-base plug fuses.

**Effective ground fault path:** An intentionally constructed, permanent, low-impedance, electrically conductive path designed and intended to carry current under ground-fault conditions from the point of a ground fault on a wiring system to the electrical supply source. It facilitates the operation of the overcurrent protective device or ground-fault detectors on a high-impedance grounded system.

**Efficacy:** The light output of a light source divided by the total power input to that source. It is expressed in lumens per watt (LPW).

**Elbow:** A 90° bend or fitting in a section of conduit or cable tray.

**Equipment:** A general term including material, fittings, devices, appliances, fixtures, apparatus, and the like used as a part of, or in connection with, an electrical installation.

**Equipment bonding jumper:** The connection between two or more portions of the equipment grounding conductor.

**Equipment grounding conductor:** The conductive path installed to connect the normally noncurrent-carrying metal parts of equipment together and to the system grounded conductor, or to the grounding electrode conductor, or both.

**Expansion joints:** Plates used at intervals along a straight run of cable tray to allow space for thermal expansion or contraction of the tray.

**Explosion-proof:** Designed and constructed to withstand an internal explosion without creating an external explosion or fire.

**Fault current:** The current that exists when an unintended path is established between an ungrounded conductor and ground.

**Field poles:** The stationary portion of a DC motor that produces the magnetic field.

**Fish line:** Light cord used in conjunction with vacuum/blower power fishing systems that attaches to the conduit piston to be pushed or pulled through the conduit. Once through, a pulling rope is attached to one end and pulled back through the conduit for use in pulling conductors.

**Fish tape:** A flat iron wire or fiber cord used to pull conductors or a pulling rope through conduit.

**Fittings:** Devices used to assemble and/or change the direction of cable tray systems.

**Frame size:** A method used to classify circuit breakers according to given current ranges.

**Frequency:** The number of cycles an alternating electric current, sound wave, or vibrating object undergoes per second.

**Fuse:** A protective device that opens a circuit when the fusible element is severed by heating due to a fault current or overcurrent passing through it.

**Fuse link:** The fusible part of a cartridge fuse.

**Gain:** The amount of pipe saved by bending on a radius as opposed to right angles. Because conduit bends in a radius and not at right angles, the length of conduit needed for a bend will not equal the total determined length. Gain is the difference between the right angle distances A and B and the shorter distance C—the length of conduit actually needed for the bend.

**Grooming:** The act of separating the braid in a coaxial conductor.

**Ground:** A conducting connection, whether intentional or accidental, between an electrical circuit or equipment and the earth.

**Ground current:** Current in the earth or grounding connection.

**Grounded:** Connected to ground or to a conductive body that extends the ground connection.

**Grounded conductor:** A system or circuit conductor that is intentionally grounded.

**Ground grids:** System of grounding electrodes interconnected by bare cables buried in the earth to provide lower resistance than a single grounding electrode.

**Ground mats:** System of bare conductors, on or below the surface of the earth, connected to a ground or ground grid to provide protection from dangerous touch voltage.

**Ground resistance:** The ohmic resistance between a grounding electrode and a remote or reference grounding electrode that are spaced such that their mutual resistance is essentially zero.

**Ground rod:** A metal rod or pipe used as a grounding electrode.

**Grounding clip:** A spring clip used to secure a bonding conductor to an outlet box.

**Grounding conductor:** A conductor used to connect equipment or the grounded circuit of a wiring system to a grounding electrode or electrodes.

**Grounding connections:** Connections used to establish a ground; they consist of a grounding conductor, a grounding electrode, and the earth surrounding the electrode.

**Grounding electrode:** A conducting object through which a direct connection to earth is established.

**Grounding electrode conductor:** A conductor used to connect the system grounded conductor or the equipment to a grounding electrode or to a point on the grounding electrode system.

**Handhole:** An enclosure used with underground systems to provide access for installation and maintenance.

**Hertz (Hz):** A unit of frequency; one hertz equals one cycle per second.

**Horsepower:** The rated output capacity of the motor. It is based on breakdown torque, which is the maximum torque a motor will develop without an abrupt drop in speed.

**Hours:** The duty cycle of a motor. Most fractional horsepower motors are marked continuous for around-the-clock operation at the nameplate rating in the rated ambient conditions. Motors marked one-half are for ½-hour ratings, and those marked one are for 1-hour ratings.

**Impedance:** The opposition to current flow in an AC circuit; impedance includes resistance (R), capacitive reactance ($X_C$), and inductive reactance ($X_L$). Impedance is measured in ohms.

**Incandescence:** The self-emission of radiant energy in the visible light spectrum resulting from thermal excitation of atoms or molecules such as occurs when an electric current is passed through the filament in an incandescent lamp.

**Incident light:** The light emitted from a self-luminous object such as the sun or an electric source.

**Inductance:** The creation of a voltage due to a time-varying current; also, the opposition to current change, causing current changes to lag behind voltage changes. The unit of measure for inductance is the henry (H).

**Inductive load:** A load in which the voltage leads the current. Electromagnet coils such as those used in relays and starters are inductive loads.

**Inside diameter (ID):** The inside diameter of conduit. All electrical conduit is measured in this manner. The outside dimensions, however, will vary with the type of conduit used.

**Instantaneous:** Contacts that make or break a circuit immediately upon coil energization or de-energization.

**Insulating tape:** Adhesive tape that has been manufactured from a nonconductive material and is used for covering wire joints and exposed parts.

**Interlocked armor cable:** Mechanically protected cable; usually a helical winding of metal tape formed so that each convolution locks mechanically upon the previous one (armor interlock).

**Intermittent duty:** Operation for alternate intervals of (1) load and no load; or (2) load and rest; or (3) load, no load, and rest.

**Junction box:** An enclosure where one or more raceways or cables enter, and in which electrical conductors can be, or are, spliced.

**Kicks:** Bends in a piece of conduit, usually less than 90°, made to change the direction of the conduit.

**Ladder tray:** A type of cable tray that consists of two parallel channels connected by rungs, similar in appearance to the common straight ladder.

**Latch coil:** The coil of a latching relay that, when energized, causes the relay to assume the switched or latched position.

**Leg length:** The distance from the end of the straight section of conduit to the bend, measured to the centerline or to the inside or outside of the bend or rise.

**Lug:** A device for terminating a conductor to facilitate the mechanical connection.

**Lumen (lm):** The basic measurement of light. One lumen is defined as the amount of light cast upon one square foot of the inner surface of a hollow sphere with a one-foot radius with a light source of one candela in its center.

**Lumen maintenance:** A measure of how a lamp maintains its light output over time. It may be expressed either numerically or as a graph of light output versus time.

**Lumens per watt (LPW):** A measure of the efficiency, or, more properly, the efficacy of a light source. The efficacy is calculated by taking the lumen output of a lamp and dividing by the lamp wattage. For example, a 100W lamp producing 1,750 lumens has an efficacy of 17.5 lumens per watt.

**Luminaire (fixture):** A complete lighting unit consisting of a lamp (or lamps) and ballasts (where applicable), together with the parts designed to distribute the light, position and protect lamps, and connect them to the power supply.

**Luminance:** The luminous intensity of any surface in a given direction per unit area of that surface as viewed from that direction. Measured in candela/m². The term luminance is commonly used to express brightness.

**Main bonding jumper:** The connection between the grounded circuit conductor and the equipment grounding conductor at the service.

**Make:** The closing of open contacts to establish an electrical circuit.

**Mechanical advantage (MA):** The force factor of a crimping tool that is multiplied by the hand force to give the total force.

**Mechanically held relay:** A latching relay in which the switched or latched position is maintained by the interference of a mechanical part that moves to a non-interfering position when the unlatch coil is energized, allowing the relay to return to the normal or unlatched position.

**Micro:** Prefix designating one-millionth of a unit. For example, one microfarad is one-millionth of a farad.

**Mogul:** A type of conduit body with a raised cover to provide additional space for large conductors.

**Molded-case circuit breaker:** A circuit breaker enclosed in an insulating housing.

**Neutral:** The conductor connected to the neutral point of a system that is intended to carry current under normal conditions.

**Ninety-degree bend:** A bend in a piece of conduit that changes its direction by 90°.

**Nonrenewable fuse:** A fuse that must be replaced after it interrupts a circuit.

**Normally closed (N.C.) contacts:** A combination of a stationary contact and a movable contact, which are engaged when the coil is de-energized. In the case of a latching relay, the contacts that are engaged after the unlatch coil is energized.

**Normally open (N.O.) contacts:** A combination of a stationary contact and a movable contact, which are disengaged when the coil is de-energized. In the case of a latching relay, the contacts that are disengaged after the unlatch coil is energized.

**Offsets:** Two equal bends made to avoid an obstruction blocking the run of the conduit.

**One-shot shoe:** A large bending shoe that is designed to make 90° bends in conduit.

**Outside diameter (OD):** The size of any piece of conduit measured on the outside diameter.

**Overcurrent:** Any current in excess of the rated current of equipment or the ampacity of a conductor. It may result from an overload, short circuit, or ground fault.

**Overcurrent protection:** De-energizing a circuit whenever the current exceeds a predetermined value; the usual devices are fuses, circuit breakers, or magnetic relays.

**Overload:** Operation of equipment in excess of the normal, full-load rating, or of a conductor in excess of rated ampacity, which, after a sufficient length of time, will cause damage or dangerous overheating. A fault, such as a short circuit or ground fault, is not an overload.

**Peak voltage:** The peak value of a sinusoidally varying (cyclical) voltage or current is equal to the root-mean-square (rms) value multiplied by the square root of two (1.414). AC voltages are usually expressed as rms values; that is, 120 volts, 208 volts, 240 volts, 277 volts, 480 volts, etc., are all rms values. The peak voltage, however, differs. For example, the peak value of 120 volts (rms) is actually $120 \times 1.414 = 169.71$ volts.

**Periodic duty:** Intermittent operation at a substantially constant load for a short and definitely specified time.

**Pipe racks:** Structural frames used to support the piping that interconnects equipment in outdoor industrial facilities.

**Plug fuse:** A type of fuse that is held in position by a screw thread contact instead of spring clips, as is the case with a cartridge fuse.

**Pole:** That portion of a device associated exclusively with one electrically separated conducting path of the main circuit or device.

**Pressure connector:** A connector applied using pressure to form a cold weld between the conductor and the connector.

**Pull box:** A sheet metal box-like enclosure used in conduit runs to facilitate the pulling of cables from point to point in long runs, or to provide for the installation of conduit support bushings needed to support the weight of long riser cables, or to provide for turns in multiple-conduit runs.

**Radian:** An angle at the center of a circle, subtending (opposite to) an arc of the circle that is equal in length to the radius.

**Radius:** The relative size of the bent portion of a pipe.

**Raintight:** Constructed or protected so that exposure to a beating rain will not result in the entrance of water under specified test conditions.

**Reactance:** The imaginary part of impedance. Also, the opposition to alternating current (AC) due to capacitance ($X_C$) and/or inductance ($X_L$).

**Reducing connector:** A connector used to join two different size conductors.

**Reflected:** The bouncing back of light waves or rays by a surface.

**Reflected light:** Light reflected from an object that does not have self-luminating properties but reflects light provided from another source.

**Refracted/refraction:** The bending of a light wave or ray as it passes obliquely from one medium to another of different density, or through layers of different density in the same medium.

**Resistivity:** Resistance between opposite faces of a unit cube. Expressed in ohm-centimeters or ohms per cubic centimeter.

**Revolutions per minute (rpm):** The approximate full-load speed at the rated power line frequency. The speed of a motor is determined by the number of poles in the winding. A four-pole, 60Hz motor runs at an approximate speed of 1,725 rpm. A six-pole, 60Hz motor runs at an approximate speed of 1,140 rpm.

**Rise:** The length of the bent section of conduit measured from the bottom, centerline, or top of the straight section to the end of the bent section.

**Root-mean-square (rms):** The square root of the average of the square of the function taken throughout the period. The rms value of a sinusoidally varying voltage or current is the effective value of the voltage or current.

**Rotation:** For single-phase motors, the standard rotation, unless otherwise noted, is counterclockwise facing the lead or opposite shaft end. All motors can be reconnected at the terminal board for opposite rotation unless otherwise indicated.

**Segment bend:** Any bend formed by a series of bends of a few degrees each, rather than a single one-shot bend.

**Segmented bending shoe:** A smaller type of shoe designed for bending segmented bends only (always less than 15°).

**Self-inductance:** A magnetic field induced in the conductor carrying the current.

**Separately derived system:** A premises wiring system whose power is derived from a source of electric energy or equipment other than a service. Such systems have no direct electrical connection, including a solidly connected ground circuit conductor, to supply conductors originating in another system.

**Setscrew grip:** A cable grip, usually with built-in clevis, in which the cable ends are inserted in holes and secured with one or more setscrews.

**Shading coil:** A shorted turn surrounding a portion of the pole of an AC electromagnet, producing by mutual inductance a delay in the change of the magnetic field in that part, thereby tending to prevent chatter and reduce hum.

**Sheave:** A pulley-like device used in cable pulls in both conduit and cable tray systems.

**Shielding:** The metal covering of a cable that reduces the effects of electromagnetic noise.

**Short circuit:** The current that exists when an unintended path is created between any two components in a circuit.

**Soap:** Slang for wire-pulling lubricant.

**Solid-state relay:** A semiconductor switch having no moving parts. These devices control power to the load and are analogous to relay contacts.

**Splice:** The electrical and mechanical connection between two pieces of cable.

**Springback:** The amount, measured in degrees, that a bent conduit tends to straighten after pressure is released on the bender ram. For example, a 90° bend, after pressure is released, will pull back about 2° to 88°.

**Starter:** A circuit component of certain fluorescent lighting fixtures that is used to heat the lamp electrodes before the lamps are lighted.

**Step voltage:** The potential difference between two points on the earth's surface separated by a distance of one pace, or about three feet.

**Strand:** A group of wires, usually stranded or braided.

**Stub-up:** Another name for the rise in a section of conduit.

**Sweep bend:** A 90° bend with a radius larger than that produced by a standard one-shot shoe.

**Swivel plates:** Devices used to make vertical offsets in cable tray.

**Synchronous speed:** When the speed of the rotor is equal to the speed of the stator. The speed is determined by multiplying 120 times the frequency divided by the number of poles.

**System grounding:** Intentional connection of one of the circuit conductors of an electrical system to ground potential.

**Take-up (comeback):** The amount that must be subtracted from the desired stub length to make the bend come out correctly using a point of reference on the bender or bending shoe.

**Tee:** A section of cable tray that branches off the main section in two other directions.

**Terminal:** A device used for connecting cables.

**Termination:** The connection of a cable.

**Thermal protector:** A protective device for assembly as an integral part of a motor or motor compressor that, when properly applied, protects the motor against dangerous overheating due to overload or failure to start.

**Touch voltage:** The potential difference between a grounded metallic structure and a point on the earth's surface equal to the normal maximum horizontal reach—approximately three feet.

**Trapeze mounting:** A method of supporting cable tray using metal channel, such as Unistrut®, Kindorf®, etc., supported by two threaded rods, and giving the appearance of a swing or trapeze.

**Tray cover:** A flat piece of metal, fiberglass, or plastic designed to provide a solid covering that is needed in some locations where conductors in the tray system may be damaged.

**Troffer:** A recessed lighting fixture installed with the opening flush with the ceiling.

**Trough:** A type of cable tray consisting of two parallel channels (side rails) having a corrugated, ventilated bottom or a corrugated, solid bottom.

**Ungrounded conductors:** Conductors in an electrical system that are not intentionally grounded.

**Unistrut®:** A brand of metal channel used as the bottom bracket for hanging cable trays. Double Unistrut® adds strength and stability to the trays and also provides a means of securing future runs of conduit.

**Unlatch coil:** The coil of a latching relay that, when energized, causes the relay to assume the normal or unlatched position.

**Varying duty:** Operation at varying loads and/or intervals of time.

**Wall mounting:** A method of supporting cable tray systems using supports secured directly to the wall.

**Waterproof:** See *Weatherproof.*

**Watertight:** Constructed so that moisture will not enter the enclosure under specified test conditions.

**Weatherproof:** Constructed or protected so that exposure to the weather will not interfere with successful operation.

**Wye:** A section of cable tray that branches off the main section in one direction.

# Index